Combinatorics is a subject of increasing importance, owing to its links with computer science, statistics and algebra. This is a textbook aimed at second-year undergraduates to beginning graduates. It stresses common techniques (such as generating functions and recursive construction) which underlie the great variety of subject matter, and the fact that a constructive or algorithmic proof is more valuable than an existence proof.

The book is divided into two parts, the second at a higher level and with a wider range than the first. Historical notes are included and give a wider perspective on the subject. More advanced topics are given as projects, and there are a number of exercises, some with solutions given.

Combinatorics:
Topics, Techniques, Algorithms

COMBINATORICS:
TOPICS, TECHNIQUES, ALGORITHMS

PETER J. CAMERON

Queen Mary & Westfield College, London

Published by the Press Syndicate of the University of Cambridge
The Pitt Building, Trumpington Street, Cambridge CB2 1RP
40 West 20th Street, New York, NY 10011-4211, USA
10 Stamford Road, Oakleigh, Melbourne 3166, Australia

First edition published 1994
Reprinted with corrections 1996

Printed in Great Britain at the University Press, Cambridge

Library of Congress cataloging in publication data

Cameron, Peter J. (Peter Jephson), 1947–
Combinatorics: topics, techniques, algorithms/Peter J. Cameron.
  p. cm.
Includes index.
ISBN 0 521 45133 7.–ISBN 0 521 45761 0 (pbk.)
1. Combinatorial analysis.   I. Title.
QA164.C346   1994
511′.6–dc20   94–4680 CIP

A catalogue record for this book is available from the British Library

ISBN 0 521 45133 7 hardback
ISBN 0 521 45761 0 paperback

Contents

Preface

Ive got to work the E qwations and the low cations
Ive got to comb the nations of it.

> Russell Hoban, *Riddley Walker* (1980)

We have not begun to understand the relationship between combinatorics and
conceptual mathematics.

> J. Dieudonné, *A Panorama of Pure Mathematics* (1982)

If anything at all can be deduced from the two quotations at the top of this page,
perhaps it is this: Combinatorics is an essential part of the human spirit; but it is
a difficult subject for the abstract, axiomatising Bourbaki school of mathematics to
comprehend. Nevertheless, the advent of computers and electronic communications
have made it a more important subject than ever.

This is a textbook on combinatorics. It's based on my experience of more than
twenty years of research and, more specifically, on teaching a course at Queen Mary
and Westfield College, University of London, since 1986. The book presupposes
some mathematical knowledge. The first part (Chapters 2–11) could be studied by
a second-year British undergraduate; but I hope that more advanced students will
find something interesting here too (especially in the Projects, which may be skipped
without much loss by beginners). The second half (Chapters 12–20) is in a more
condensed style, more suited to postgraduate students.

I am grateful to many colleagues, friends and students for all kinds of contributions, some of which are acknowledged in the text; and to Neill Cameron, for the
illustration on p. 128.

I have not provided a table of dependencies between chapters. Everything is
connected; but combinatorics is, by nature, broad rather than deep. The more
important connections are indicated at the start of the chapters.

> Peter J. Cameron
> 17 March 1994

This reprinting has given me the opportunity to correct a number of errors and
to make some small imporovements. I have been helped by many colleagues and
friends who sent me lists of misprints and mistakes. But two of these have made
contributions which go far beyond mere proofreading, Rosemary Bailey and Peter
Johnson[1]; to them, special thanks are due. There are also two new pictures by Neill
Cameron (over the page).

> Peter J. Cameron
> 18 September 1995

---

[1] See pages 103 and 186 for two of their contributions.

Euler's officers in Königsberg (pp. 2, 97, 167)

Kirkman arrives at Hilbert's hotel (pp. 2, 119, 310)

1. What is Combinatorics?

Combinatorics is the slums of topology.

J. H. C. Whitehead (attr.)[1]

I have to admit that he was not bad at combinatorial analysis — a branch, however, that even then I considered to be dried up.

Stanislaw Lem, *His Master's Voice* (1968)

Combinatorics is special. Most mathematical topics which can be covered in a lecture course build towards a single, well-defined goal, such as Cauchy's Theorem or the Prime Number Theorem. Even if such a clear goal doesn't exist, there is a sharp focus (finite groups, perhaps, or non-parametric statistics). By contrast, combinatorics appears to be a collection of unrelated puzzles chosen at random.

Two factors contribute to this. First, combinatorics is broad rather than deep. Its tentacles stretch into virtually all corners of mathematics. Second, it is about techniques rather than results. As in a net,[2] threads run through the entire construction, appearing unexpectedly far from where we last saw them. A treatment of combinatorics which neglects this is bound to give a superficial impression.

This feature makes the teacher's job harder. Reading, or lecturing, is inherently one-dimensional. If we follow one thread, we miss the essential interconnectedness of the subject.

I have attempted to meet this difficulty by various devices. Each chapter begins with a list of topics, techniques, and algorithms considered in the chapter, and cross-references to other chapters. Also, some of the material is set in smaller type and can be regarded as optional. This usually includes a 'project' involving a more difficult proof or construction (where the arguments may only be sketched, requiring extra work by the reader). These projects could be used for presentations by students. Finally, the book is divided into two parts; the second part treats topics in greater depth, and the pace hots up a bit (though, I hope, not at the expense of intelligibility).

As just noted, there are algorithms scattered throughout the book. These are not computer programs, but descriptions in English of how a computation is performed. I hope that they can be turned into computer programs or subroutines by readers with programming experience. The point is that an explicit construction of an object usually tells us more than a non-constructive existence proof. (Examples will be given to illustrate this.) An algorithm resembles a theorem in that it requires a proof (not of the algorithm itself, but of the fact that it does what is claimed of it).

---

[1] This attribution is due to Graham Higman, who revised Whitehead's definition to 'Combinatorics is the mews of algebra.'

[2] '*Net*. Anything reticulated or decussated at equal distances, with interstices between the intersections.' Samuel Johnson, *Dictionary of the English Language* (1775).

But what is combinatorics? Why should you read further?

Combinatorics could be described as the art of arranging objects according to specified rules. We want to know, first, whether a particular arrangement is possible at all, and if so, in how many different ways it can be done. If the rules are simple (like picking a cricket team from a class of schoolboys), the existence of an arrangement is clear, and we concentrate on the counting problem. But for more involved rules, it may not be clear whether the arrangement is possible at all. Examples are Kirkman's schoolgirls and Euler's officers, described below.

Sample problems

In this section, I will give four examples of combinatorial questions chosen to illustrate the nature of the subject. Each of these will be discussed later in the book.

Derangements

> Given n letters and n addressed envelopes, in how many ways can the letters be placed in the envelopes so that no letter is in the correct envelope?

DISCUSSION. The total number of ways of putting the letters in the envelopes is the number of *permutations* of n objects,[3] which is $n!$ (factorial n). We will see that the fraction of these which are all incorrectly addressed is very close to $1/e$, where $e = 2.71828\ldots$ is the base of natural logarithms — a surprising result at first sight. In fact, the exact number of ways of mis-addressing all the letters is the nearest integer to $n!/e$ (see Exercise 1).

Kirkman's schoolgirls

> Fifteen schoolgirls walk each day in five groups of three. Arrange the girls' walks for a week so that, in that time, each pair of girls walks together in a group just once.

DISCUSSION. If it is possible at all, seven days will be required. For any given girl must walk once with each of the other fourteen; and each day she walks with two others. However, showing that the walks are actually possible requires more argument. The question was posed and solved by Kirkman in 1847. The same question could be asked for other numbers of girls (see Exercise 2). Only in 1967 did Ray-Chaudhuri and Wilson show that solutions exist for any number of girls congruent to 3 modulo 6.

Euler's officers

> Thirty-six officers are given, belonging to six regiments and holding six ranks (so that each combination of rank and regiment corresponds to just one officer). Can the officers be paraded in a 6×6 array so that, in any line (row or column) of the array, each regiment and each rank occurs precisely once?

---

[3] Permutations will be described in Chapter 3.

DISCUSSION. Euler posed this problem in 1782; he believed that the answer was 'no'. This was not proved until 1900, by Tarry. Again, the problem can be generalised, to n^2 officers, where the number of regiments, ranks, rows and columns is n (we assume $n > 1$) — see Exercise 3. There is no solution for $n = 2$. Euler knew solutions for all n not congruent to 2 modulo 4, and guessed that there was no solution for $n \equiv 2$ (mod 4). However, he was wrong about that. Bose, Shrikhande and Parker showed in 1960 that there is a solution for all n except $n = 2$ and $n = 6$.

A Ramsey game

> This two-player game requires a sheet of paper and pencils of two colours, say red and blue. Six points on the paper are chosen, with no three in line. Now the players take a pencil each, and take turns drawing a line connecting two of the chosen points. The first player to complete a triangle of her own colour loses. (Only triangles with vertices at the chosen points count.)
> *Can the game ever result in a draw?*

DISCUSSION. We'll see that a draw is not possible; one or other player will be forced to create a triangle. Ramsey proved a wide generalisation of this fact. His theorem is sometimes stated in the form 'Complete disorder is impossible.'

How to use this book

1. The book is divided into two parts: Chapters 2–11 and Chapters 12–20. In the second part, along with some new material, we revisit many of the topics from the first part and treat them from a more advanced viewpoint; also, as I mentioned earlier, the pace is a little faster in the second part. In any case, a first course can be devised using only the first part of the book. (The second–third year undergraduate course at Queen Mary and Westfield College includes a selection of material from Chapters 3 (Sections 3.1, 3.2, 3.3, 3.5, 3.7, 3.11, 3.12), 4 (Sections 4.1, 4.3, 4.4, 4.5), 5, 6, 7, 8 and 10; other courses treat material from Chapters 9, 11, 14–17.)

2. Chapter 3 plays a special rôle. The material here is central to combinatorics: subsets, partitions, and permutations of finite sets. Within the other chapters, you are encouraged to dabble, taking or leaving sections as you choose; but I recommend reading all of Chapter 3 (except perhaps the Projects, see below).

3. A number of sections are designated as Projects. These are to be regarded as less central and possibly more difficult than the others. The word suggests that they could be worked through by individuals outside class time, and then made the subject of presentations to the class.

4. Each chapter after this one begins with a box containing 'topics, techniques, algorithms and cross-references'. This is designed to give you some indication of the scope of the chapter. Roughly speaking, topics are specific results or constructions; techniques are of wider applicability, indicating general methods which may be illustrated in specific cases in the chapter; algorithms are self-explanatory; and

cross-references pinpoint at least some occurrences of the material in other chapters. These are usually backward references, but the multidimensional nature of the subject means that this is not always so. You should use these as pointers to places where you might find help if you are stuck on something. The index can also be used for this purpose.

5. The exercises are a mixed bunch; but, by and large, I have tended to avoid 'drill' and give more substantial problems. You will certainly learn more if you work conscientiously through them. But I have tried not to assume that you have done all the problems. When (as often happens) the result of an exercise is needed in a later chapter, I have usually supplied a proof (or, failing that, a hint). Indeed, hints are strewn liberally through the exercises, and some example solutions are given (rather more briefly than I would expect from students!) at the end of the book.

6. The last chapter does two jobs. First, it treats (somewhat sketchily) some further topics not mentioned earlier; second, it gives pointers to further reading in various parts of combinatorics. I have included a small collection of unsolved problems here, to indicate the sort of thing that research in combinatorics might involve. But beware: these problems are unsolved; this means that somebody has given some thought to them and failed to solve them, so they are probably more difficult than the exercises in other chapters.

7. The numbering is as follows. Chapter A is divided into sections, of which a typical one is Section A.B. Within a section, theorems (and similar statements such as propositions, lemmas, corollaries, facts, algorithms, and numbered equations) have numbers of the form A.B.C. On the other hand, diagrams are just numbered within the chapter, as A.D, for example; and exercises are typically referred to as 'exercise E of Chapter A'. Some theorems or facts are displayed in a box for easy reference. But don't read too much into the difference between displayed and undisplayed theorems, or between theorems and propositions; it's a matter of taste, and consistency is not really possible.

8. An important part of combinatorics today is the algorithmic side: I can prove that some object exists; how do I construct it? I have described algorithms for a wide range of constructions. No knowledge of computers or programming languages is assumed. The description of the algorithms makes use of words like 'While ...', 'Repeat ... until ...', and so on. These are to be interpreted as having their usual English meaning. Of course, this meaning has been taken over by programming languages; if you are fluent in Pascal, you will I hope find my descriptions quite congenial. If you are a competent programmer and have access to a computer, you are advised at several places to implement these algorithms.

What you need to know

The mathematical results that I use are listed here. You don't need everything all at once; the more advanced parts of algebra, for example, are only required later in the book, so you could study algebra and combinatorics at the same time. If all else fails, I have tried to arrange things so that you can take on trust what you don't know. Topics in square brackets are treated in the book, but you may feel the need of more explanation from a course or textbook in that subject. As you see, combinatorics connects with all of mathematics; you will see material from many other areas being used here.

- *Basic pure mathematics*: Sets and functions, ordered n-tuples and cartesian products; integers, factorisation, modular arithmetic; [equivalence and order relations].
- *Linear algebra*: Vector spaces, subspaces; linear transformations, matrices; row operations, row space; eigenvalues of real symmetric matrices.
- *Abstract algebra*: [Elementary group theory; finite fields].
- *Number theory*: [Quadratic residues; two and four squares theorems].
- *Analysis*: Basic operations (limits, differentiation, etc.); [power series].[4]
- *Topology*: [Definition of metric and topological space; surfaces; Jordan curve theorem].
- *Probability*: Basic concepts (for finite spaces only) [except in Chapter 19].
- *Set theory*: See Chapter 19.

Exercises

1. For $n = 3, 4, 5$, calculate the number of ways of putting n letters into their envelopes so that every letter is incorrectly addressed. Calculate the ratio of this number to $n!$ in each case.

2. Solve Kirkman's problem for nine schoolgirls, walking for four days.

3. Solve Euler's problem for nine, sixteen and twenty-five officers. Show that no solution is possible for four officers.

4. Test the assertion that the Ramsey game cannot end in a draw by playing it with a friend. Try to develop heuristic rules for successful play.

---

[4] As will be explained in Section 4.2, our treatment of power series is formal and does not involve questions of convergence.

2. On numbers and counting

One of them is all alone and ever more shall be so
Two of them are lily-white boys all clothed all in green Oh
Three of them are strangers o'er the wide world they are rangers
Four it is the Dilly Hour when blooms the Gilly Flower
Five it is the Dilly Bird that's seldom seen but heard
Six it is the ferryman in the boat that o'er the River floats Oh
Seven are the Seven Stars in the Sky, the Shining Stars be Seven Oh
Eight it is the Morning's break when all the World's awake Oh
Nine it is the pale Moonshine, the Shining Moon is Nine Oh
Ten Forgives all kinds of Sin, from Ten begin again Oh

English traditional folksong
from Bob Stewart, *Where is Saint George?* (1977)

TOPICS: Natural numbers and their representation; induction; useful functions; rates of growth; counting labelled and unlabelled structures; Handshaking Lemma

TECHNIQUES: Induction; double counting

ALGORITHMS: Odometer Principle; [Russian peasant multiplication]

CROSS-REFERENCES:

This chapter is about counting. In some sense, it is crucial to what follows, since counting is so basic in combinatorics. But this material is part of mathematical culture, so you will probably have seen most of it before.

2.1. Natural numbers and arithmetic

Kronecker is often quoted as saying about mathematics, 'God made the integers; the rest is the work of man.' He was referring to the natural numbers (or counting numbers), which are older than the earliest archæological evidence. (Zero and the negative numbers are much more recent, having been invented (or discovered) in historical time.)[1] Since much of combinatorics is concerned with counting, the natural numbers have special significance for us.

---

[1] See Georges Ifrah, *From One to Zero: A Universal History of Numbers* (1985), for an account of the development of numbers and their representation.

As each new class of numbers was added to the mathematical repertoire, it was given a name reflecting the prejudice against its members, or the 'old' numbers were given a friendly, reassuring name. Thus, zero and negative integers are contrasted with the 'natural' positive integers. Later, quotients of integers were 'rational', as opposed to the 'irrational' square root of 2; and later still, all numbers rational and irrational were regarded as 'real', while the square root of -1 was 'imaginary' (and its friends were 'complex').

The natural numbers are the first mathematical construct with which we become familiar. Small children recite the names of the first few natural numbers in the same way that they might chant a nursery rhyme or playground jingle. This gives them the concept that the numbers come in a sequence. They grasp this in a sophisticated way. The rhyme[2]

> One, two,
> Missed a few,
> Ninety-nine,
> A hundred

expresses confidence that the sequence of numbers stretches at least up to 100, and that the speaker could fill in the gap if pressed.

Order or progression is thus the most basic property of the natural numbers.[3] How is this expressed mathematically? First we must stop to consider how natural numbers are represented. The simplest way to represent the number n is by a sequence of n identical marks. This is probably the earliest scheme mankind adopted. It is well adapted for tallying: to move from one number to the next, simply add one more mark. However, large numbers are not easily recognisable. After various refinements (ranging from grouping the marks in sets of five to the complexities of Roman numerals), positional notation was finally adopted.

This involves the choice of a base b (an integer greater than 1), and b *digits* (distinguishable symbols for the integers $0, 1, 2, \ldots, b-1$). (Early attempts at positional notation were bedevilled because the need for a symbol for zero was not recognised.) Now any natural number N is represented by a finite string of digits. Logically the string is read from right to left; so we write it as $x_{n-1} \ldots x_1 x_0$, where each x_i is one of our digits. By convention, the leftmost digit is never zero. The algorithm for advancing to the next number is called the *Odometer Principle*. It is based on the principle of trading in b counters in place i for a single counter in place $i+1$, and should be readily understood by anyone who has watched the odometer (or mileage gauge) of a car.

---

[2] I have heard the feminist version of this: 'One, two, Mrs. Few, ...'

[3] 'The operations of arithmetic are based on the tacit assumption that *we can always pass from any number to its successor,* and this is the essence of the ordinal concept.' Tobias Dantzig, *Number: the Language of Science* (1930).

(2.1.1) Odometer Principle

to find the successor of a natural number to base b

Start by considering the rightmost digit.

- *If the digit we are considering is not $b - 1$, then replace it by the next digit in order, and terminate the algorithm.*
- *If we are considering a blank space (to the left of all the digits), then write in it the digit 1, and terminate the algorithm.*
- *If neither of the above holds, we are considering the digit $b - 1$. Replace it with the digit 0, move one place left, and return to the first bullet point.*

For example, if the base b is 2 and the digits are 0 and 1, the algorithm (starting with 1) generates successively 10, 11, 100, 101, 110, ...

Now it can be proved by induction that the string $x_{n-1} \ldots x_1 x_0$ represents the positive integer

$$x_{n-1} b^{n-1} + \cdots + x_1 b + x_0$$

(see Exercise 2).

Often the number 0 is included as a natural number. (This is most usually done by logicians, who like to generate the whole number system out of zero, or nothing. But it conflicts with our childhood experience: I have never heard a child say 'nought, one, two, ...'[4], and we don't count that way.) This is done by modifying our representation so that the digit 0 represents the number 0. This is the one allowed exception to the rule that the left-most digit cannot be 0; the alternative, representing 0 by a blank space, would be confusing.

The odometer of a car actually works slightly differently. It works with a fixed number of digits which are initially all zero, so that the 'blank space' case of the algorithm cannot arise. If there are k digits, then the integers $0, \ldots, b^k - 1$ are generated in turn, and then the odometer returns to 0 and the process repeats.

Now that we have a representation of positive integers, and understand how to move to the next integer, we should explore the arithmetic operations (ambition, distraction, uglification and derision).[5] Algorithms for these are taught in primary school.[6] I will not consider the details here. It is a good exercise to program a computer to perform these algorithms[7], or to investigate how many elementary

---

[4] A possible exception occurs when one child has been appointed to be first, and another wishes to claim precedence, as in 'Zero the hero'. But this is closer to the historical than the logical approach.

[5] Lewis Carroll, *Alice's Adventures in Wonderland* (1865).

[6] These algorithms were known to the Babylonians in 1700 B.C.

[7] Most programming languages specify the 'maximum integer' to be something like 32767 or 2147483647. Often, the answer to a counting problem will be much larger than this. To find it by computer, you may have to write routines for arithmetic operations on integers with many digits. If you need to do this, write your routines so that you can re-use them!

operations are required to add or multiply two n-digit numbers (where elementary operations might consist of referring to one's memory of the multiplication tables, or writing down a digit).

2.2. Induction

Induction is a very powerful principle for proving assertions about the natural numbers. It is applied in various different forms, some of which are described in this section. We also see that it is a consequence of our most basic intuition about the natural numbers.

The *Principle of Induction* asserts the following:

(2.2.1) Principle of Induction

Let $P(n)$ be a proposition or assertion about the natural number n. Suppose that $P(1)$ is true. Suppose also that, if $P(n)$ is true, then $P(n+1)$ is also true. Then $P(n)$ is true for all natural numbers n.

Why is this true? As we saw, the basic property of the natural numbers, recognised even by children, is that we can count up to any natural number n starting from 1 (given sufficient patience!) Now, with the assumptions of the Principle, $P(1)$ is true, so $P(2)$ is true, so (miss a few here) so $P(n-1)$ is true, so $P(n)$ is true.

As this argument suggests, if you are reading a mathematical argument, and the author puts in a few dots or the words 'and so on', there is probably a proof by induction hiding there. Consider, for example, the function f satisfying $f(1) = 2$ and $f(n+1) = 2f(n)$ for all natural numbers n. Then

$$f(2) = 4 = 2^2, f(3) = 8 = 2^3, \quad \ldots \quad f(n) = 2^n.$$

The dots hide a proof by induction. Let $P(n)$ be the assertion that $f(n) = 2^n$. Then $P(1)$ holds; and, assuming that $P(n)$ holds, we have

$$P(n+1) = 2P(n) = 2 \cdot 2^n = 2^{n+1},$$

so $P(n+1)$ also holds. So the Principle of Induction justifies the conclusion. The point is that very simple arguments by induction can be written out with three dots in place of the detailed verification, but this verification could be supplied if necessary. We'll see more examples of this later.

Now I give some alternative forms of the Principle of Induction and justify their equivalence. The first one is transparent. Suppose that $P(n)$ is an assertion, for which we know that $P(27)$ is true, and that if $P(n)$ holds then so does $P(n+1)$. Then we conclude that $P(n)$ holds for all $n \geq 27$. (To prove this formally, let $Q(n)$ be the assertion that $P(n+26)$ is true, and verify the hypotheses of the Principle of Induction for $Q(n)$.)

For the next variation, let $P(n)$ be a proposition about natural numbers. Suppose that, for every natural number n, *if $P(m)$ holds for all natural numbers m less than*

n, *then* $P(n)$ holds. Can we conclude that $P(n)$ holds for all n? On the face of it, this seems a much stronger principle, since the hypothesis is much weaker. (Instead of having to prove $P(n)$ from just the information that $P(n-1)$ holds, we may assume the truth of $P(m)$ for *all* smaller m.) But it is true, and it follows from the Principle as previously stated.

We let $Q(n)$ be the statement '$P(m)$ holds for all $m < n$'. Now it is clear that $Q(n+1)$ implies $P(n)$, so we will have succeeded if we can prove that $Q(n)$ holds for all n. We prove this by induction.

First, $Q(1)$ holds: for there are no natural numbers less than 1, so the assertion P holds for all of them (vacuously).

Now suppose that $Q(n)$ holds. That is, $P(m)$ holds for all $m < n$. By assumption, $P(n)$ also holds. Now $P(m)$ holds for all $m < n+1$ (since the numbers less than $n+1$ are just n and the numbers less than n)[8]. In other words, $Q(n+1)$ holds.

Now the Principle of Induction shows that $Q(n)$ holds for all n.

The final re-formulation gives us the technique of 'Proof by Minimal Counterexample'. Suppose that $P(n)$ is a proposition such that it is *not* true that $P(n)$ holds for all natural numbers n. Then there is a least natural number n for which $P(n)$ is false; in other words, $P(m)$ is true for all $m < n$ but $P(n)$ is false. For suppose that no such n exists; then the truth of $P(m)$ for all $m < n$ entails the truth of $P(n)$, and as we have seen, this suffices to show that $P(n)$ is true for all n, contrary to assumption.

This argument shows that any non-empty set of natural numbers contains a minimal element. (If S is the set, let $P(n)$ be the assertion $n \notin S$.)

2.3. Some useful functions

I assume that you are familiar with common functions like polynomials, the function $|x|$ (the *absolute value* or *modulus*), etc.

Floor and Ceiling. The *floor* of a real number x, written $\lfloor x \rfloor$, is the greatest integer not exceeding x. In other words, $\lfloor x \rfloor$ is the integer m such that $m \le x < m+1$. If x is an integer, then $\lfloor x \rfloor = x$. This function is sometimes written $[x]$; but the notation $\lfloor x \rfloor$ suggests 'rounding down'. It is the number of the floor of a building on which x would be found, if the height of x above the ground is measured in units of the distance between floors. (The British system of floor numbering is used, so that the ground floor is number 0.)

The *ceiling* is as you would probably expect: $\lceil x \rceil$ is the smallest integer not less than x. So, if x is not an integer, then $\lceil x \rceil = \lfloor x \rfloor + 1$; if x is an integer, its floor and ceiling are equal. In any case, you can check that

$$\lceil x \rceil = -\lfloor -x \rfloor.$$

Factorial. The factorial function is defined on positive integers by the rule that $n!$ is the product of all the integers from 1 to n inclusive. It satisfies the condition

$$n! = n \cdot (n-1)! \qquad (*)$$

---

[8] Let p be an integer less than $n+1$. Then $p < n$ or $p = n$ or $p > n$; and the last case is impossible, since there is no integer between n and $n+1$.

for $n > 1$. In fact, we can consistently define $0! = 1$; then $(*)$ holds for all $n > 0$. In fact, the conditions $0! = 1$ and $(*)$ actually define $n!$ for all integers $n \geq 0$. (This is proved by induction: $0!$ is defined; if $n!$ is defined then so is $(n+1)!$; so $n!$ is defined for all n.)

Exponential and logarithm.   These two functions are familiar from elementary calculus. We will often use the power series expansions of them. The equation

$$e^x = \sum_{n=0}^{\infty} \frac{x^n}{n!} = 1 + x + \frac{x^2}{2!} + \dots$$

is valid for all real numbers x. On the other hand, the function $\log x$ can't be expanded as a series of powers of x, since $\log 0$ is undefined. Instead, we have

$$\log(1+x) = \sum_{n=1}^{\infty} \frac{(-1)^{n-1} x^n}{n} = x - \frac{x^2}{2} + \dots,$$

which is valid for all x with $|x| < 1$ (and in fact also for $x = 1$).

The exponential function grows more rapidly than any power of x; this means that $e^x > x^c$ for all sufficiently large x (depending on c). In fact, for $x > (c+1)!$, we have

$$e^x > \frac{x^{c+1}}{(c+1)!} > x^c.$$

On the other hand, the logarithm function grows more slowly than any power of x.

We will often write $\exp(x)$ instead of e^x.

2.4. Orders of magnitude

People use the phrase 'the combinatorial explosion' to describe a counting function which grows very rapidly. This is a common phenomenon, and it means that, while we may be able to give a complete description of all the objects being counted for small values of the parameter, soon there will be far too many for this to be possible, and maybe even far too many for an exact count; we may have to make do with fairly rough estimates for the counting function. I will consider now what such rough estimates might look like. In this section, some results from later chapters will be anticipated. If you are unfamiliar with these, take them on trust until we meet them formally.

Let X be a set with n elements, say $X = \{1, 2, \dots, n\}$. The *number of subsets* of X is 2^n. This is the most familiar example of an *exponential function*, or *function with exponential growth*. A function f which has (precisely) exponential growth has the property that

$$f(n+1) = cf(n)$$

for some $c > 1$. (If $c = 1$, the function is constant; if $c < 1$, then $f(n) \to 0$ as $n \to \infty$. In these cases, the term 'exponential *growth*' is not really appropriate![9]) A function

---

[9] Economists define a recession as a period when the exponential constant for the GDP is less than 1.004. Sometimes you have to run in order to stand still.

f satisfying the above equation is given by $f(n) = ac^n$, where a is a constant (and is equal to the value of $f(0)$).

We also say that a function f has 'exponential growth' if it is roughly the same size as an exponential function. So the function $f(n) = 2^n + n$ has exponential growth, since the term n is dwarfed by 2^n for large n. Formally, the function f is said to have *exponential growth* if $f(n)^{1/n}$ tends to a limit $c > 1$ as $n \to \infty$. This means that, for any positive number ϵ, $f(n)$ lies between $(c - \epsilon)^n$ and $(c + \epsilon)^n$ for all sufficiently large n. The number c is called the *exponential constant* for f.

Of course, a function may grow more slowly than exponentially. Examples include
- *polynomial growth* with degree c, like the function $f(n) = n^c$;
- *fractional exponential growth* with exponent c, like the function e^{n^c}, where $0 < c < 1$.

These functions arise in real combinatorial counting problems, as we will see. But many functions grow faster than exponentially. Here are two examples.

The number of *permutations* of the set X is equal to $n! = n(n - 1) \ldots 1$, the product of the integers from 1 to n inclusive. We have

$$2^{n-1} \leq n! \leq n^{n-1},$$

because (ignoring the factor 1) there are $n - 1$ factors, each lying between 2 and n. In fact it is easy to see that the growth is not exponential. We will find better estimates in the next chapter.

Now let $\mathcal{P}(X)$, the *power set* of X, denote the set of all subsets of X. We will be considering subsets of $\mathcal{P}(X)$, under the name *families of sets*. How many families of sets are there? Clearly the number is 2^{2^n}. This number grows much faster than exponentially, and much faster than the factorial function. A function like this is called a *double exponential*.

For comparing the magnitudes of functions like these, it is often helpful to consider the logarithm of the function, rather than the function itself. The logarithm of an exponential function is a (roughly) linear function. The logarithm of $n!$ is fairly well approximated by $n \log n$; and the logarithm of a double exponential is exponential. Other possibilities are functions whose logarithms are polynomial.

Of course, this is only the beginning of a hierarchy of growth rates; but for the most part we won't have to consider anything worse than a double exponential.

In connection with growth rates, there is a convenient analytic notation. We write $O(f(n))$ (read 'big Oh $f(n)$') to mean a (possibly unknown) function $g(n)$ such that, for all sufficiently large n, $|g(n)| \leq cf(n)$ for some constant c. This is typically used in the form

$$\phi(n) = F(n) + O(f(n)),$$

where ϕ is a combinatorial counting function and F, f are analytic functions where f grows more slowly than F; this has the interpretation that the order of magnitude of ϕ is similar to that of F. For example, in Section 3.6, we show that

$$\log n! = n \log n - n + O(\log n).$$

We write $o(f(n))$ (and say 'small oh $f(n)$') to mean a function $g(n)$ such that $g(n)/f(n) \to 0$ as $n \to \infty$; that is, g is of smaller order of magnitude than f.

There are several variants. For example, Ω is the opposite of O; that is, $\Omega(f(n))$ is a function $g(n)$ with $|g(n)| \geq cf(n)$ for some constant $c > 0$. Also, $f(n) \sim g(n)$ means that $f(n)/g(n) \to 1$ as $n \to \infty$ (assuming that $g(n) \neq 0$ for large enough n): roughly, f and g have the same order of magnitude.

2.5. Different ways of counting

In combinatorics (unlike real life[10]), when we are asked to count something, there are very many different answers which can be regarded as correct. Consider the simple problem of choosing three items from a set of five. Before we can work out the right answer, the problem must be specified more precisely. Are the objects in the set identical (five electrons, say, or five red billiard balls), or all different (the ace, two, three, four, and five of spades, for example)? Does the order of selection matter? (That is, do we just put in a hand and pull three objects out, or do we draw them one at a time and record the order?) And are we allowed to choose the same object more than once (say, by recording the result of each draw and returning the object to the urn), or not? There are various intermediate cases, like making words using the letters of a given word, where a letter may be repeated but not more often than it occurs in the original word.

Almost always, we assume that the objects are distinguishable, like the five spade cards. Under this assumption, the problem will be solved under the four possible combinations of the other assumptions in Chapter 3. What if they are indistinguishable? In this case, there is obviously only one way to select three red billiard balls from a set of five: any three red billiard balls are identical to any other three.

What difference does indistinguishability make? If the underlying objects are distinguishable, we can assume that they carry labels bearing the numbers $1, 2, \ldots, n$. In this case, we say that the configurations we are counting are *labelled*. If the n underlying objects are indistinguishable, we are counting *unlabelled* things. An example will illustrate the difference.

Suppose that we are interested in n towns; some pairs of towns are joined by a direct road, others not. We are not concerned with the geographical locations, only in whether the towns are connected or not. (This is described by the structure known as a *graph*.[11] See Chapter 11 for more about graphs.) Figure 1 shows the eight labelled graphs for $n = 3$. If the towns are indistinguishable, then the second,

---

[10] According to folklore, it is impossible to count the Rollright Stones consistently.

[11] This usage of the term is quite different from the sense in the phrase 'the graph of $y = \sin x$'. Some people distinguish the two meanings by different pronunciation, with a short a for the sense used here.

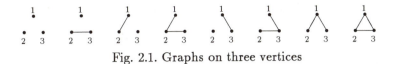

Fig. 2.1. Graphs on three vertices

third and fifth graphs are identical, as are the fourth, sixth and seventh. So there are just four unlabelled graphs with $n = 3$.

In general, let $f(n)$ and $g(n)$ denote the numbers of labelled and unlabelled configurations, respectively, with n underlying objects. Then two labelled configurations will be regarded as identical as unlabelled configurations if and only if there is a *permutation* of $\{1, 2, \ldots, n\}$ which carries one to the other. (For example, the cyclic permutation $1 \mapsto 2 \mapsto 3 \mapsto 1$ carries the second graph in Fig. 1 to the fifth.) So at most $n!$ labelled configurations collapse into a single unlabelled one, and we have

$$f(n)/n! \le g(n) \le f(n).$$

Now there are two possibilities for the 'order of magnitude' behaviour.

If $f(n)$ grows much more rapidly than $n!$, then the left and right hand sides of this equation are not so very far apart, and we have a reasonable estimate for $g(n)$. For example, we saw that there are 2^{2^n} families of subsets of the n-element set X. The number of permutations is insignificant by comparison, so it doesn't matter very much whether the elements of X are distinguishable or not, that is, whether we count labelled or unlabelled families.

But if this doesn't occur, then more care is needed. There are just 2^n subsets of the n-element set X, and this function grows more slowly than $n!$. In this case, we can count unlabelled sets another way. If all elements of X are indistinguishable, then the only thing we can tell about a subset of X is its cardinality; two subsets containing the same number of elements are equivalent under a permutation. So the number of unlabelled subsets is $n + 1$, since the cardinality of a subset can take any one of the $n + 1$ values $0, 1, 2, \ldots, n$.

This theme can be refined, using the concepts of *permutation group* and *cycle index*. These are more advanced topics, and will be treated in Chapter 15.

2.6. Double counting

We come now to a deceptively simple but enormously important counting principle:

> *If the same set is counted in two different ways, the answers are the same.*

This is analogous to finding the sum of all the entries in a matrix by adding the row totals, and then checking the calculation by adding the column totals.

The principle is best illustrated by applications (of which there will be many later) — here is one:

> **(2.6.1) Handshaking Lemma**
> *At a convention, the number of delegates who shake hands an odd number of times is even.*

To show this, let $D_1, \ldots, D_n$ be the delegates. We apply double counting to the set of ordered pairs (D_i, D_j) for which D_i and D_j shake hands with each other at the convention. Let x_i be the number of times that D_i shakes hands, and y the total number of handshakes that occur. On the one hand, the number of pairs is $\sum_{i=1}^{n} x_i$, since for each D_i the number of choices of D_j is equal to x_i. On the other hand, each handshake gives rise to two pairs (D_i, D_j) and (D_j, D_i); so the total is $2y$. Thus

$$\sum_{i=1}^{n} x_i = 2y.$$

But, if the sum of n numbers is even, then evenly many of the numbers are odd. (If we add an odd number of odd numbers and any number of even numbers, the answer will be odd.)

The double counting principle is usually applied to counting ordered pairs. For lovers of formalism, here is a general result, which encapsulates most of the applications we will make of it.

(2.6.2) Proposition. Let $A = \{a_1, \ldots, a_m\}$ and $B = \{b_1, \ldots, b_n\}$ be sets. Let S be a subset of $A \times B$. Suppose that, for $i = 1, \ldots, m$, the element a_i is the first component of x_i pairs in S, while, for $j = 1, \ldots, n$, the element b_j is the second component of y_j pairs in S. Then

$$|S| = \sum_{i=1}^{m} x_i = \sum_{j=1}^{n} y_j.$$

Often it happens that x_i is constant (say x) and y_j is also constant (say y). Then we have

$$mx = ny.$$

2.7. Appendix on set notation

The basic notation for sets is listed here. If A and B are sets, then we write $x \in A$ if x is an element of A, $x \notin A$ otherwise. Also

$|A|$ (the *cardinality* of A) is the number of elements in A;

$A \cup B$ (the *union*) is the set of elements in A or B (or both);

$A \cap B$ (the *intersection*) is the set of elements in both A and B;

$A \setminus B$ (the *difference*) is the set of elements in A but not B;

$A \triangle B$ (the *symmetric difference*) is the set of elements in just one of the two sets;

$A \subseteq B$ if every element of A belongs to B;

$A = B$ if A and B have exactly the same elements.

So, for example,

$$A \triangle B = (A \setminus B) \cup (B \setminus A) = (A \cup B) \setminus (A \cap B),$$
$$|A \cup B| + |A \cap B| = |A| + |B|.$$

The notation $\{x : P\}$ means the set of all elements x having property P. So, for example,

$$A \cup B = \{x : x \in A \text{ or } x \in B\}.$$

Similarly, $\{x, y\}$ is the set consisting of the elements x and y only. If $x \neq y$, it is called an *unordered pair*, since $\{x, y\} = \{y, x\}$. By contrast, the *ordered pair* (x, y) has the property that $(x, y) = (u, v)$ if and only if $x = u$ and $y = v$. (We permit $x = y$ here.) This is familiar from Cartesian coordinates of points in the Euclidean plane.

The *Cartesian product* $A \times B$ is the set of all ordered pairs (a, b), with $a \in A$ and $b \in B$. Similarly for more than two factors. For example, we write A^n for the set of ordered n-tuples of elements of A, for any positive integer n. We have

$$|A \times B| = |A| \cdot |B|,$$
$$|A^n| = |A|^n.$$

Until last century, a function was something described by a formula (typically a polynomial or a power series); it was the ambiguity in this definition which led to the modern version. A *function* f from A to B is a subset of $A \times B$ with the property that, for any $a \in A$, there is a unique $b \in B$ such that $(a, b) \in f$. If $(a, b) \in f$, we write $f(a) = b$.[12] Usually there is a rule for calculating $b = f(a)$ from a, but this is not part of the definition.

If $A = \{a_1, a_2, \ldots, a_n\}$ with $a_1, \ldots, a_n$ all distinct, then any function $f : A \to B$ can be specified by giving the n-tuple of values $(f(a_1), f(a_2), \ldots, f(a_n))$. Thus the number of functions from A to B is $|B|^{|A|}$. Motivated by this, the set of functions from A to B is sometimes written B^A, so that $|B^A| = |B|^{|A|}$.

The *power set* $\mathcal{P}(A)$ is the set of all subsets of A. Any subset X of A is specified by its *characteristic function*, the function $f_X : A \to \{0, 1\}$ defined by

$$f_X(a) = \begin{cases} 1 & \text{if } a \in X; \\ 0 & \text{if } a \notin X. \end{cases}$$

(Two subsets are equal if and only if their characteristic functions are equal.) So there are as many subsets of A as there are functions from A to $\{0, 1\}$; that is, $|\mathcal{P}(A)| = 2^{|A|}$.

---

[12] This definition is very familiar, despite appearances. You probably visualise 'the function $y = x^2$' in terms of its graph in the Euclidean plane with coordinates (x, y); and the graph consists of precisely those ordered pairs (x, y) for which $y = x^2$. In other words, the graph *is* the function!

2.8. Exercises

1. Criticise the following proof that 1 is the largest natural number.

> *Let n be the largest natural number, and suppose than $n \neq 1$. Then $n > 1$, and so $n^2 > n$; thus n is not the largest natural number.*

2. Prove by induction that the Odometer Principle with base b does indeed give the representation $x_{n-1} \ldots x_1 x_0$ for the natural number

$$N = x_{n-1} b^{n-1} + \cdots + x_1 b + x_0.$$

3. (a) Prove by induction that

$$n! > \left(\frac{n}{e}\right)^n$$

for $n \geq 1$. (You may use the fact that $(1 + \frac{1}{n})^n < e$ for all n.)

   (b) Use the arithmetic–geometric mean inequality[13] to show that $n! < \left(\frac{n+1}{2}\right)^n$ for $n > 1$, and deduce that

$$n! < e\left(\frac{n}{2}\right)^n$$

for $n \geq 1$.

4. (a) Prove that $\log x$ grows more slowly than x^c for any positive number c.

   (b) Prove that, for any $c, d > 1$, we have $c^x > x^d$ for all sufficiently large x.

5. (a) We saw that there are $2^{2^3} = 256$ labelled families of subsets of a 3-set. How many unlabelled families are there?

   (b) Prove that the number $F(n)$ of unlabelled families of subsets of an n-set satisfies $\log_2 F(n) = 2^n + O(n \log n)$.

6. Verify that the numbers of graphs are given in Table 1 for $n \leq 5$.

n	2	3	4	5
labelled	2	8	64	1024
unlabelled	2	4	11	34

Table 2.1. Graphs

7. Suppose that an urn contains four balls with different colours. In how many ways can three balls be chosen? As in the text, we may be interested in the order of choice, or not; and we may return balls to the urn, allowing repetitions, or not. Verify the results of Table 2.

	order important	order unimportant
repetition allowed	64	20
repetition not allowed	24	4

Table 2.2. Selections

---

[13] The arithmetic–geometric mean inequality states that the arithmetic mean of a list of positive numbers is greater than or equal to their geometric mean, with equality only if all the numbers are equal. Can you prove it? (HINT: Do the special case when all but one of the numbers are equal by calculus, and then the general case by induction.)

8. A *Boolean function* takes n arguments, each of which can have the value TRUE or FALSE. The function takes the value TRUE or FALSE for each choice of values of its arguments. Prove that there are 2^{2^n} different Boolean functions. Why is this the same as the number of families of sets?

9. Logicians define a natural number to be the set of all its predecessors: so 3 is the set $\{0, 1, 2\}$. Why do they have to start counting at 0?

10. A function f has *polynomial growth* of degree d if there exist positive real numbers a and b such that $an^d < f(n) < bn^d$ for all sufficiently large n. Suppose that f has polynomial growth, and g has exponential growth with exponential constant greater than 1 (as defined in the text). Prove that $f(n) < g(n)$ for all sufficiently large n. If $f(n) = 10^6 n^{10^6}$ and $g(n) = (1.000001)^n$, how large is 'sufficiently large'?

11. Let $\mathcal{B}$ be a set of subsets of the set $\{1, 2, \ldots, v\}$, containing exactly b sets. Suppose that
  - every set in $\mathcal{B}$ contains exactly k elements;
  - for $i = 1, 2, \ldots, v$, the element i is contained in exactly r members of $\mathcal{B}$.
Prove that $bk = vr$.
  Give an example of such a system, with $v = 6$, $k = 3$, $b = 4$, $r = 2$.

12. The 'Russian peasant algorithm' for multiplying two natural numbers m and n works as follows.[14]

(2.7.3) Russian peasant multiplication

to multiply two natural numbers m and n

Write m and n at the head of two columns.
REPEAT the sequence
  - *halve the last number in the first column (discarding the remainder) and write it under this number;*
  - *double the last number in the second column and write it under this number;*
UNTIL *the last number in the first column is 1.*

For each even number in the first column, delete the adjacent entry in the second column. Now add the remaining numbers in the second column. Their sum is the answer.

For example, to calculate 18×37:

---

[14] No tables needed, except two times!

18	~~37~~
9	74
4	~~148~~
2	~~296~~
1	592

666

Table 2.3. Multiplication

PROBLEMS. (i) Prove that this method gives the right answer.

(ii) What is the connection with the primary school method of long multiplication? HINT FOR (i) AND (ii): Express m (and n) to the base 2.

(iii) Suppose we change the algorithm by squaring (instead of doubling) the numbers in the second column, and, in the last step, multiplying (rather than adding) the undeleted numbers. Prove that the number calculated is n^m. How many multiplications does this method require?

13. According to the Buddha,

> *Scholars speak in sixteen ways of the state of the soul after death. They say that it has form or is formless; has and has not form, or neither has nor has not form; it is finite or infinite; or both or neither; it has one mode of consciousness or several; has limited consciousness or infinite; is happy or miserable; or both or neither.*

How many different possible descriptions of the state of the soul after death do you recognise here?

14. The library of Babel[15] consists of interconnecting hexagonal rooms. Each room contains twenty shelves, with thirty-five books of uniform format on each shelf. A book has four hundred and ten pages, with forty lines to a page, and eighty characters on a line, taken from an alphabet of twenty-five orthographical symbols (twenty-two letters, comma, period and space). Assuming that one copy of every possible book is kept in the library, how many rooms are there?

15. COMPUTER PROJECT. Develop a suite of subroutines for performing arithmetic on integers of arbitrary size, regarded as strings of digits. (You should deal with input and output, arithmetic operations — note that division should return a quotient and a remainder — and comparisons. You might continue with exponentiation and factorials, as well as various combinatorial functions to be defined later.)

---

[15] Jorge Luis Borges, *Labyrinths* (1964).

3. Subsets, partitions, permutations

The emphasis on mathematical methods seems to be shifted more towards combinatorics and set theory — and away from the algorithm of differential equations which dominates mathematical physics.

J. von Neumann & O. Morgenstern,
Theory of Games and Economic Behaviour (1944).

The process is directed always towards analysing and separating the material into a collection of discrete counters, with which the detached intellect can make, observe and enjoy a series of abstract, detailed, artificial patterns of words and images (you may be reminded of the New Criticism)...

Elizabeth Sewell, 'Lewis Carroll and T. S. Eliot as Nonsense Poets'
in Neville Braybrooke (ed.), *T. S. Eliot* (1958).

TOPICS: Subsets, binomial coefficients, Pascal's Triangle, Binomial Theorem; [congruences of binomial coefficients]; permutations, ordered and unordered selections, cycle decomposition of a permutation; estimates for the factorial function; relations; [finite topologies; counting trees]; partitions, Bell numbers

TECHNIQUES: Binomial coefficient identities; use of double counting; estimates via integration

ALGORITHMS: Sequential and recursive generation of combinatorial objects; [Prüfer's algorithm]

CROSS-REFERENCES: Odometer Principle; double counting (Chapter 2); recurrence relations (Chapter 4)

This chapter is about the central topic of 'classical' combinatorics, what is often referred to as 'Permutations and Combinations'. Given a set with n elements, how many ways can we choose a selection of its elements, with or without respect to the order of selection, or divide it up into subsets? We'll define the various numbers involved, and prove some of their properties; but these echo through subsequent chapters.

3.1. Subsets

How many subsets does a set of n elements have?

The number of subsets is 2^n. There are several different ways to see this. Perhaps most easily, for each of the n elements of the set, there are two choices in building a subset (viz., put the element in, or leave it out); all combinations of these choices are possible, giving a total of 2^n.

Implicitly, this argument sets up a bijection between the subsets of a set X and the functions from X to $\{0,1\}$. The function f_Y corresponding to the subset Y is defined by the rule

$$f_Y(x) = \begin{cases} 1 & \text{if } x \in Y \\ 0 & \text{if } x \notin Y. \end{cases}$$

Conversely, a function f corresponds to the set $Y = \{x \in X : f(x) = 1\}$. The function f_Y is called the *characteristic function* or *indicator function* of Y.

If $X = \{0, 1, \ldots, n-1\}$, then we can represent a function $f : X \to \{0,1\}$ by the n-tuple $(f(0), f(1), \ldots, f(n-1))$ of its values. Thus subsets of X correspond to n-tuples of zeros and ones.

We can take this one step further, and regard the n-tuple as the base 2 representation of an integer

$$N = f(n-1)2^{n-1} + \ldots + f(1)2 + f(0),$$

as described in Chapter 2. Each n-tuple corresponds to a unique integer; the smallest is 0 (corresponding to the empty set), and the largest is $2^{n-1} + \ldots + 2 + 1 = 2^n - 1$ (corresponding to the whole set X), and every integer between represents a unique subset. So the number of subsets is equal to the number of integers between 0 and $2^n - 1$ (inclusive), namely 2^n.

Note that this method gives a convenient numbering of the subsets of the set $\{0, \ldots, n-1\}$: the kth subset X_k corresponds to the integer k, where $0 \le k \le 2^n - 1$. The set X_k is easily recovered by writing k to base 2. The numbering has some further virtues. For example, the set X_k depends only on k, and not on the particular value of n used; replacing n by a larger value doesn't change it. So we get a unique set X_k of non-negative integers corresponding to each non-negative integer k. For another nice property, see Exercise 2.

Yet another proof of the formula for the number $F(n)$ of subsets of an n-set is obtained by noting that we can find all subsets of $\{1, \ldots, n+1\}$ by taking all subsets of $\{1, \ldots, n\}$ and extending each in the two possible ways — either do nothing, or add the element $n+1$. So $F(n+1) = 2F(n)$. This is a *recurrence relation*, by which the value of F is determined by its values on smaller arguments. Recurrence relations form the subject of the next chapter.

3.2. Subsets of fixed size

Let n and k be non-negative integers, with $0 \le k \le n$. The *binomial coefficient* $\binom{n}{k}$ is defined to be the number of k-element subsets of a set of n elements. (The number obviously doesn't depend on which n-element set we use.) This number is often written as ^nC_k, and is read 'n choose k'. It is called a *binomial coefficient* (for reasons to be elaborated later).

(3.2.1) Formula for binomial coefficients

$$\binom{n}{k} = \frac{n(n-1)\ldots(n-k+1)}{k(k-1)\ldots 1} = \frac{n!}{k!(n-k)!}$$

Note that $\binom{n}{0} = 1$ (the empty set) and $\binom{n}{n} = 1$ (the whole set) — the proposed formula is correct in these cases, in view of the convention that $0! = 1$ (see Section 2.3).

As suggested by the name, we prove this by counting choices. Given a set X of n elements, in how many ways can we choose a set of k of them? Clearly there are n possible choices for the 'first' element, $(n-1)$ choices for the 'second', ... , and $(n-k+1)$ choices for the 'k^{th}'; in total, $n(n-1)\ldots(n-k+1)$. But we put the terms 'first', 'second', etc., in quotes because a subset has no distinguished first, second, ... element. In other words, if the same k elements were chosen in a different order, the same subset would result. So we must divide this number by the number of orders in which the k elements could have been chosen. Arguing exactly as before, there are k choices for which one is 'first', $(k-1)$ for which is 'second', and so on. Division gives the middle expression in the box. Now the third expression is equal to the second because $n(n-1)\ldots(n-k+1) = n!/(n-k)!$; the denominator cancels all the factors from $n-k$ on in the numerator.

Once we have a formula, there are two possible ways to prove assertions or identities about binomial coefficients. There is a combinatorial proof, arguing from the definition (we will interpret $\binom{n}{k}$ as the number of ways of choosing a team of k players from a class of n pupils); and there is an algebraic proof, from the formula. We give a few simple ones.

(3.2.2) Fact.

$$\binom{n}{k} = \binom{n}{n-k}.$$

FIRST PROOF. Choosing a team of k from a class of n is equivalent to choosing the $n-k$ people to leave out.

SECOND PROOF. It's obvious from the last formula in the box.

(3.2.3) Fact.

$$k\binom{n}{k} = n\binom{n-1}{k-1}.$$

FIRST PROOF. We choose a team of k and designate one team member as captain. There are $\binom{n}{k}$ possible teams and, for each team, there are k choices for the captain.

Alternatively, we could choose the captain first (in n possible ways), and then the remainder of the team ($k - 1$ from the remaining $n - 1$ class members).

Note that this is an application of the 'double counting' principle described in Section 2.6.

SECOND PROOF. Try it yourself!

You will find that the SECOND PROOFs above probably come more naturally to you. For this reason, I'll concentrate on the combinatorial style of proof for the next couple of results. Remember that the algebraic proof is not always appropriate or even possible — sometimes we won't have a formula for the numbers in question, or the formula is too complex. (See the discussion of Stirling numbers in Section 5.3 for examples of this.)

(3.2.4) Fact.

$$\binom{n+1}{k} = \binom{n}{k-1} + \binom{n}{k}.$$

PROOF. We have a class of $n+1$ pupils, one of whom is somehow 'distinguished', and wish to pick a team of k. We could either include the distinguished pupil (in which case we must choose the other $k - 1$ team members from the remaining n pupils), or leave him out (when we have to choose the whole team from the remaining n).

(3.2.5) Fact.

$$\sum_{k=0}^{n} \binom{n}{k} = 2^n.$$

PROOF. This one is easy — there are 2^n subsets altogether (of arbitrary size).

(3.2.6) Fact.

$$\sum_{k=0}^{n} \binom{n}{k}^2 = \binom{2n}{n}.$$

PROOF. The right-hand side is the number of ways of picking a team of n from a class of $2n$. Now suppose that, of the $2n$ pupils, n are girls and n are boys. In how many ways can we pick a team of k girls and $n - k$ boys? Obviously this number is $\binom{n}{k}\binom{n}{n-k}$, which is equal to $\binom{n}{k}^2$, by Fact 3.2.2. The result now follows.

The definition of the binomial coefficient $\binom{n}{k}$ actually makes sense for any non-negative integers n and k: if $k > n$, then there are no k-subsets of an n-set, and $\binom{n}{k} = 0$. The (first) formula gives the right answer, since if $k > n$ then one of the factors in the numerator is zero. (This cannot be assumed, since the argument we gave is only valid if $k \leq n$.) However, the second formula makes no sense (unless, very dubiously, we assume that the factorial of a negative integer is infinite!).

Facts 3.2.2–4 above remain valid with this more general interpretation. (You should check this.)

Sometimes it is convenient to widen the definition still further. For example, if $k < 0$, we should define $\binom{n}{k} = 0$, in order that Fact 3.2.2 should hold in general. We'll see in Chapter 4 that it is possible to relax the requirement that n is a non-negative integer even further. The most general definition, using the formula, works for any *real number* n and any *integer* k: we set

$$\binom{n}{k} = \begin{cases} \dfrac{n(n-1)\dots(n-k+1)}{k!} & \text{if } k \geq 0; \\ 0 & \text{if } k < 0. \end{cases}$$

3.3. The Binomial Theorem and Pascal's Triangle

Fact 3.2.5 above can be generalised to the celebrated *Binomial Theorem*.[1] A *binomial* is a polynomial with two terms; the Binomial Theorem states that, if a power of a binomial is expanded, the coefficients in the resulting polynomial are the binomial coefficients (from which, obviously, they get their name).

(3.3.1) Binomial Theorem

$$(1 + t)^n = \sum_{k=0}^{n} \binom{n}{k} t^k.$$

FIRST PROOF. It's clear that $(1 + t)^n$ is a polynomial in t of degree n. To find the coefficient of t^k, consider the product

$$(1 + t)(1 + t)\dots(1 + t) \qquad (n \text{ factors}).$$

The expansion is obtained by choosing either 1 or t from each factor in all possible ways, multiplying the chosen terms, and adding all the results. A term t^k is obtained when t is chosen from k of the factors, and 1 from the other $n - k$ factors. There are $\binom{n}{k}$ ways of choosing these k factors; so the coefficient of t^k is $\binom{n}{k}$, as claimed.

SECOND PROOF. The theorem can be proved by induction on n. It is trivially true for $n = 0$. Assuming the result for n, we have

$$(1 + t)^{n+1} = (1 + t)^n \cdot (1 + t)$$
$$= \left(\sum_{k=0}^{n} \binom{n}{k} t^k \right) \cdot (1 + t);$$

the coefficient of t^k on the right is $\binom{n}{k-1} + \binom{n}{k}$ (the first term coming from $t^{k-1} \cdot t$ and the second from $t^k \cdot 1$); and

$$\binom{n}{k-1} + \binom{n}{k} = \binom{n+1}{k},$$

by Fact 3.2.4.

---

[1] Proved by Sir Isaac Newton in about 1666.

The Binomial Theorem allows the possibility of completely different proofs of properties of binomial coefficients, some of which are quite difficult to prove in other ways. Here are a couple of examples. First, a proof of Fact 3.2.3.

Differentiate the Binomial Theorem with respect to t:

$$n(1+t)^{n-1} = \sum_{k=1}^{n} k \binom{n}{k} t^{k-1}.$$

The coefficients of t^{k-1} on the left and right of this equation are $n\binom{n-1}{k-1}$ and $k\binom{n}{k}$ respectively.

(3.3.2) Fact. *For $n > 0$, the numbers of subsets of an n-set of even and of odd cardinality are equal (viz., 2^{n-1}).*

PROOF. Put $t = -1$ in the Binomial Theorem to obtain

$$0 = (1-1)^n = \sum_{k=0}^{n} \binom{n}{k}(-1)^k,$$

hence

$$\sum_{\substack{0 \le k \le n \\ k \text{ even}}} \binom{n}{k} = \sum_{\substack{0 \le k \le n \\ k \text{ odd}}} \binom{n}{k}.$$

But the two sides of this equation are just the numbers of subsets of even, resp. odd, cardinality.

If n is odd, then k is even if and only if $n-k$ is odd; so complementation sets up a bijection between the subsets of even and odd size, proving the result. However, in general, a different argument is required. The map $X \mapsto X \triangle \{n\}$ (that is, if $n \in X$, then remove it; otherwise put n into X) is a bijection on subsets of $\{1, \ldots, n\}$ which changes the cardinality by 1, and hence reverses the parity; so there are equally many sets of either parity.

The argument can be refined to calculate the number of sets whose size lies in any particular congruence class. I illustrate by calculating the number of sets of size divisible by 4. I assume that n is a multiple of 8. (The answer takes different forms depending on the congruence class of n mod 8.)

(3.3.3) Proposition. *If n is a multiple of 8, then the number of sets of size divisible by 4 is $2^{n-2} + 2^{(n-2)/2}$.*

For example, if $n = 8$, the number of such sets is $\binom{8}{0} + \binom{8}{4} + \binom{8}{8} = 2^6 + 2^3$.

PROOF. We let A be the required number, and B the number of sets whose size is congruent to 2 (mod 4). By Fact 3.3.2, $A + B = 2^{n-1}$.

Now substitute $t = \mathrm{i}$ in the Binomial Theorem. Note that $1 + \mathrm{i} = \sqrt{2}\mathrm{e}^{\mathrm{i}\pi/4}$, and so (since n is a multiple of 8), $(1+\mathrm{i})^n = 2^{n/2}$. Thus

$$2^{n/2} = \sum_{k=0}^{n} \binom{n}{k} \mathrm{i}^k.$$

Take the real part of the right-hand side, noting that $\mathrm{i}^k = 1, \mathrm{i}, -1, -\mathrm{i}$ according as $k \equiv 0, 1, 2$ or 3 (mod 4). We obtain $A - B = 2^{n/2}$. From this and the expression for $A + B$ above, we obtain the value of A (and that of B).

REMARK. By taking the imaginary part of the equation, we find the numbers of sets with size congruent to 1, or to 3, mod 4.

The binomial coefficients are often written out in the form of a triangular array, known as *Pascal's Triangle*:[2]

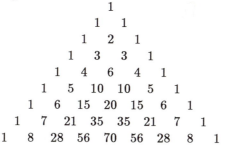

$$
\begin{array}{ccccccccc}
 & & & & 1 & & & & \\
 & & & 1 & & 1 & & & \\
 & & 1 & & 2 & & 1 & & \\
 & 1 & & 3 & & 3 & & 1 & \\
1 & & 4 & & 6 & & 4 & & 1 \\
\end{array}
$$

```
              1
            1   1
          1   2   1
        1   3   3   1
      1   4   6   4   1
    1   5  10  10   5   1
  1   6  15  20  15   6   1
1   7  21  35  35  21   7   1
1   8  28  56  70  56  28   8   1
```

Thus, $\binom{n}{k}$ is the kth element in the nth row, where both the rows and the elements in them are numbered starting at zero. Fact 3.2.4 shows that each internal element of the triangle is the sum of the two elements above it (i.e., above and to the left and right). Moreover, the borders of the triangle are filled with the number 1 (since $\binom{n}{0} = \binom{n}{n} = 1$). With these two rules, it is very easy to continue the triangle as far as necessary. This suggests that Pascal's Triangle is an efficient tool for calculating binomial coefficients. (See Exercise 8.)

3.4. Project: Congruences of binomial coefficients

A popular school project is to examine the patterns formed by the entries of Pascal's Triangle modulo a prime. For example, the first eight rows mod 2 are as follows:

```
              1
            1   1
          1   0   1
        1   1   1   1
      1   0   0   0   1
    1   1   0   0   1   1
  1   0   1   0   1   0   1
1   1   1   1   1   1   1   1
```

If T consists of the first 2^n rows, then the first 2^{n+1} rows look like $\begin{smallmatrix} T & & \\ T & 0 & T \end{smallmatrix}$.

Thus the pattern has a 'self-similarity' of the kind more usually associated with fractals than with combinatorics! A similar pattern holds for congruence modulo other primes, except that the copies of T are multiplied by the entries of the p-rowed Pascal triangle.

---

[2] Not surprisingly, this object was known long before Pascal. I owe to Robin Wilson the information that it appears in the works of the Majorcan theologian Ramon Llull (1232–1316). Llull also gives tables of combinations and mechanical devices for generating them, complete graphs, trees, etc. However, combinatorics for him was only a tool in his logical system, and logic was firmly subservient to theology. In his first major work, a commentary on Al-Ghazali, he says, 'We will speak briefly of Logic, since we should speak of God.'

The mathematical formulation is remarkably simple. It was discovered by Lucas in the nineteenth century.

(3.4.1) Lucas' Theorem

Let p be prime, and let $m = a_0 + a_1 p + \ldots + a_k p^k$, $n = b_0 + b_1 p + \ldots + b_k p^k$, where $0 \leq a_i, b_i < p$ for $i = 0, \ldots, k-1$. Then

$$\binom{m}{n} \equiv \prod_{i=0}^{k} \binom{a_i}{b_i} \pmod{p}.$$

NOTE. We assume here the usual conventions for binomial coefficients, in particular, $\binom{a}{b} = 0$ if $a < b$.

PROOF. It suffices to show that, if $m = cp + a$ and $n = dp + b$, where $0 \leq a, b < p$, then

$$\binom{m}{n} \equiv \binom{c}{d} \binom{a}{b} \pmod{p}.$$

For $a = a_0$, $b = b_0$, and $c = a_1 + \ldots + a_k p^{k-1}$, $d = b_1 + \ldots + b_k p^{k-1}$; and then induction finishes the job.

This assertion can be proved directly, but there is a short proof using the Binomial Theorem. The key is the fact that, if p is prime, then

$$(1 + t)^p \equiv 1 + t^p \pmod{p}.$$

This is because each binomial coefficient $\binom{p}{i}$, for $1 \leq i \leq p - 1$, is a multiple of p, so all intermediate terms in the Binomial Theorem vanish mod p. (For $\binom{p}{i} = p!/i!(p-i)!$, and p divides the numerator but not the denominator.) Thus (congruence mod p):

$$(1 + t)^m = (1 + t)^{cp} (1 + t)^a$$
$$\equiv (1 + t^p)^c (1 + t)^a$$
$$= \sum_{i=0}^{c} \binom{c}{i} t^{pi} \cdot \sum_{j=0}^{a} \binom{a}{j} t^j.$$

Since $0 \leq a, b < p$, the only way to obtain a term in $t^n = t^{dp+b}$ in this expression is to take the term $i = d$ in the first sum and the term $j = b$ in the second; this gives

$$\binom{m}{n} \equiv \binom{c}{d} \binom{a}{b} \pmod{p},$$

as required.

3.5. Permutations

There are two ways of regarding a permutation, which I will call 'active' and 'passive'. Let X be a finite set. A *permutation* of X, in the active sense, is a one-to-one mapping from X to itself. For the passive sense, we assume that there is a natural ordering of the elements of X, say $\{x_1, x_2, \ldots, x_n\}$. (For example, X

might be $\{1, 2, \ldots, n\}$.) Then the passive representation of the permutation π is the ordered n-tuple $(\pi(x_1), \pi(x_2), \ldots, \pi(x_n))$.[3]

In the preceding paragraph, I wrote $\pi(x)$ for the result of applying the function π to the element x. However, in the algebraic theory of permutations, we often have to *compose* permutations, i.e., apply one and then the other. In order that the result of applying first π_1 and then π_2 can be called $\pi_1\pi_2$, it is more natural to denote the image of x under π as $x\pi$. Then

$$x(\pi_1\pi_2) = (x\pi_1)\pi_2,$$

which looks like a kind of associative law![4]

As is (I hope) familiar to you, the set of all permutations of $\{1, \ldots, n\}$, equipped with the operation of composition, is a group. It is known as the *symmetric group* of degree n, denoted by S_n (or sometimes $\mathrm{Sym}(n)$). The symmetric groups form one of the oldest and best-loved families of groups.

From now on, we take $X = \{1, 2, \ldots, n\}$.

A permutation π can be represented in so-called *two-line notation* as

$$\begin{pmatrix} 1 & 2 & \ldots & n \\ 1\pi & 2\pi & \ldots & n\pi \end{pmatrix}.$$

The top row of this symbol can be in any order, as long as $x\pi$ is directly under x for all x. If the top row is in natural order, then the bottom row is the passive form of the permutation.

(3.5.1) Proposition. *The number of permutations of an n-set is $n!$.*

PROOF. Take the top row of the two-rowed symbol to be $(1\ 2\ \ldots\ n)$. Then there are n choices for the first element in the bottom row; $n - 1$ choices for the second (anything except the first chosen element); and so on.

Note that this formula is correct when $n = 0$: the only permutation of the empty set is the 'empty function'.

There is another, shorter, representation of a permutation, the *cycle form*. A *cycle*, or *cyclic permutation*, is a permutation of a set X which maps

$$x_1 \mapsto x_2 \mapsto \ldots \mapsto x_n \mapsto x_1,$$

where $x_1, \ldots, x_n$ are all the elements of X in some order. It is represented as $(x_1\ x_2\ \ldots\ x_n)$ (not to be confused with the passive form of a permutation!) The cycle is not unique: we can start at any point, so $(x_i\ \ldots\ x_n\ x_1\ \ldots\ x_{i-1})$ represents the same cycle.

---

[3] In the nineteenth century, it was more usual to refer to a passive permutation as a *permutation*, synonymous with 'rearrangement'. An active permutation was called a *substitution*.

[4] We say that permutations *act on the right* if they compose according to this rule.

(3.5.2) Proposition. *Any permutation can be written as the composition of cycles on pairwise disjoint subsets. The representation is unique, apart from the order of the factors, and the starting-points of the cycles.*

The proof of this theorem is algorithmic. Let π be a permutation of X.

(3.5.3) Decomposition into disjoint cycles

WHILE *there is a point of X not yet assigned to a cycle,*
- *choose any such point x;*
- *let m be the least positive integer such that $x\pi^m = x$;*
- *construct the cycle $(x\ x\pi\ \ldots\ x\pi^{m-1})$.*

RETURN *the product of all cycles constructed.*

PROOF. In the algorithm, we use the notation π^m for the composition of m copies of π. We first have to show that the construction makes sense, that is, $(x\ x\pi\ \ldots\ x\pi^{m-1})$ really is a cycle. This could only fail if the sequence of elements contains a repetition. But, if $x\pi^i = x\pi^j$, where $0 \le i < j < m$, then (because π is one-to-one) it holds that $x = x\pi^{j-i}$; but this contradicts the choice of m as least integer such that $x\pi^m = x$.

Next, we establish that the cycles use disjoint sets of points. Suppose that $x\pi^i = y\pi^j$, and suppose that x is chosen before y. If $y\pi^m = y$, then $x\pi^{i+m-j} = y\pi^m = y$, contradicting the fact that y (when chosen) doesn't already lie in a cycle.

It is clear that any point of X lies in one of the chosen cycles. Finally, the composition of all these cycles is equal to π. For, given a point z, there is a unique y and i such that $z = y\pi^i$. Then the cycle containing y agrees with π in mapping z to $y\pi^{i+1}$, and all the other cycles have no effect on z.

EXAMPLE. The permutation $\left(\begin{smallmatrix} 1 & 2 & 3 & 4 & 5 & 6 \\ 3 & 6 & 4 & 1 & 5 & 2 \end{smallmatrix}\right)$, in cycle notation, is $(1\ 3\ 4)(2\ 6)(5)$. This is just one of 36 different expressions: there are $3! = 6$ ways to order the three cycles, and $3 \cdot 2 \cdot 1 = 6$ choices of starting points.

3.6. Estimates for factorials

Since many kinds of combinatorial objects (for example, binomial coefficients) can be expressed in terms of factorials, it is often important to know roughly how large $n!$ is. In Exercise 3 of the last chapter, upper and lower bounds were found by *ad hoc* methods. In this section, a more systematic approach will yield better estimates. I will prove:

(3.6.1) Theorem.

$$n \log n - n + 1 \le \log n! \le n \log n - n + (\log(n+1) + 2 - 2\log 2).$$

From this, it follows that

$$\log n! = n \log n - n + O(\log n).$$

This is weaker than an asymptotic estimate for $n!$ itself: the exponentials of the upper and lower bounds are $e(n/e)^n$ and $\frac{1}{4}(n+1)e^2(n/e)^n$, which differ by a factor of $(n+1)e/4$. A more precise estimate (not proved here) is:

(3.6.2) Stirling's Formula

$$n! = \sqrt{2\pi n}\left(\frac{n}{e}\right)^n\left(1 + O\left(\frac{1}{n}\right)\right).$$

PROOF OF THEOREM. The main tool is shown in the pictures of Fig. 3.1.   Since

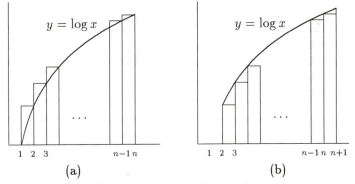

Fig. 3.1. Sums and integrals

$y = \log x$ is an increasing function of x for all positive x (its derivative, $1/x$, is positive), the tops of the rectangles in Fig. 3.1(a) all lie above the curve $y = \log x$, and those in Fig. 3.1(b) lie below the curve. In other words,

$$\int_1^n \log x \,\mathrm{d}x \le \sum_{i=2}^n \log i \le \int_2^{n+1} \log x \,\mathrm{d}x.$$

The term in the middle is $\log n!$. So

$$n \log n - n + 1 \le \log n! \le (n+1)\log(n+1) - (n+1) - 2\log 2 + 2.$$

The lower bound is exactly what is needed. For the upper bound, note that

$$\log(n+1) - \log n = \int_n^{n+1}\frac{\mathrm{d}x}{x} < \frac{1}{n},$$

so $n\log(n+1) < n\log n + 1$. Combining this with the upper bound, we obtain

$$\log n! \le \log(n+1) + n\log n - n - 2\log 2 + 2.$$

If you are interested, you could regard the proof of Stirling's Formula as a project.[5] A lower bound only slightly weaker than Stirling's is given in Exercise 12.

Exercise 13 gives an example of the use of Stirling's Formula to estimate a binomial coefficient. A weaker result can be obtained much more easily:

---

[5] An accessible proof can be found in Alan Slomson, *Introduction to Combinatorics* (1991).

(3.6.3) Proposition.

$$2^{2n}/(2n+1) \leq \binom{2n}{n} \leq 2^{2n}.$$

PROOF. Immediate from the fact that the $2n+1$ binomial coefficients $\binom{2n}{i}$, for $i = 0, \ldots, 2n$, have sum 2^{2n}, and the middle one is the largest.

3.7. Selections

> In how many ways can one select k objects from a set of size n?

The answer differs according to the terms of the problem, as we saw in Chapter 2. Specifically, is the order in which the objects are chosen significant (a *permutation*) or not (a *combination*)? and is the same object permitted to feature more than once in the selection, or not? (The term 'permutation' is used in a more general sense than in the last section: this is what might more accurately be called a 'partial permutation'.)

(3.7.1) Theorem. *The number of selections of k objects from a set of n objects is given by the following table:*

	Permutations and combinations	
	Order significant	*Order not significant*
Repetitions allowed	n^k	$\binom{n+k-1}{k}$
Repetitions not allowed	$n(n-1)\cdots(n-k+1)$	$\binom{n}{k}$

PROOF. For the column 'order significant', these are straightforward. If repetitions are allowed, there are n choices for each of the k objects; if repetitions are not allowed, there are n choices for the first, $n-1$ for the second, $n-k+1$ for the kth.

For 'order not significant', if repetitions are not allowed, we are counting the k-subsets of an n-set, which we already know how to do. The final entry is a bit harder.

(3.7.2) Lemma. *The number of choices of k objects from n with repetitions allowed and order not significant is equal to the number of ways of choosing n non-negative integers whose sum is k.*

PROOF. Given a choice of k objects from the set $a_1, \ldots, a_n$, let x_i be the number of times that the object a_i gets chosen. Then $x_i \geq 0$, $\sum_{i=1}^{n} x_i = k$. Conversely, given $(x_1, \ldots, x_n)$, form a selection by choosing object a_i just x_i times.

(3.7.3) Lemma. *The number of n-tuples of non-negative integers $x_1, \ldots, x_n$ with $x_1 + \ldots + x_n = k$ is $\binom{n+k-1}{n-1} = \binom{n+k-1}{k}$.*

PROOF. Consider the following correspondence. Put $n + k - 1$ spaces in a row, and fill $n - 1$ of them with markers. Let x_1 be the number of spaces before the first marker; x_i the number of spaces between the $(i-1)^{\text{st}}$ and i^{th} marker, for $2 \leq i \leq n - 1$; and x_n the number of spaces after the $(n-1)^{\text{st}}$ marker. Then $x_i \geq 0$, $\sum x_i = (n + k - 1) - (n - 1) = k$. Conversely, given $x_1, \ldots, x_n$, put markers after x_1 spaces, after x_2 more spaces, $\ldots$, after x_{n-1} more spaces (so that x_n spaces remain).

EXAMPLE. Suppose that $n = 3$, $k = 4$. The pattern of spaces and markers

$$\square \; \square \; \boxtimes \; \square \; \boxtimes \; \square$$

corresponds to the values $x_1 = 2$, $x_2 = 1$, $x_3 = 1$. Conversely, the values $(x_1, x_2, x_3) = (0, 0, 4)$ correspond to the pattern

$$\boxtimes \; \boxtimes \; \square \; \square \; \square \; \square .$$

Now the number of ways of choosing the positions of the markers is $\binom{n+k-1}{n-1} = \binom{n+k-1}{k}$, as claimed.

REMARK. Using the extended definition of binomial coefficients, the number of selections with repetitions allowed and order not significant can be written

$$(-1)^k \binom{-n}{k}.$$

A common puzzle is to find as many words as possible which can be formed from the letters of a given word. Of course, the crucial feature of this problem is that the words formed should belong to some given human language (i.e., they should be found in a standard dictionary). There are two possible strategies for this problem. We could either form all potential words (all permutations of whatever length), and look each one up in the dictionary; or go through the entire dictionary, and check whether each word uses a subset of the given letters. In order to decide which strategy is more efficient, we need to answer a theoretical question (how many permutations are there?) and some practical ones (how many words are there in the dictionary, and how fast can we look them up?)

We will solve a special case of the theoretical question. Assume that the n given letters are all distinct. We will call any ordered selection without repetition from these letters a *word* (without judging its legality — note in particular that we include the 'empty word' with no letters, which doesn't appear in any dictionary[6]).

(3.7.4) Proposition. *For $n > 0$, the number of ordered selections without repetition from a set of n objects is $\lfloor e \cdot n! \rfloor$, where e is the base of natural logarithms.*

---

[6] If it did, how would you look it up?

PROOF. The number $f(n)$ in question is obtained by summing the number of ordered selections of k elements for $k = 0, \ldots, n$:

$$\sum_{k=0}^{n} \frac{n!}{k!} = n! \sum_{k=0}^{n} \frac{1}{k!}.$$

From the familiar Taylor series for e^x, we see that

$$\sum_{k=0}^{\infty} \frac{1}{k!} = e.$$

So

$$e \cdot n! - f(n) = \frac{1}{n+1} + \frac{1}{(n+1)(n+2)} + \cdots$$
$$< \frac{1}{n+1} + \frac{1}{(n+1)^2} + \cdots$$
$$= \frac{1}{n} \leq 1;$$

so $f(n) = \lfloor e \cdot n! \rfloor$.

If the allowed letters contain repetitions, the problem is harder. It is possible to derive a general formula; but it is probably easier to argue *ad hoc* in a particular case, as the next example shows.

EXAMPLE. How many words can be made from the letters of the word FLEECE?

We count words according to the number of occurrences of the letter E. If there is at most one E, we can invoke the previous result: there are $24 + 24 + 12 + 4 + 1 = 65$ such words (including the empty word). If there are two Es, let us imagine first that they are distinguishable; then there are $120 + 3 \cdot 24 + 3 \cdot 6 + 2 = 212$ possibilities. (For example, with four letters altogether, we choose two of the remaining three letters in $\binom{3}{2} = 3$ ways, and arrange the resulting four in $4! = 24$ ways.) Since the two Es are in fact indistinguishable, we have to halve this number, giving 106 words. Finally, with three distinguishable Es, there would be $720 + 3 \cdot 120 + 3 \cdot 24 + 6 = 1158$ possibilities, and so there are $1158/6 = 193$ words of this form. So the total is $65 + 106 + 193 = 364$ words.

3.8. Equivalence and order

A *relation* on a set X is normally regarded as a property which may or may not hold between any two given elements of X. Typical examples are 'equal', 'less than', 'divides', etc. The definition comes as a surprise at first: a *relation* on X is a subset of X^2 (the set of ordered pairs of elements of X). What is the connection? Of course, a relation in the familiar sense is completely determined by the set of pairs which satisfy it; and conversely, given any set of pairs, we could imagine a property which was true for those pairs and false for all others.

This dual interpretation causes a small problem of notation. In general, if $R \subseteq X^2$ is a relation, we could write $x \, R \, y$ to have the same meaning as $(x, y) \in R$. This is consistent with the usual notations $x = y$, $x < y$, $x|y$, etc. But we don't reverse the procedure and write $(x, y) \in =$, $(x, y) \in <$, etc.!

Here are some important properties which a relation R may or may not have:

- R is *reflexive* if, for all $x \in X$, we have $(x, x) \in R$.
- R is *irreflexive* if, for all $x \in X$, we have $(x, x) \notin R$. (This is not the same as saying 'R is not reflexive'.)
- R is *symmetric* if, for all $x, y \in X$, $(x, y) \in R$ implies $(y, x) \in R$.
- R is *antisymmetric* if, whenever $(x, y) \in R$ and $(y, x) \in R$ both hold, then $x = y$.
- R is *transitive* if, for all $x, y, z \in X$, $(x, y) \in R$ and $(y, z) \in R$ together imply $(x, z) \in R$.

For example, the relation of equality is reflexive, symmetric and transitive; the relation 'less than or equal' is reflexive, antisymmetric and transitive; the relation 'less than' is irreflexive, antisymmetric and transitive; and the relation of adjacency in a *graph* (as described in Section 2.5) is irreflexive and symmetric.

Note that there are two ways of modelling an order relation: as 'less than' (irreflexive) or as 'less than or equal' (reflexive).

We proceed to define some important classes of relations in terms of these properties.

An *equivalence relation* is a reflexive, symmetric and transitive relation. It turns out that equivalence relations describe partitions of a set. Let R be an equivalence relation on X. For $x \in X$, the *equivalence class* containing x is the set $R(x) = \{y \in X : (x, y) \in R\}$. A *partition* of X is a family of pairwise disjoint, non-empty subsets whose union is X — thus, every point of X lies in exactly one of the sets.

(3.8.1) Theorem. *Let R be an equivalence relation on X. Then the equivalence classes of R form a partition of X. Conversely, given any partition of X, there is a unique equivalence relation on X whose equivalence classes are the parts of the partition.*

PROOF. Let R be an equivalence relation on X. Each equivalence class is non-empty, and their union is X; for, by reflexivity, each point $x \in X$ lies in the class $R(x)$, and conversely, $R(x)$ contains x.

Also, the equivalence classes are pairwise disjoint: for suppose that two classes $R(x)$, $R(y)$ have a common point z. We will show that $R(x) = R(y)$. By definition, $(x, z), (y, z) \in R$. By symmetry, $(z, y) \in R$; then, by transitivity, $(x, y) \in R$. Now, to prove two sets equal, we have to show that each set contains the other. So suppose that $w \in R(y)$. Then $(y, w) \in R$. Since $(x, y) \in R$, transitivity implies that $(x, w) \in R$, or $w \in R(x)$. So $R(y) \subseteq R(x)$. The reverse implication is similar.

For the converse, suppose that the sets $Y_1, Y_2, \ldots$ form a partition of X. Define a relation R by the rule that $(x, y) \in R$ if there is an index i such that x and y both lie in Y_i. It is not difficult to prove that R is an equivalence relation. For example, to show reflexivity, take $x \in X$; by assumption there is a (unique) i such that $x \in Y_i$; so $(x, x) \in R$. The other two properties are an exercise, as is the fact that R is the unique equivalence relation with equivalence classes $Y_1, Y_2, \ldots$.

Thus, for example, the number of partitions of a set is equal to the number of equivalence relations on that set. We will study these numbers (the *Bell numbers*) in Section 3.11.

We turn now to order relations. As mentioned above, there are two ways to model an order relation: we use the reflexive one (taking 'less than or equal') as the prototype.

A relation R on X is a *partial order* if it is reflexive, antisymmetric, and transitive.

Note that there may be some pairs of elements which are not comparable at all (i.e., neither $(x, y) \in R$ nor $(y, x) \in R$ hold). A relation R is said to satisfy *trichotomy* if, for any $x, y \in R$, one of the cases $(x, y) \in R$, $x = y$, or $(y, x) \in R$ holds. Then a relation R is a *total order* if it is a partial order which satisfies trichotomy. We commonly omit the word 'total' here; an *order* is a total order.

(3.8.2) Proposition. *The number of orders of an n-set is equal to $n!$.*

REMARK. In fact we show that, given any order on an n-set, its elements can be numbered $x_1, \ldots, x_n$ so that $(x_i, x_j) \in R$ if and only if $i \leq j$; and there is a unique way of doing this. In other words, the axiomatic definition of order agrees with our expectations!

PROOF. We show first that there is a 'last' element of X, an element x such that, if $(x, y) \in R$, then $y = x$. Suppose that no such x exists. Then, for any x, there exists $y \neq x$ such that $(x, y) \in R$. Start with $x = x_1$, and choose $x_2, x_3, \ldots$ according to this principle (so that $(x_i, x_{i+1}) \in R$ for all i). By transitivity, $(x_i, x_j) \in R$ for all $i \leq j$, and $x_i \neq x_{i+1}$ for all i. Now X is finite, so the sequence eventually bites its tail; that is, there exists $i < j$ so that $x_i = x_j$. Then $(x_{j-1}, x_j) \in R$, and $(x_j, x_{j-1}) = (x_i, x_{j-1}) \in R$ since $i \leq j - 1$. By antisymmetry, $x_j = x_{j-1}$, contrary to the construction.

Now there cannot be more than one 'last' element, since, for any x and y, either $(x, y) \in R$ or $(y, x) \in R$ by trichotomy.

Call the last element x_n; then, by trichotomy, $(x, x_n) \in R$ for all $x \in X$.

Arguing by induction, there is a unique way to label the remaining elements as $x_1, \ldots, x_{n-1}$, in accordance with the assertion. The proposition is proved.

We see that orders on X are equinumerous with permutations of X; indeed, our representation of an order looks like the 'passive' form of a permutation. But there is no 'canonical' bijection between orders and permutations; we need one distinguished order to set up this correspondence. (Then any order R corresponds to the permutation which takes the distinguished order into R.)

In the next section, we will consider a generalisation of (partial) orders. A relation R is a *partial preorder* (or *pre-partial order*) if it is reflexive and transitive — we relax the condition of antisymmetry. Exercise 19 outlines a proof that, if R is a partial preorder on X, then there is a natural way to define an equivalence relation on X so that the set of equivalence classes is partially ordered. (We set $x \equiv y$ if both $x \, R \, y$ and $y \, R \, x$ hold: think of such x and y as being indistinguishable. Now the truth of the relation $x \, R \, y$ is unaffected if either x or y is replaced by a point which is indistinguishable from it; so R induces a relation on the equivalence classes which is still reflexive and transitive, and is also antisymmetric.)

A partial preorder satisfying trichotomy is called a *preorder*.

3.9. Project: Finite topologies

Topology is the study of continuity. The term suggests doughnuts, Möbius bands, and such like. There is, however, an abstract definition of a topology, and it applies to finite as well as infinite spaces. We are going to translate the meaning of 'finite topology' into something simpler and more combinatorial.

A *topology* consists of a set X, and a set $\mathcal{T}$ of subsets of X, satisfying the following axioms:
- $\emptyset \in \mathcal{T}$ and $X \in \mathcal{T}$;
- the union of any collection of sets in $\mathcal{T}$ is in $\mathcal{T}$;
- the intersection of any two sets in $\mathcal{T}$ is in $\mathcal{T}$.

Sets in $\mathcal{T}$ are said to be *open*. The idea is that, if x is a point and U an open set containing x, the points of U are in some sense 'close' to x. (Indeed, U is often called a *neighbourhood* of x.)

REMARK. It follows by induction from the third axiom that the intersection of any finite number of members of $\mathcal{T}$ is a member of $\mathcal{T}$. If X is finite, the second axiom need only deal with finite unions, and so it too can be simplified to the statement that the union of any two sets in $\mathcal{T}$ is in $\mathcal{T}$; then the axioms are 'self-dual'. This is not the case in general!

(3.9.1) Theorem. *Let X be finite. Then there is a one-to-one correspondence between the topologies on X, and the partial preorders (i.e., reflexive and transitive relations) on X.*

Thus, describing finite topologies (sets of sets) reduces to the simpler task of describing partial preorders (sets of pairs). No such correspondence holds for infinite sets!

PROOF. The correspondence is simple to describe; the verification less so.

CONSTRUCTION 1. Let $\mathcal{T}$ be a topology on X. Define a relation R by the rule that $(x, y) \in R$ if every open set containing x also contains y. It is trivial that R is reflexive and transitive; that is, R is a partial preorder.

CONSTRUCTION 2. Let R be a partial preorder on X. Call a subset U of X *open* if, whenever $x \in U$, we have $R(x) \subseteq U$, where

$$R(x) = \{y : (x, y) \in R\}.$$

Let $\mathcal{T}$ be the set of all open sets. We have to verify that $\mathcal{T}$ is a topology. The first axiom requires no comment. For the second axiom, let $U_1, U_2, \ldots$ be open, and $x \in \bigcup_i U_i$; then $x \in U_j$ for some j, whence

$$R(x) \subseteq U_j \subseteq \bigcup_i U_i.$$

For the third axiom, let U and V be open and $x \in U \cap V$. Then $R(x) \subseteq U$ and $R(x) \subseteq V$, and so $R(x) \subseteq (U \cap V)$; thus $U \cap V$ is open.

All this argument is perfectly general. It is the fact that we have a bijection which depends on the finiteness of X. We have to show that applying the two constructions in turn brings us back to our starting point.

Suppose first that R is a partial preorder, and $\mathcal{T}$ the topology derived from it by Construction 2. Suppose that $(x, y) \in R$. Then $y \in R(x)$, so every open set containing x also contains y. Conversely, suppose that every open set containing x also contains y. The set $R(x)$ is itself open (this uses the transitivity of R: if $z \in R(x)$, then $R(z) \subseteq R(x)$), and so $y \in R(x)$; thus $(x, y) \in R$. Hence the partial preorder derived from $\mathcal{T}$ by Construction 1 coincides with R. (We still haven't used finiteness!)

Conversely, let $\mathcal{T}$ be a topology, and R the partial preorder obtained by Construction 1. If $U \in \mathcal{T}$ and $x \in U$, then $R(x) \subseteq U$; so U is open in the sense of Construction 2. Conversely, suppose that U is open, that is, $x \in U$ implies $R(x) \subseteq U$. Now each set $R(x)$ is the intersection of all members of $\mathcal{T}$ containing x. (This follows from the definition of R in Construction 1.) But there are only finitely many such open sets (here, at last, we use the fact that X is finite!); and the intersection of finitely many open sets is open, as we remarked earlier; so $R(x)$ is open. But, by hypothesis, U is the union of the sets $R(x)$ for all points $x \in U$; and a union of open sets is open, so U is open, as required.

In the axiomatic development of topology, the next thing one meets after the definition is usually the so-called 'separation axioms'. A topology is said to satisfy the axiom T_0 if, given any two distinct points x and y, there is an open set containing one but not the other; it satisfies axiom T_1 if, given distinct x and y, there is an open set containing x but not y (and *vice versa*).

These two axioms for finite topologies have a natural interpretation in terms of the partial preorder R. Axiom T_1 asserts that R never holds between distinct points x and y; that is, R is the trivial relation of equality. Construction 2 in the proof of the theorem then shows that every subset is open. (This is called the *discrete topology*.) It follows that any stronger separation axiom (in particular, the so-called 'Hausdorff axiom' T_2) also forces the topology to be discrete.

Axiom T_0 translates into the condition that the relation R is antisymmetric; thus, it is a partial order. So there is a one-to-one correspondence between T_0 topologies on the finite set X and partial orders on X.

3.10. Project: Cayley's Theorem on trees

As we saw at the end of Section 3.8, the number of orderings of an n-set is equal to the number of permutations of the same set, namely $n!$. This seems too trivial to be of any use at all, but in fact it forms the basis of a conceptual proof of a very famous theorem of Cayley:[7]

(3.10.1) Cayley's Theorem on trees

The number of labelled trees on n vertices is n^{n-2}.

The definitions will be given somewhat briefly; graphs (and trees in particular) are discussed in more detail in Chapter 11. A *graph* consists of a set of *vertices* and a set of *edges*, each edge consisting of a pair of vertices. The edge is regarded as joining the two vertices. Graphs were mentioned in Chapter 2, where we also introduced the distinction between labelled and unlabelled graphs. Here, we will be counting labelled graphs; that is, the vertex set is always $\{1, 2, \ldots, n\}$, and two graphs are the same precisely when they have the same set of edges.

A *path* in a graph is a sequence of vertices, all distinct except perhaps the first and the last, with the property that consecutive vertices in the sequence are adjacent (joined by an edge). A graph is *connected* if any two vertices are the ends of a path. A *circuit* is a path (having more than two vertices) such that the first and last vertices are equal. A *tree* is a connected graph containing no circuit. Cayley's Theorem asserts that there are n^{n-2} trees on n vertices.

We prove this theorem by counting slightly different structures called vertebrates. A *vertebrate* is a tree with two distinguished vertices called the *head* and the *tail*, which may or may not be equal. There is a path from the head to the tail, and it is unique (or else there would be a circuit); this path is called the *backbone*. If $T(n)$ is the number of trees on n vertices, then the number of vertebrates is $n^2 T(n)$ (each of the head and tail is chosen from a set of size n). So it is enough to prove that there are exactly n^n vertebrates on the set $N = \{1, \ldots, n\}$.

An *endofunction* on N is simply a function from N to itself. In fact, what we show is:

(3.10.2) Proposition. *The numbers of vertebrates and endofunctions on N are equal.*

---

[7] The proof outlined here is adapted from an argument by André Joyal (1981).

Obviously there are n^n endofunctions; so this will prove Cayley's Theorem. It would suffice to find a bijection between vertebrates and endofunctions. But there is no 'natural' bijection, so we have to do something a bit more complicated. First, one more small piece of notation. A *rooted tree* is a tree with a single distinguished vertex (called, naturally, the *root*.)

FACT 1. A vertebrate on N is uniquely specified by the following data:
- a non-empty subset K of N (say $\{m_1, m_2, \ldots, m_k\}$, where $1 \le m_1 < m_2 < \ldots < m_k \le n$);
- an ordering of K;
- a partition of $N \setminus K$ into $|K| = k$ subsets $S_1, \ldots, S_k$;
- a rooted tree on $S_i \cup \{m_i\}$ with root m_i, for $i = 1, \ldots, k$.

How do we construct a vertebrate from these data? The set K is the backbone, and the order describes the sequence of vertices from the head (the first vertex in K according to the chosen order) to the tail (the last vertex in K). Now the tree on $S_i \cup \{m_i\}$ is hooked on to the backbone at m_i. Any vertebrate arises in a unique way from this construction.

FACT 2. An endofunction on N is uniquely specified by the following data:
- a non-empty subset K of N (say $\{m_1, m_2, \ldots, m_k\}$, where $1 \le m_1 < m_2 < \ldots < m_k \le n$);
- a permutation of K;
- a partition of $N \setminus K$ into $|K| = k$ subsets $S_1, \ldots, S_k$;
- a rooted tree on $S_i \cup \{m_i\}$ with root m_i, for $i = 1, \ldots, k$.

From these data, we build an endofunction as follows. The points in K are mapped according to the given permutation. Each vertex of S_i is mapped one step closer to the root m_i (along the unique path joining them).

It is not quite so clear that any endofunction arises uniquely by this construction. Let f be an endofunction on N. A point m is *periodic* if $f^t(m) = m$ for some $t > 0$. (The notation f^t means the result of applying the function t times.) Let K be the set of periodic points. Then f induces a permutation of K. Moreover, any point p of N is *ultimately periodic*, in the sense that $f^s(p) \in K$ for some s. Let S_i be the set of points in $N \setminus K$ for which the images under f first enter K at the point m_i. If we join y to $f(y)$ by an edge for each $y \in S_i \cup \{m_i\}$, the graph obtained contains no cycles, and so is a tree with root m_i.

Considering the data required to specify the two types of object, and recalling that there are equally many orders and permutations, it is clear that there are also equally many vertebrates and endofunctions.

3.11. Bell numbers

The *Bell number* B_n is the number of partitions of an n-set (or the number of equivalence relations on an n-set; this is the same thing, by Theorem 3.8.1).

For example, $B_3 = 5$; the five partitions of $\{1, 2, 3\}$ are

$$\{\{1, 2, 3\}\}$$
$$\{\{1, 2\}, \{3\}\}$$
$$\{\{1, 3\}, \{2\}\}$$
$$\{\{2, 3\}, \{1\}\}$$
$$\{\{1\}, \{2\}, \{3\}\}$$

Similarly, $B_2 = 2$, $B_1 = 1$. What is B_0? Since the parts of a partition are non-empty by definition, a partition of the empty set cannot have any parts at all, and must be the empty set. But the empty set is indeed a partition of itself! So $B_0 = 1$. (You may regard this as a convention if you like.)

Subsets, permutations, and partitions of a set are the fundamental objects with which combinatorics deals. Strangely, unlike the other cases, there is no convenient short formula for the Bell numbers. They can be calculated by a 'recurrence relation' which expresses the value of B_n in terms of smaller Bell numbers and binomial coefficients, as follows.

(3.11.1) Recurrence for Bell numbers

For $n \geq 1$,

$$B_n = \sum_{k=1}^{n} \binom{n-1}{k-1} B_{n-k}.$$

PROOF. Take $X = \{1, \ldots, n\}$, and consider a partition of X. It has a unique part containing n, say $\{n\} \cup Y$, where Y is a subset of the $(n-1)$-set $\{1, \ldots, n-1\}$. The remaining parts form a partition of the set $\{1, \ldots, n-1\} \setminus Y$. These data (the subset Y, and the partition) determine the original partition uniquely. If $|Y| = k-1$, then there are $\binom{n-1}{k-1}$ choices of Y, and B_{n-k} choices of a partition of the remaining points. Multiplying, and summing over all possible values of k (from 1 to n), gives the result.

For example,

$$B_4 = \binom{3}{0} B_3 + \binom{3}{1} B_2 + \binom{3}{2} B_1 + \binom{3}{3} B_0$$
$$= 5 + (3 \cdot 2) + (3 \cdot 1) + 1$$
$$= 15.$$

3.12. Generating combinatorial objects

Combinatorial problems have a tendency to grow in size explosively as the size of the set increases. It often happens that a few small values can be done by hand, and then we have to resort to the computer to settle a few more cases.[8] If the problem involves checking all objects of some kind (subsets, permutations, etc.), then we need an algorithm to generate all of these.

Usually the simplest algorithm (conceptually) involves recursion, based on the way in which the objects are built up from smaller ones. For example, here is a recursive algorithm for generating the power set of $\{1, \ldots, n\}$. Note how the algorithm resembles the proof of the recurrence relation $F(n+1) = 2F(n)$, $F(0) = 1$ for the counting function.

---

[8] After this, brainwork is the only way.

(3.12.1) Recursive algorithm: Power set of $\{1, \ldots, n\}$

If $n = 0$, return $\{\emptyset\}$.

Otherwise,

- *generate the power set of $\{1, \ldots, n-1\}$;*
- *make a new copy of each subset and adjoin the element n to it;*
- *return the set of all sets created.*

In symbols: $\mathcal{P}(\emptyset) = \{\emptyset\}$;

$$\mathcal{P}(\{1, \ldots, n\}) = \{Y, Y \cup \{n\} : Y \in \mathcal{P}(\{1, \ldots, n-1\})\}$$

for $n > 0$.

In a similar way, the recurrence relations

- $\dbinom{n}{k} = \dbinom{n-1}{k-1} + \dbinom{n-1}{k}$
- $n! = n(n-1)!$
- the recurrence relation for Bell numbers

suggest recursive algorithms for k-subsets, permutations, and partitions.

However, there are disadvantages to this simple approach. The main one is that, even for moderate values of n, the set of all subsets (or all permutations) is so large that the computer's memory will not hold it. What we have to do is, rather than creating all the objects in one step, generate them one at a time, process each one, and then throw it away when the next one is generated.[9] The algorithm will have the following general form. There are two parts. The first step generates the 'first' object. The second step takes any object and tries to calculate the 'next' one; if there is no 'next' one (so that the current object is the last), it should report this fact. Then the structure of a program will be like this:

Generate first object.
REPEAT
- process current object;
- generate next object
UNTIL there's no next object.

One very important observation is that this set-up presupposes that the objects come in some order. But the order is not specified, except in the progression from each object to the next.[10] So these algorithms implicitly define an ordering of the relevant objects.

---

[9] If you are writing programs implementing the following algorithms, a good 'minimal processing' is to count the objects; this provides an additional check on the correctness of the program.

[10] Compare the remarks about the order of the natural numbers in Chapter 2.

For subsets of a set, we use the *Odometer Principle* from Chapter 2. Re-writing the algorithm given there, we get:

(3.12.2) Algorithm: Subsets of $\{1, \ldots, n\}$

FIRST SUBSET *is $\emptyset$.*

NEXT SUBSET *after Y:*

- *Find the last element i not in Y (working back from the end).*
- *If there's no such element, then Y was the last subset.*
- *Remove from Y all elements after i, and add i to Y. Return this set.*

This displays the principle correctly. In practice, it would be more efficient to combine the steps. Thus, we take a pointer i, initialised to n. While $i \in Y$, we remove i from Y and decrease i by 1. If we fall off the bottom (i.e., reach $i = 0$), then Y was the last set. Otherwise, add the final value of i to the set Y and return the result.

Note that, if we represent a subset Y of X by its 'characteristic function', the sequence $(a_1, \ldots, a_n)$ with

$$a_i = \begin{cases} 1 & \text{if } i \in Y, \\ 0 & \text{if } i \notin Y, \end{cases}$$

and interpret this as the base 2 representation of an integer $N = a_1 2^{n-1} + \ldots + a_n$, then the algorithm proceeds through the integers from 0 to $2^n - 1$ in order. This ordering of the subsets of a set was discovered by Shao Yung (1160), who proposed it as an alternative to the traditional order of the sixty-four I Ching hexagrams attributed to King Wên (ca. 1150 BC); and independently and much later by Leibniz (1703).[11] So we could simplify the algorithm, using the computer's inbuilt arithmetic. We define the set corresponding to a non-negative integer N by writing N to the base 2 and interpreting the result as a characteristic function. If we denote the set corresponding to N by $Y(N)$, then the FIRST SUBSET is $Y(0)$; and the NEXT SUBSET after $Y(N)$ is $Y(N+1)$. (The 'next subset' procedure fails if $N = 2^n - 1$.)

This procedure has an additional advantage, in that it gives us 'random access' to the subsets of a set: we can easily produce the Nth set $Y(N)$ for any N with $0 \leq N \leq 2^n - 1$. However, for other cases considered below, it is harder to do this.

Consider the problem of generating all the k-subsets of a set. Here, there are two essentially different 'natural' orders in which the subsets could be generated, exemplified by the case $n = 5$, $k = 3$:[12]

$$123, 124, 125, 134, 135, 145, 234, 235, 245, 345;$$

$$123, 124, 134, 234, 125, 135, 235, 145, 245, 345.$$

---

[11] It is said that, after Leibniz' discovery, he was informed of the Chinese precedence by a Jesuit missionary, Fr. Joachim Bouvet. But Leibniz went further, using the binary representation for arithmetic where Shao Yung was concerned only with the progression. For further discussion, see S. N. Afriat, *The Ring of Linked Rings* (1982).

[12] In fact, reversing the order of the numbers $\{1, \ldots, n\}$ and the order of the subsets takes the first ordering to the second. But, to a computer, this would look like time reversal: not an easy trick!

The first ordering is generated by a fragment of program which (in BASIC) would look like this (for $k = 3$, n arbitrary):

```
FOR i = 1 TO n − 2
   FOR j = i + 1 TO n − 1
      FOR k = j + 1 TO n
         process {i, j, k}
      NEXT k
   NEXT j
NEXT i
```

(Hopefully this is clear even to non-programmers.) This seems a natural way to do it. But it only works in this form if k is small and fixed. Also, the other order has a subtle advantage. Observe that the 3-subsets of $\{1, \ldots, 4\}$ occur first, in their 'natural' order, followed by the subsets containing 5 (which are obtained by adjoining 5 to the 2-subsets of $\{1, \ldots, 4\}$ in their natural order). This is in accord with the recursive version discussed earlier. Anyway, the following algorithm does the job (producing the second order above):

(3.12.3) Algorithm: k**-subsets of** $\{1, \ldots, n\}$

FIRST SUBSET *is* $\{1, \ldots, k\}$.

NEXT SUBSET *after* $Y = \{y_1, \ldots, y_k\}$, *where* $y_1 < \ldots < y_k$:

- *Find the first i such that $y_i + 1 \notin Y$;*
- *increase y_i by 1, set $y_j = j$ for $j < i$, and return the new set Y;*
- *this fails if $i = k$, $y_k = n$, in which case $Y = \{n − k + 1, \ldots, n\}$ is the last set.*

The two 'natural' orders of k-sets can be characterised as follows. The first is the so-called *lexicographic* order. This means that, if we regard the symbols $1, \ldots, n$ as letters of an alphabet, and regard each k-set as a word by writing its elements in alphabetical order, then the words occur in lexicographic order (the order in which they would be found in a dictionary). The second order is *reverse lexicographic*: we turn k-sets into words as above, but then reverse each word before putting them in dictionary order.

Lexicographic order or something similar is usually the most natural for problems of this kind. The next algorithm, for permutations, uses lexicographic order, where a permutation is taken in passive form. (That is, we regard a permutation as an n-tuple $(x_1, \ldots, x_n)$, where $x_1, \ldots, x_n$ are $1, \ldots, n$ in some order.) Here is the algorithm.

(3.12.4) Algorithm: Permutations of $\{1,\ldots,n\}$

FIRST PERMUTATION *is given by* $x_i = i$ *for* $i = 1,\ldots,n$.

NEXT PERMUTATION *after* $(x_1,\ldots,x_n)$:

- *Find the largest* j *for which* $x_j < x_{j+1}$ *(working back from the end).*
- *If no such* j *exists, then the current permutation is the last.*
- *Interchange the value of* x_j *with the least* x_k *greater than* x_j *with* $k > j$; *then reverse the sequence of values of* $x_{j+1},\ldots,x_n$; *return this permutation.*

Here is an example. Suppose the current permutation is (436521). The algorithm first locates $j = 2$, $x_j = 3$. We assert that the current permutation is the last (in lexicographic order) of the form $(43\ldots)$, and should be followed by the first of the form $(45\ldots)$, namely (451236). To obtain this, we find $k = 4$, $x_k = 5$. (Since the values after x_j are decreasing, this can be located by working back from the end until we first find a value greater than x_j.) Then we interchange the entries in the second and fourth positions, giving (456321); and reverse the entries in positions 3 to 6, giving (451236), as required.

It is much harder to give an algorithm of this kind for partitions of a set. This is related to the non-existence of a simple formula for the Bell number.

3.13. Exercises

1. A restaurant near Vancouver offered Dutch pancakes with 'a thousand and one combinations' of toppings. What do you conclude?

2. Using the numbering of subsets of $\{0,1,\ldots,n-1\}$ defined in Section 3.1, prove that, if $X_k \subseteq X_l$, then $k \leq l$ (but not conversely).

3. Prove that, for fixed n, the greatest binomial coefficient $\binom{n}{k}$ occurs when $k = \lfloor n/2 \rfloor$ or $\lceil n/2 \rceil$.

4. Prove the following identities:

(a) $\binom{n}{k}\binom{k}{l} = \binom{n}{l}\binom{n-l}{k-l}$.

(b) $\sum_{i=0}^{k} \binom{m}{i}\binom{n}{k-i} = \binom{m+n}{k}$

     (recall the convention that $\binom{n}{k} = 0$ if $k < 0$ or $k > n$).

(c) $\sum_{i=0}^{k} \binom{n+i}{i} = \binom{n+k+1}{k}$.

(d) $\sum_{k=1}^{n} k\binom{n}{k} = n2^{n-1}$.

(e) $\sum_{k=0}^{n} (-1)^k \binom{n}{k}^2 = \begin{cases} 0 & \text{if } k \text{ is odd;} \\ (-1)^m \binom{2m}{m} & \text{if } n = 2m. \end{cases}$

5. Following the method in the text, calculate the number of subsets of an n-set of size congruent to m (mod 3) $(m = 0, 1, 2)$ for each value of n (mod 6).

6. Let k be a given positive integer. Show that any non-negative integer N can be written uniquely in the form

$$N = \binom{x_k}{k} + \binom{x_{k-1}}{k-1} + \ldots + \binom{x_1}{1},$$

where $0 \le x_1 < \ldots < x_{k-1} < x_k$. [HINT: Let x be such that $\binom{x}{k} \le N < \binom{x+1}{k}$. Then any possible representation has $x_k = x$. Now use induction and the fact that $N - \binom{x}{k} < \binom{x}{k-1}$ (Fact 3.2.5) to show the existence and uniqueness of the representation.]
   Show that the order of k-subsets corresponding in this way to the usual order of the natural numbers is the same as the reverse lexicographic order generated by the algorithm in Section 3.11. [HINT: $\sum_{j=0}^{i} \binom{n-j}{i-j} = \binom{n+1}{i}$.]

7. Use the fact that $(1+t)^p \equiv 1 + t^p$ (mod p) to prove by induction that $n^p \equiv n$ (mod p) for all positive integers n.

8. A computer is to be used to calculate values of binomial coefficients. The largest integer which can be handled by the computer is 32767. Four possible methods are proposed:

(1) $\binom{n}{k} = n!/k!(n-k)!$;

(2) $\binom{n}{k} = n(n-1)\ldots(n-k+1)/k!$;

(3) $\binom{n}{0} = 1$, $\binom{n}{k} = \binom{n}{k-1} \cdot \dfrac{n-k+1}{k}$ for $k > 0$;

(4) $\binom{n}{0} = \binom{n}{n} = 1$, $\binom{n}{k} = \binom{n-1}{k-1} + \binom{n-1}{k}$ for $0 < k < n$ (i.e., Pascal's Triangle).

For which values of n and k can $\binom{n}{k}$ be calculated by each method? What can you say about the relative speed of the different methods?

9. Show that there are $(n-1)!$ cyclic permutations of a set of n points.

10. The *order* of a permutation π is the least positive integer m such that π^m is the identity permutation. Prove that the order of a cycle on n points is n. Prove that the order of an arbitrary permutation is the least common multiple of the lengths of the cycles in its cycle decomposition.

11. How many words can be made from the letters of the word ESTATE?

12. Given n letters, of which m are identical and the rest are all distinct, find a formula for the number of words which can be made.

13. Show that, for $n = 2, 3, 4, 5, 6$, the number of *unlabelled* trees on n vertices is 1, 1, 2, 3, 6 respectively.

14. The line segments from $(i, \log i)$ to $(i+1, \log(i+1))$ lie below the curve $y = \log x$. (This is because the curve is convex, i.e., its second derivative $-1/x^2$ is negative.) The area under these line segments from $i = 1$ to $i = n$ is $\log n! + \frac{1}{2}\log(n+1)$, since it consists of the rectangles of Fig. 3.1(b) together with triangles with width 1 and heights summing to $\log(n+1)$. Deduce that

$$n! \leq e\sqrt{n+1}\left(\frac{n}{e}\right)^n.$$

[REMARK. According to Stirling's Formula, the limiting ratio of this upper bound to $n!$ is $e/\sqrt{2\pi} = 1.0844\ldots.$]

15. Use Stirling's Formula to prove that

$$\binom{2n}{n} \sim 2^{2n}/\sqrt{\pi n}.$$

16. (a) Let $n = 2k$ be even, and X a set of n elements. Define a *factor* to be a partition of X into k sets of size 2. Show that the number of factors is equal to $1 \cdot 3 \cdot 5 \cdots (2k-1)$. This number is sometimes called a *double factorial*, written $(2k-1)!!$ (with !! regarded as a single symbol, the two exclamation marks suggesting the gap of two, *not* the factorial function iterated!)

(b) Show that a permutation of X interchanges *some* k-subset with its complement if and only if all its cycles have even length. Prove that the number of such permutations is $((2k-1)!!)^2$. [HINT: any pair of factors defines a partition of X into a disjoint union of cycles, and conversely. The correspondence is not one-to-one, but the non-bijectiveness exactly balances.]

(c) Deduce that the probability that a random element of S_n interchanges some $\frac{1}{2}n$-set with its complement is $O(1/\sqrt{n})$. [HINT: You will probably need two analytic facts: $1 - x < e^{-x}$ for positive x; and $\sum_{i=1}^{n}(1/i) = \log n + O(1)$.]

17. How many relations on an n-set are there? How many are (a) reflexive, (b) symmetric, (c) reflexive and symmetric, (d) reflexive and antisymmetric?

18. Given a relation R on X, define

$$R^+ = \{(x, y) : (x, y) \in R \text{ or } x = y\}.$$

Prove that the map $R \to R^+$ is a bijection between the irreflexive, antisymmetric and transitive relations on X, and the reflexive, antisymmetric and transitive relations on X. Show further that this bijection preserves the property of trichotomy.

REMARK. This exercise shows that it doesn't matter whether we use the 'less than' or the 'less than or equal' model for order relations.

19. Recall that a *partial preorder* is a relation R on X which is reflexive and transitive. Let R be a partial preorder. Define a relation S by the rule that $(x, y) \in S$ if and only if both (x, y) and (y, x) belong to R. Prove that S is an equivalence relation. Show further that R 'induces' a partial order $\overline{R}$ on the set of equivalence classes of S in a natural way: if $(x, y) \in R$, then $(\overline{x}, \overline{y}) \in \overline{R}$, where $\overline{x}$ is the S-equivalence

class containing x, etc. (You should verify that this definition is independent of the choice of representatives x and y.)

Conversely, let X be a set carrying a partition, and R' a partial order on the parts of the partition. Prove that there is a unique partial preorder on X giving rise to this partition and partial order as in the first part of the question.

Show further that the results of this question remain valid if we replace *partial preorder* and *partial order* by *preorder* and *order* respectively, where a *preorder* is a partial preorder satisfying trichotomy.

20. List the (a) partial preorders, (b) preorders, (c) partial orders, (d) orders on the set $\{1, 2, 3\}$.

21. Prove that $B_n < n!$ for all $n > 2$. [HINT: associate a partition with each permutation.]

22. Verify, theoretically or practically, the following algorithm for generating all partial permutations of $\{1, \ldots, n\}$:

(3.13.1) Algorithm: Partial permutations of $\{1, \ldots, n\}$
FIRST PARTIAL PERMUTATION *is the empty sequence.*
NEXT OBJECT *after* $(x_1, \ldots, x_m)$:
- *If the length m of the current sequence is less than n, extend it by adjoining the least element it doesn't contain.*
- *Otherwise, proceed as in the algorithm for permutations, up to the point where x_j and x_k are interchanged; then, instead of reversing the terms after x_j, remove them from the sequence.*

---

23. Verify the following recursive procedure for generating the set of partitions of a set X.

(3.13.2) Recursive algorithm: Partitions of X
If $X = \emptyset$, *then* $\emptyset$ *is the only partition.*
If $X \neq \emptyset$, *then*
- *select an element* $x \in X$;
- *generate all subsets of* $X \setminus \{x\}$;
- *for each subset* Y, *generate all partitions of* $X \setminus (\{x\} \cup Y)$, *and adjoin to each the additional part* $\{x\} \cup Y$.

---

24. Let $A = (a_{ij})$ and $B = (b_{ij})$ be $(n+1) \times (n+1)$ matrices (with rows and columns indexed from 0 to n) defined by $a_{ij} = \binom{i}{j}$, $b_{ij} = (-1)^{i+j} \binom{i}{j}$ (where $\binom{i}{j} = 0$ if $i < j$). Prove that $B = A^{-1}$. [HINT: let V be the vector space of polynomials of degree at most n, with basis $1, t, t^2, \ldots, t^n$. Show that A represents the linear transformation $f(t) \mapsto f(t+1)$. What transformation is represented by B?]

25. PROJECT. A couple of harder binomial identities. Prove:

(a) $\displaystyle\sum_{k=0}^{n}\binom{2n+1}{2k+1}\binom{m+k}{2n}=\binom{2m}{2n}.$

(b) $\displaystyle\sum_{k=0}^{n}(-1)^{k}\binom{n}{k}^{3}=\begin{cases}(-1)^{m}(3m)!/(m!)^{3} & \text{if } n=2m; \\ 0 & \text{if } n \text{ is odd}.\end{cases}$

26. PROJECT. According to (3.10.1) and (3.7.1), the number of labelled trees on n vertices is equal to the number of $(n-2)$-tuples of numbers chosen from the set $\{1,2,\ldots,n\}$. Is there a direct proof of this, establishing a one-to-one correspondence between the two sets? The following argument was found by Prüfer. Verify that the procedures below give mutually inverse bijections. A *Prüfer code* is an $(n-2)$-tuple from $\{1,\ldots,n\}$; and a *leaf* of a tree is a vertex lying on only one edge.

(a) Labelled tree to Prüfer code:
- START with the empty sequence P.
- REPEAT the following operation:

  Take the leaf with smallest label. Append the label of its unique neighbour to P and delete the leaf and the edge containing it from the tree.

  UNTIL only one edge remains.
- The sequence P is the Prüfer code of the tree.

(b) Prüfer code P to tree:
- Take $\{1,\ldots,n\}$ to be the vertex set of the tree. Set $S=\{1,\ldots,n\}$.
- REPEAT the following operation:

  Let p be the first element of the sequence P, and s the smallest element of S not occurring in P. Join p to s with an edge. Remove p from P and s from S.

  UNTIL the sequence P is empty.
- At this point, two elements remain in S; join them with an edge. This completes the tree.

Show further that the number of edges containing a vertex in a labelled tree is one more than the number of occurrences of its label in the Prüfer code of the tree.

27. PROJECT. A *forest* is a graph without cycles. Prove that the number $F(n)$ of forests on the set $\{1,\ldots,n\}$ satisfies the recurrence relation

$$F(n)=\sum_{k=1}^{n}\binom{n-1}{k-1}k^{k-2}F(n-k).$$

Calculate the ratio of $F(n)$ to the number n^{n-2} of trees for small n. What can you say about this ratio in the limit?

4. Recurrence relations and generating functions

The way begets one; one begets two; two begets three; three begets the myriad creatures.

Lao Tse, *Tao Te Ching* (ca. 500 BC)

TOPICS: Fibonacci, Catalan and Bell numbers, derangements, [finite fields, sorting, binary trees, 'Twenty Questions']

TECHNIQUES: Recurrence relations, solution of linear recurrence relations with constant coefficients, generating functions and their manipulation, [the ring of formal power series]

ALGORITHMS: Computation of Fibonacci numbers, [QUICKSORT]

CROSS-REFERENCES: Derangements (Chapter 5), set partitions (Chapter 3)

A *recurrence relation* expresses the value of a function f at the natural number n in terms of its values at smaller natural numbers. We saw a simple example of this already: the number $F(n)$ of subsets of an n-set satisfies $F(n+1) = 2F(n)$. This relation, together with the initial value $F(0) = 1$, determines the value of F for every natural number. In this chapter, we examine recurrence relations in more detail.

An important technique, often associated with recurrence relations but useful in its own right, is that of *generating functions*. These are power series whose coefficients form the number sequence in question. We show how generating functions can be used either to solve recurrence relations explicitly, or to derive some information about the (unknown) solution. The techniques look suspiciously like analysis![1]

To begin, here is an introductory example of a proof by generating function. Let $F(n)$ be the number of subsets of an n-set. We saw several times already that $F(n) = 2^n$; now we will evaluate $F(n)$ by yet another method, seemingly more complicated but in fact of very general applicability. Set

$$\phi(t) = \sum_{n=0}^{\infty} F(n)t^n.$$

---

[1] To Newton, 'analysis' meant manipulation of power series. See V. I. Arnol'd, *Huygens & Barrow, Newton & Hooke* (1990).

(Don't worry for the moment about whether this power series converges.) Now

$$2t\phi(t) = \sum_{n=0}^{\infty} 2F(n)t^{n+1}$$

$$= \sum_{n=0}^{\infty} F(n+1)t^{n+1}$$

$$= \phi(t) - 1,$$

the last equality holding because the sum is identical with the definition of $\phi(t)$ (with $n+1$ replacing n) except that the first term $F(0)t^0 = 1$ is missing. Thus

$$\phi(t) = \frac{1}{1-2t}.$$

The right-hand side is the sum of a geometric progression:

$$\phi(t) = \sum_{n=0}^{\infty} (2t)^n.$$

Comparing this with the original series, we conclude that $F(n) = 2^n$. (If two power series are equal, then all their coefficients coincide.)

Incidentally, we now see that the power series converges for all t with $|t| < \frac{1}{2}$; so our manipulations are justified by analysis. We will return to this question of justification later. First, however, we do a less trivial example.

4.1. Fibonacci numbers

PROBLEM. *In how many ways can the non-negative integer n be written as a sum of ones and twos (in order)?*

Let F_n be this number. Then, for example, $F_4 = 5$, since

$$4 = 1+1+1+1 = 2+1+1 = 1+2+1 = 1+1+2 = 2+2.$$

Similarly, we find that $F_1 = 1, F_2 = 2, F_3 = 3$. By convention, we take $F_0 = 1$: the only solution for $n = 0$ is the empty sequence.

Suppose that $n \geq 2$. Any expression for n as a sum of ones and twos must end with either a 1 or a 2. If it ends with 1, then the preceding terms sum to $n - 1$; if it ends with a 2, they sum to $n - 2$. So we have

$$F_n = F_{n-1} + F_{n-2}.$$

The numbers $F_0, F_1, F_2, \ldots$ are called the *Fibonacci numbers*.

This is an example of a *recurrence relation*, more specifically, a three-term linear recurrence relation with constant coefficients. The meaning of these terms is, I hope, obvious. But, in general, a $(k+1)$-term *recurrence relation* expresses any value $F(n)$ of a function in terms of the k preceding values $F(n-1), F(n-2), \ldots, F(n-k)$; it is *linear* if it has the form

$$F(n) = a_1(n)F(n-1) + a_2(n)F(n-2) + \ldots + a_k(n)F(n-k),$$

where $a_1, \ldots, a_k$ are functions of n; and it is *linear with constant coefficients* if $a_1, \ldots, a_k$ are constants. We will see examples later of recurrence relations in which the value of $F(n)$ depends on all the preceding values, in a highly non-linear way; so this one is very special.

> FACT: *A function satisfying a $(k+1)$-term recurrence relation is uniquely determined by its values on the first k natural numbers.*

(The first k natural numbers could be $0, \ldots, k-1$ or $1, \ldots, k$, depending on context.)

For, if we know $F(1), \ldots, F(k)$ (say), then these values determine $F(k+1)$, and then the values $F(2), \ldots, F(k+1)$ determine $F(k+2)$, and so on. The words *and so on* are a signal that we are using induction. Formally, if two functions F and G satisfy the same recurrence relation and agree on the first k natural numbers, then one proves by induction that they agree everywhere.

This is rather like the situation with differential equations, where we expect a kth order differential equation and k initial conditions to determine a solution uniquely. However, our situation is very much simpler in one way: the existence and uniqueness follows immediately from the Principle of Induction, without the need for any hard analysis. For any recurrence relation whatever, it is usually obvious just what sort of initial values are required to determine the solution uniquely.

We turn to methods for solving the recurrence relation:

(4.1.1) Fibonacci Recurrence Relation

For $n \geq 2$,
$$F_n = F_{n-1} + F_{n-2}$$

Two methods will be given; both of them generalise.

FIRST METHOD. Since the recurrence relation is linear, if we can find any solutions, we can take linear combinations of them to generate new solutions. (Again this is like what happens with differential equations.) Specifically, let F and G satisfy the recurrence relation above, and let $H_n = aF_n + bG_n$. Then

$$
\begin{aligned}
H_n &= aF_n + bG_n \\
&= a(F_{n-1} + F_{n-2}) + b(G_{n-1} + G_{n-2}) \\
&= (aF_{n-1} + bG_{n-1}) + (aF_{n-2} + bG_{n-2}) \\
&= H_{n-1} + H_{n-2}.
\end{aligned}
$$

We try to fit the initial conditions by choice of a and b.

Try a solution of the form $F_n = \alpha^n$. (The justification for this will be that it works!) We require

$$\alpha^n = \alpha^{n-1} + \alpha^{n-2},$$
$$\alpha^{n-2}(\alpha^2 - \alpha - 1) = 0.$$

So, if $\alpha^2 - \alpha - 1 = 0$, the recurrence holds for all n.

The roots of this equation are $\alpha = \frac{1}{2}(1 + \sqrt{5})$, $\beta = \frac{1}{2}(1 - \sqrt{5})$. So we have a general solution of the form

$$F_n = a\left(\frac{1 + \sqrt{5}}{2}\right)^n + b\left(\frac{1 - \sqrt{5}}{2}\right)^n.$$

To fit the initial conditions (which are $F_0 = 1$, $F_1 = 1$ in our case), we require

$$a + b = 1,$$

$$a\left(\frac{1 + \sqrt{5}}{2}\right) + b\left(\frac{1 - \sqrt{5}}{2}\right) = 1,$$

whence $a + b = 1$, $a - b = \frac{1}{\sqrt{5}}$, giving

$$a = \left(\frac{\sqrt{5} + 1}{2\sqrt{5}}\right), \qquad b = \left(\frac{\sqrt{5} - 1}{2\sqrt{5}}\right).$$

and so:

(4.1.2) Fibonacci numbers

$$F_n = \left(\frac{1}{\sqrt{5}}\right)\left(\frac{1 + \sqrt{5}}{2}\right)^{n+1} - \left(\frac{1}{\sqrt{5}}\right)\left(\frac{1 - \sqrt{5}}{2}\right)^{n+1}.$$

REMARKS. 1. $\left(\frac{1+\sqrt{5}}{2}\right) \approx 1.618\ldots$, and $\left(\frac{1-\sqrt{5}}{2}\right) \approx -0.618\ldots$. So the function grows exponentially; for $n > 0$, its value is the nearest integer to $\left(\frac{1}{\sqrt{5}}\right)\left(\frac{1+\sqrt{5}}{2}\right)^{n+1}$.
2. Note that we could easily find values of a and b to fit any given initial values.
3. We'll see that, for some purposes, the explicit formula is less useful than the recurrence relation.

SECOND METHOD. We now solve the recurrence relation using the technique of generating functions. We let $\phi(t)$ be the power series

$$\phi(t) = \sum_{n \geq 0} F_n t^n,$$

where t is an indeterminate.

We have

$$t\phi(t) = \sum_{n \geq 0} F_n t^{n+1} = \sum_{n \geq 1} F_{n-1} t^n,$$

$$t^2\phi(t) = \sum_{n \geq 0} F_n t^{n+2} = \sum_{n \geq 2} F_{n-2} t^n.$$

(Be clear about what is happening here. To get from the second term to the third in each equation, we have used the fact that n is only a 'dummy variable' whose actual name is not important. So, for example, in the first equation, we substitute $m = n+1$, and then replace the dummy variable m by n. If this confuses you, write out the first few terms of both sums.)

Now $F_n = F_{n-1} + F_{n-2}$, so it is 'almost true' that $\phi(t) = (t + t^2)\phi(t)$. Certainly, the coefficients of t^2 and all higher powers will be the same on both sides of this equation, but we might have to adjust the constant term and the term in t. Remember that $F_0 = 1, F_1 = 1$.

The coefficient of t is F_1 on the left and F_0 on the right, so these agree. The constant term is F_0 on the left and 0 on the right, so we have to add 1 to the right-hand side to obtain equality. Thus,

$$\phi(t) = 1 + (t + t^2)\phi(t),$$

whence

$$\phi(t) = \frac{1}{1 - t - t^2}.$$

Now the value of F_n is the coefficient of t^n in the Taylor series for this function. This is most easily found by a partial fraction expansion. Let $1 - t - t^2 = (1 - \alpha t)(1 - \beta t)$. Thus, α and β are roots of $x^2 - x - 1 = 0$; so $\alpha = \left(\frac{1+\sqrt{5}}{2}\right), \beta = \left(\frac{1-\sqrt{5}}{2}\right)$. (The same as before — no coincidence!) If we let

$$\frac{1}{(1 - \alpha t)(1 - \beta t)} = \frac{a}{1 - \alpha t} + \frac{b}{1 - \beta t},$$

then

$$1 = a(1 - \beta t) + b(1 - \alpha t),$$

so $a + b = 1$, $a\beta + b\alpha = 0$. These equations can be solved for a and b (with the same solution as before!).

Now

$$\phi(t) = \frac{a}{1 - \alpha t} + \frac{b}{1 - \beta t}$$
$$= a(1 + \alpha t + \alpha^2 t^2 + \ldots) + b(1 + \beta t + \beta^2 t^2 + \ldots);$$

equating coefficients of t^n, we find that

$$F_n = a\alpha^n + b\beta^n.$$

4.2. Aside on formal power series

Once we have found the power series in the above argument, we can use the theory of power series to show that it converges for $|t| < 1/\alpha$, and so the manipulations above are justified analytically. But in fact there is a theory of *formal power series*, according to which it is legitimate to do such manipulations without any regard to questions of convergence. This is important in cases where either the series don't converge for any non-zero value of t, or we are unable to find out enough about it to resolve the question of convergence. In this section, I'll outline the algebraic formalism for this. If you feel comfortable with the arguments of the last section, there is no need to read what follows.

A *formal power series* over a field F should be thought of as an expression

$$\sum_{n \geq 0} a_n t^n = a_0 + a_1 t + a_2 t^2 + \ldots,$$

but more formally it is an infinite sequence $(a_0, a_1, a_2, \ldots)$ of elements of F. (In fact, the definition will work over an arbitrary ring.) The set of all formal power series has operations of addition and multiplication defined on it, under which it forms a ring. Also, we can differentiate (and we have a *differential ring*). There are additional operations defined only for certain formal power series, such as infinite sums and products, and substitution; we will define these informally as required.

The *addition* and *multiplication* are exactly what you would expect: you add and multiply 'term-by-term'. That is,

$$\left(\sum_{n \geq 0} a_n t^n \right) + \left(\sum_{n \geq 0} b_n t^n \right) = \sum_{n \geq 0} (a_n + b_n) t^n,$$

and

$$\left(\sum_{n \geq 0} a_n t^n \right) \cdot \left(\sum_{n \geq 0} b_n t^n \right) = \sum_{n \geq 0} c_n t^n$$

where

$$c_n = \sum_{i=0}^{n} a_i b_{n-i}.$$

It can be checked with some effort that these operations are associative and commutative, and that the distributive law holds.

For example, we can sum geometric progressions:

$$\sum_{n \geq 0} (ct)^n = \frac{1}{1 - ct}.$$

This is easily verified by showing that $(1 - ct) \left(\sum_{n \geq 0} (ct)^n \right) = 1$.

Another very important operation on formal power series is *differentiation*:

$$\frac{\mathrm{d}}{\mathrm{d}t} \left(\sum_{n \geq 0} a_n t^n \right) = \sum_{n \geq 1} (n a_n) t^{n-1}.$$

The standard rules of elementary calculus for differentiating sums and products hold in this situation. The standard functions of analysis are *defined* as formal power series by their usual Taylor series: for example,

$$\exp(t) = \sum_{n \geq 0} \frac{t^n}{n!},$$

$$\log(1 + t) = \sum_{n \geq 1} \frac{(-1)^{n-1} t^n}{n}.$$

They satisfy the usual differential equations: $\frac{\mathrm{d}}{\mathrm{d}t} \exp(t) = \exp(t)$, $\frac{\mathrm{d}}{\mathrm{d}t} \log(1 + t) = 1/(1 + t)$.

We can add infinitely many formal power series as long as we are never required to add infinitely many field elements. So, for example, if $g(t) = \sum_{n \geq 1} b_n t^n$ is a formal power series whose constant term is zero, then $(g(t))^n$ has no term involving powers of t less than t^n. Thus it makes sense to evaluate

$$\sum_{n \geq 0} a_n (g(t))^n,$$

since a given term, say the term in t^m, only contains contributions from expressions $a_n(g(t))^n$ for $n \leq m$. The resulting formal power series is obtained by *substitution* of g into f, where $f(t) = \sum_{n \geq 0} a_n t^n$. We see that we can substitute one formal power series into another, provided the first has constant term zero.

Substitution behaves as one would expect: for example,

$$\exp(\log(1+t)) = 1+t,$$
$$\log(1+(\exp(t)-1)) = t.$$

(Note that $\log(1+t)$ and $\exp(t) - 1$ do have zero constant term.) Furthermore, if f and g have constant term 0, then $\exp(f)$ and $\exp(g)$ are defined, and

$$\exp(f) \cdot \exp(g) = \exp(f+g).$$

One notable example of a formal power series is provided by the *Binomial Theorem* for a general exponent. In our situation, the following statement is a definition (of $(1+t)^r$), not a theorem:[2]

(4.2.1) Binomial Theorem

For any real number r,

$$(1+t)^r = \sum_{n \geq 0} \binom{r}{n} t^n.$$

(Here the 'binomial coefficient' $\binom{r}{n}$ is defined by

$$\binom{r}{n} = \frac{r(r-1)\ldots(r-n+1)}{n!};$$

if r is a positive integer, this agrees with the usual definition, and it vanishes for $n > r$.)

Now it can be verified that the 'law of exponents' holds:

$$(1+t)^r \cdot (1+t)^s = (1+t)^{r+s}.$$

For $r = -1$, this agrees with our calculation of the sum of a geometric progression above (with $c = 1$). Moreover, we can define $((1+t)^r)^s$ by substitution, since $(1+t)^r$ has the form $1 + f(t)$ where f has constant term zero; and we find that

$$((1+t)^r)^s = (1+t)^{rs}.$$

Finally, we have

$$\frac{d}{dt}(1+t)^r = r(1+t)^{r-1}.$$

(This follows from the easily-checked identity $r\binom{n}{r} = n\binom{n-1}{r-1}$.)

One more important operation on formal power series is *infinite product*. Let $f_1, f_2, \ldots$ be formal power series with constant term 0. Then the product

$$\prod_{n \geq 1}(1 + f_n(t))$$

---

[2] Just as 'Zorn's Lemma' is an axiom of set theory, and 'Bertrand's Postulate' is a theorem.

should be defined by taking, in all possible ways, either 1 or f_n from the nth factor, multiplying these together, and adding the resulting terms. To avoid having to multiply infinitely many non-trivial terms, we specify that we choose 1 from all but finitely many of the factors; this gives a sum over all finite sets of natural numbers. There is still a potential problem; we have to ensure that only finitely many terms contribute to the coefficient of any given power of t. This will be true, for example, if $f_n(t)$ contains no terms of degree less than n in t. So, for example,

$$\prod_{n\geq 1}(1+t^n)$$

is defined — see Exercise 14. It can be shown that, if $\prod_{n\geq 1}(1+f_n(t)) = 1+g(t)$ is defined, then

$$\log(1+g(t)) = \sum_{n\geq 1}\log(1+f_n(t)).$$

Suppose that F is the field of real or complex numbers. Then, if the sequence $(a_0, a_1, \ldots)$ grows no faster than exponentially, its generating function will have non-zero radius of convergence,[3] and techniques of analysis can be used on it. However, for many interesting counting functions of combinatorial interest, the growth is faster than exponential, and the series must be treated formally. For example, the generating function for permutations is $\sum_{n\geq 0} n!t^n$. This diverges for all $t \neq 0$, and yet the coefficients in its inverse have combinatorial significance (see Exercise 13).

4.3. Linear recurrence relations with constant coefficients

The procedure for solving a general linear recurrence relation with constant coefficients is similar to that in the Fibonacci case. Consider the recurrence

$$F(n) = a_1 F(n-1) + a_2 F(n-2) + \ldots + a_k F(n-k).$$

Using the first method, we try a solution of the form $F(n) = \alpha^n$; we find that α must be a root of the polynomial

$$x^k = a_1 x^{k-1} + a_2 x^{k-2} + \ldots + a_k.$$

If this *characteristic equation* has all its roots distinct, then we obtain k independent solutions of the recurrence relation. Taking a linear combination of these, and fitting k initial values of F, we get k linear equations in k unknowns; these equations have a unique solution. So we have obtained the most general solution of the problem. However, if the characteristic polynomial has repeated roots, then we don't obtain enough solutions. In this case, suppose that α is a root of the characteristic equation with multiplicity d. Then it can be verified that the d functions $\alpha^n, n\alpha^n, \ldots, n^{d-1}\alpha^n$ are all solutions of the recurrence relation. Doing this for every root, we again find enough independent solutions that k initial values can be fitted.

---

[3] Recall from analysis that the *radius of convergence* of the power series $\sum_{n\geq 0} a_n t^n$ is given by

$$R = 1/\limsup_{n\to\infty}(a_n)^{1/n}.$$

The series converges for $|t| < R$ and diverges for $|t| > R$.

The justification of this is the fact that the solutions claimed can be substituted in the recurrence relation and its truth verified.

EXAMPLE. Solve the recurrence relation

$$F(n) = 3F(n-2) - 2F(n-3)$$

with initial values $F(0) = 3, F(1) = 1, F(2) = 8$.

The characteristic equation is

$$x^3 = 3x - 2,$$

with solutions $x = 1, 1, -2$. So the general solution of the recurrence relation is

$$F(n) = a(-2)^n + bn + c.$$

To fit the initial conditions, we require $a = b = 1, c = 2$, so the solution is $F(n) = (-2)^n + n + 2$.

4.4. Derangements and involutions

For linear recurrences with non-constant terms, or for non-linear recurrences, there is no general method which always works. Sometimes it is possible to solve such relations, either by guessing a solution (and verifying that it works), or by some other method. We give a couple of examples. In the first case, we solve the recurrence; in the second, we will merely derive some information about the solution.

EXAMPLE: DERANGEMENTS. A *derangement* of $1, 2, \ldots, n$ is a permutation of this set which leaves no point fixed. In Chapter 1, you were asked to calculate the number of derangements for $n \leq 5$. Now we will find the general formula. (This will be done again in Chapter 5 to illustrate a different technique, the *Principle of Inclusion and Exclusion*.)

Let $d(n)$ be the number of derangements of $\{1, \ldots, n\}$. Any derangement moves the point n to some point $i < n$. Clearly, the same number of derangements is obtained for each value of i from 1 to $n-1$; so we will find $d(n)$ by computing the number of derangements that map n to i and multiplying by $n-1$.

Let π be a derangement with $n\pi = i$. (Remember that permutations act on the right!) There are two cases:

CASE 1: $i\pi = n$. In other words, π interchanges n and i. Now it operates on the remaining $n-2$ points as a derangement. Furthermore, given any derangement of the points different from i and n, we may extend it to interchange i and n, and obtain a derangement of the entire set. So the number of derangements of this type is $d(n-2)$.

CASE 2: $i\pi \neq n$; say, $j\pi = n$ for some $j \neq i$. Now define a permutation π' of $\{1, \ldots, n-1\}$ by the rule

$$k\pi' = \begin{cases} k\pi, & \text{if } k \neq j; \\ i, & \text{if } k = j. \end{cases}$$

Then π' is a derangement. Any derangement π' of $\{1, \ldots, n-1\}$ can be 'extended' to a derangement π of $\{1, \ldots, n\}$, by reversing the construction. So there are $d(n-1)$ derangements under this case.

So we obtain

$$d(n) = (n-1)(d(n-1) + d(n-2)).$$

This is a three-term recurrence relation. The initial values are given by $d(0) = 1$, $d(1) = 0$.

(4.4.1) Theorem. *The number $d(n)$ of derangements of an n-set is given by*

$$d(n) = n! \left(\sum_{i=0}^{n} \frac{(-1)^i}{i!} \right).$$

This is the nearest integer to $n!/e$ for $n \geq 1$, where e is the base of natural logarithms.

REMARK. This demonstrates the claim made in Chapter 1, that if n letters are randomly distributed among n addressed envelopes, the probability that no letter is correctly addressed is close to $1/e$. (The problem asks for the probability that a random permutation is a derangement; this is $d(n)/n!$.)

To prove the theorem, we must show that the two sides of the equation satisfy the same recurrence relation and have the same initial values. So let $f(n) = n! \sum_{i=0}^{n} (-1)^i/i!$. Then

$$f(0) = 1 = d(0), \qquad f(1) = 1 = d(1).$$

Also

$$(n-1)(f(n-1) + f(n-2)) = (n-1) \cdot (n-1)! \sum_{i=0}^{n-1} \frac{(-1)^i}{i!}$$

$$+ (n-1) \cdot (n-2)! \sum_{i=0}^{n-2} \frac{(-1)^i}{i!}$$

$$= ((n-1) \cdot (n-1)! + (n-1) \cdot (n-2)!) \sum_{i=0}^{n-2} \frac{(-1)^i}{i!}$$

$$+ (-1)^{n-1}(n-1)$$

$$= n! \sum_{i=0}^{n-2} \frac{(-1)^i}{i!} + (-1)^{n-1} \frac{n!}{(n-1)!} + (-1)^n \frac{n!}{n!}$$

$$= n! \sum_{i=0}^{n} \frac{(-1)^i}{i!}$$

$$= f(n),$$

since $n - 1 = n!/(n-1)! - n!/n!$.

So the equality is established.

Now the Taylor series for e^{-1} is $\sum_{i=0}^{\infty} (-1)^i/i!$. Since this series has terms of alternating sign and decreasing in absolute value, the difference between the nth

term and the limit is less than the $(n+1)^{\text{st}}$ term. So

$$\left| d(n) - \frac{n!}{e} \right| = n! \left| \sum_{i=0}^{\infty} \frac{(-1)^i}{i!} - \sum_{i=0}^{n} \frac{(-1)^i}{i!} \right|$$

$$< n! \left| \frac{(-1)^{n+1}}{(n+1)!} \right|$$

$$= \frac{1}{n+1}$$

$$\leq \tfrac{1}{2} \text{ for } n \geq 1.$$

So $n!/e$ differs from the integer $d(n)$ by less than $\frac{1}{2}$. It follows that $d(n)$ is the nearest integer to $n!/e$, as claimed.

EXAMPLE: INVOLUTIONS. Here is an example of a naturally occurring sequence, with a simple recurrence relation where we won't find a simple formula either for the terms in the sequence itself or for a generating function for them (but see Exercise 18); however, we can get quite precise information just using the recurrence relation.

PROBLEM. How many permutations are there of a set of n elements having the property that all their cycles have length 1 or 2?

The cycles of a permutation refer to its expression as a product of disjoint cycles, found in the usual way. For example, $s(3) = 4$, counting the permutations $(1)(2)(3)$, $(1\ 2)(3)$, $(1\ 3)(2)$ and $(2\ 3)(1)$. Similarly, $s(2) = 2$, and $s(1) = 1$. (What is $s(0)$?)

Let $s(n)$ be the number of permutations satisfying this condition. As usual, we assume that the n-set is $\{1, 2, \ldots, n\}$, and divide the permutations into two classes:

- Those which fix the point n. These act on the set $\{1, \ldots, n-1\}$ as permutations with all cycles of length 1 or 2, so there are $s(n-1)$ of them.
- Those which don't fix n. If such a permutation moves n to i, say, then by assumption it contains a cycle $(n\ i)$, and it acts on the $n-2$ points other than n and i as a permutation with all cycles of length 1 or 2. There are $n-1$ choices for i, and for each choice, $s(n-2)$ choices for the permutation.

So we have the recurrence relation

$$s(n) = s(n-1) + (n-1)s(n-2).$$

This recurrence relation makes the calculation of further values easy. For example,

$$s(4) = 4 + 3 \cdot 2 = 10,$$
$$s(5) = 10 + 4 \cdot 4 = 26,$$
$$s(6) = 26 + 5 \cdot 10 = 76.$$

We demonstrate the following properties of the numbers $s(n)$:

(4.4.2) **Proposition.** (a) $s(n)$ is even for all $n > 1$;
(b) $s(n) > \sqrt{n!}$ for all $n > 1$.

PROOF. Both statements are proved by induction, being easily verified for $n = 2, 3$.

(i) If $s(n-1)$ and $s(n-2)$ are even, then $s(n) = s(n-1) + (n-1)s(n-2)$ is even. So induction applies.

(ii) Suppose that $s(n-1) > \sqrt{(n-1)!}$ and $s(n-2) > \sqrt{(n-2)!}$. Then

$$\begin{aligned}
s(n) &= s(n-1) + (n-1)s(n-2) \\
&> \sqrt{(n-1)!} + (n-1)\sqrt{(n-2)!} \\
&= \sqrt{(n-1)!}\,(1 + \sqrt{n-1}) \\
&> \sqrt{(n-1)!} \cdot \sqrt{n} \\
&= \sqrt{n!},
\end{aligned} \qquad (*)$$

and the induction goes through. (In $(*)$, we have used the fact that

$$1 + \sqrt{n-1} > \sqrt{n},$$

which is true because $(1 + \sqrt{n-1})^2 = n + 2\sqrt{n-1}$.)

REMARKS. 1. The second inequality is actually quite a good estimate.

2. The evenness of $s(n)$ is a special case of a general group-theoretic fact, in the case where G is the symmetric group $\mathrm{Sym}(n)$ of all permutations of $\{1,\dots,n\}$: *In a finite group G of even order n, the number of solutions of $x^2 = 1$ is even.* This is because the elements y for which $y^2 \neq 1$ come in pairs $\{y, y^{-1}\}$, and so are even in number.

4.5. Catalan and Bell numbers

In this section, we look at two important sequences of numbers. They have several, apparently accidental, common properties: both are 'named'; they start out similarly (the Catalan numbers are 1, 2, 5, 14, 42, ..., while the Bell numbers are 1, 2, 5, 15, 52, ...); and both are given by recurrence relations.

The Catalan numbers appear in many guises throughout combinatorics and computer science.[4] Here is a typical application:

> In how many ways can a sum of n terms be bracketed so that it can be calculated by adding two terms at a time?

For example, if $n = 4$, there are five possibilities:

$$\begin{aligned}
&(((a+b)+c)+d), \\
&((a+(b+c))+d), \\
&(a+((b+c)+d)), \\
&(a+(b+(c+d))), \\
&((a+b)+(c+d)).
\end{aligned}$$

---

[4] And elsewhere. Two of my colleagues, independently, asked me about the Catalan numbers which had come up in their research. One studies non-linear dynamics; the other, Lie superalgebras.

We have 'normalised' by enclosing the entire expression in an extra pair of brackets. (Note that, in an algebraic system where the operation is non-associative, these expressions could all have different values.)

Let C_n be the number of ways of bracketing a sum of n terms. To obtain a recurrence relation for C_n, note that any bracketed expression has the form (E_1+E_2), where E_1 and E_2 are bracketed expressions with (say) i and $n-i$ terms, for some i satisfying $1 \le i \le n-1$. There are C_i choices for E_i, and C_{n-i} for E_{n-i}. Summing over i, we obtain our first example of a non-linear recurrence relation:

(4.5.1) Recurrence relation for Catalan numbers

For $n > 1$,

$$C_n = \sum_{i=1}^{n-1} C_i C_{n-i}$$

Let $F(t) = \sum_{n \ge 1} C_n t^n$ be the generating function. (By convention, we take $C_0 = 0$; also, $C_1 = 1$, which is the start of the recurrence.) The recurrence relation shows that the terms in t^2 and higher powers of t in $F(t)^2$ are equal to those of $F(t)$. However, because the constant term is zero, $F(t)^2$ has no term in t. Thus, we have

$$F(t) = t + F(t)^2.$$

Re-writing this as a quadratic equation and solving, we obtain

$$F(t) = \frac{1}{2}\left(1 \pm (1 - 4t)^{1/2}\right).$$

Because $F(0) = 0$, we must choose the minus sign in the solution. Now, from the Binomial Theorem, we can read off the coefficient of t^n:

$$C_n = -\frac{1}{2}\binom{1/2}{n}(-4)^n$$

$$= -\frac{1}{2} \cdot \frac{1}{2} \cdot \frac{-1}{2} \cdot \frac{-3}{2} \cdots \frac{-(2n-3)}{2} \cdot (-4)^n/n!$$

$$= (2n-2)!/(n-1)!n!$$

(In the above expression, there are $n+1$ twos in the denominator, and $4^n/2^{n+1} = 2^{n-1}$. Then the product of all odd numbers from 1 to $2n-3$ is equal to $(2n-2)!/2^{n-1}(n-1)!$. Moreover, there are altogether $2n-2$ minus signs.) Thus:

(4.5.2) Catalan numbers

$$C_n = \frac{1}{n}\binom{2n-2}{n-1}.$$

See the Exercises for other combinatorial interpretations of Catalan numbers.

We encountered the Bell numbers briefly in Chapter 3. The Bell number B_n is the number of partitions of a set of size n. We proved there that it satisfies the recurrence relation

$$B_n = \sum_{i=1}^{n} \binom{n-1}{i-1} B_{n-i},$$

with the convention that $B_0 = 1$. This recurrence is linear, but involves all the preceding terms, rather than a fixed number.

There is no simple closed formula for B_n, but there is a nice expression for its generating function, which we now derive. This is a type of generating function we haven't met before. The *exponential generating function*, or e.g.f., of the sequence $(a_0, a_1, \ldots)$ is the formal power series

$$\sum_{n \geq 0} \frac{a_n t^n}{n!}.$$

The name comes from the fact that the e.g.f. of the all-1 sequence is just the ordinary exponential function $\exp(t)$. We will see in Part 2 that the exponential generating function is well suited to counting *labelled* objects, in the sense introduced in Chapter 2. Note that, if $F(t) = \sum_{n \geq 0} a_n t^n / n!$, then the derivative is $\frac{d}{dt} F(t) = \sum_{n \geq 1} a_n t^{n-1} / (n-1)!$; this is the e.g.f. of the sequence with the first term deleted.

Let $F(t) = \sum_{n \geq 0} B_n t^n / n!$ be the e.g.f. of the Bell numbers. Take the recurrence relation, multiply by $t^{n-1}/(n-1)!$, and sum over n, to obtain

$$\frac{d}{dt} F(t) = \sum_{n \geq 1} \frac{B_n t^{n-1}}{(n-1)!}$$

$$= \sum_{n \geq 1} \left(\sum_{i=1}^{n} \binom{n-1}{i-1} B_{n-i} \right) \frac{t^{n-1}}{(n-1)!}$$

$$= \sum_{n \geq 1} \sum_{i=1}^{n} \frac{t^{i-1}}{(i-1)!} \cdot \frac{B_{n-i} t^{n-i}}{(n-i)!}$$

$$= \left(\sum_{j \geq 0} \frac{t^j}{j!} \right) \cdot \left(\sum_{k \geq 0} \frac{B_k t^k}{k!} \right)$$

$$= \exp(t) F(t).$$

(In the penultimate line, we changed dummy variables to $j = i - 1$ and $k = n - i$; as n runs from 1 to ∞, and i from 1 to n, j and k independently take all non-negative integer values.)

Now we have

$$\frac{d}{dt}(\exp(-\exp(t)) F(t)) = 0,$$

so $F(t) = c \exp(\exp(t))$ for some constant c. Using the fact that $F(0) = 1$, we find that $c = \exp(-1)$; so

(4.5.3) E.g.f. for Bell numbers.

$$\sum_{n \geq 0} \frac{B_n t^n}{n!} = \exp(\exp(t) - 1).$$

4.6. Computing solutions to recurrence relations

In principle, nothing could be simpler than computing, say, Fibonacci numbers from their recurrence relation. By the way it works, knowing that $F_0 = F_1 = 1$, we find $F_2 = 1 + 1 = 2$, then F_3, and so on. For example,

$$F_{1000} = 70,330,367,711,422,815,821,835,254,877,183,549,770,181,$$
$$269,836,358,732,742,604,905,087,154,537,118,196,933,579,$$
$$742,249,494,562,611,733,487,750,449,241,765,991,088,186,$$
$$363,265,450,223,647,106,012,053,374,121,273,867,339,111,$$
$$198,139,373,125,598,767,690,091,902,245,245,323,403,501$$

takes just 999 additions to compute.[5]

However, there is an important point to consider. It is tempting to program the calculation exactly as the sequence is defined; that is, to define a function F on the natural numbers by the rules

- $F(0) = F(1) = 1$;
- $F(n) = F(n-1) + F(n-2)$ for $n > 1$.

But this is *not* wise. Let us trace the calculation of $F(4)$. We find that $F(4) = F(2) + F(3)$. First, we evaluate $F(2) = F(0) + F(1) = 1 + 1 = 2$. Next, we evaluate $F(3) = F(1) + F(2)$. Now the computer does not realise that $F(2)$ has already been calculated; it throws away its rough working. So we have to repeat the computation $F(2) = F(0) + F(1) = 1 + 1 = 2$ before we can find $F(3) = 1 + 2 = 3$ and finally $F(4) = 2 + 3 = 5$. For larger arguments, the amount of repeated labour grows exponentially (see Exercise 7).

So it is important to tell the computer to remember earlier results. For this, define an array of numbers $(F_0, F_1, \ldots, F_{1000})$ (if the largest Fibonacci number we'll need is F_{1000}), with the first two entries equal to 1, and each subsequent entry equal to the sum of the two before it.

This consideration applies to any sequence of numbers defined by a recurrence relation of any sort. In the specific case of Fibonacci numbers, if we only need one number F_n rather than the whole sequence $F_0, \ldots, F_n$, it's possible to economise on storage space. We only need to remember two numbers, say x and y (and a counter n). Start with $x = y = 1$ and $n = 1$. Now, in a single step,

- increase n by 1;
- calculate $x + y$, and replace either x or y by this number according as n is even or odd.

The last number written (viz., x or y depending on the parity of n) is the nth Fibonacci number.

It is possible to calculate F_n faster than this, using only $c \log n$ arithmetic operations, using the 'Russian peasant multiplication' trick. Exercise 8 gives details.

---

[5] On the other hand, if we were to use the formula, we would be faced with the need to calculate $((\sqrt{5} + 1)/2\sqrt{5})((\sqrt{5} + 1)/2)^{1000}$ to such high accuracy that the final answer is guaranteed to have an error of less than 0.5 — a much more difficult task!

4.7. Project: Finite fields and QUICKSORT

In this section, we will work through two more elaborate applications of recurrence relations and generating functions. We prove the existence of irreducible polynomials over finite fields; and we calculate the average number of comparisons needed to sort a list using QUICKSORT.[6]

IRREDUCIBLE POLYNOMIALS AND FINITE FIELDS.

If p is a prime, the integers modulo p form a field: addition, subtraction, multiplication and division (except by zero) are defined, and the commutative, associative and distributive laws hold. What other finite fields exist?

This question was answered by Galois in the nineteenth century.[7] He proved the following result:

> **(4.7.1) Galois' Theorem**
> *The number of elements in a finite field is a prime power; and, for any prime power q, there is a unique field with q elements.*

The field with q elements is called the *Galois field* of order q, denoted by $\mathrm{GF}(q)$. Thus, if p is prime, then $\mathrm{GF}(p) = \mathbb{Z}/(p)$, the integers mod p. Suppose that $q = p^n$. Then a field of order q is constructed from a polynomial

$$f(x) = x^n + b_{n-1}x^{n-1} + \ldots + b_1 x + b_0$$

over $\mathbb{Z}/(p)$, which has degree n, is *monic* (leading coefficient 1), and is irreducible: the elements of the field are the p^n expressions

$$c_0 + c_1 \alpha + \ldots + c_{n-1}\alpha^{n-1}$$

for $c_0, c_1, \ldots, c_{n-1} \in \mathbb{Z}/(p)$; addition and multiplication are defined in the obvious way, but setting $f(\alpha) = 0$ where necessary to reduce the degree of any expression to $n-1$ or less. (Compare the construction of the complex numbers as the set of objects of the form $a + bi$ for $a, b \in \mathbb{R}$, where $i^2 = -1$; note that the polynomial $x^2 + 1$ is irreducible over $\mathbb{R}$.)

The point of this brief discussion is that the existence of finite fields will follow if we can show that there is an irreducible polynomial of any possible degree over $\mathbb{Z}/(p)$. We will prove this in the most naïve way possible, by counting the polynomials. We need one algebraic fact: *a monic polynomial over a field can be factorised into monic irreducible factors, uniquely up to the order of the factors.*

Fix a prime power q, and let F be a field of order q. (For Galois' Theorem, take $q = p$ prime, and $F = \mathbb{Z}/(p)$.) Let a_n be the number of monic irreducible polynomials of degree n over F. The total number of monic polynomials of degree n is q^n, since each of the n coefficients $b_{n-1}, \ldots, b_1, b_0$ can be chosen arbitrarily from F.

---

[6] I am indebted to Colin McDiarmid for the second example.

[7] Évariste Galois was killed in a duel at the age of 20. The night before the duel, he had written all his recent mathematical discoveries in a hastily scrawled letter to a friend; this document can be regarded as the foundation of modern algebra, though its influence was not felt until its publication by Liouville fifteen years later. The theorem on finite fields, however, is one of the few pieces of his work published during his lifetime.

Now an arbitrary polynomial has a unique factorisation into irreducibles. Consider those polynomials which have m_1 factors of degree 1, m_2 of degree 2, and so on. We must have $m_1 + 2m_2 + \ldots = n$. The m_i factors of degree i are chosen from the set of a_i irreducibles of degree i; repetition is allowed, and the order of the factors is not important. By (3.7.1), there are $\binom{a_i + m_i - 1}{m_i}$ choices for these factors, and hence

$$\prod_{i \geq 1} \binom{a_i + m_i - 1}{m_i}$$

polynomials with a factorisation of this shape. So, counting all monic polynomials of degree n, we have

$$\sum_{m_1 + 2m_2 + \ldots = n} \prod_{i \geq 1} \binom{a_i + m_i - 1}{m_i} = q^n. \qquad (*)$$

This is a recurrence relation (albeit a highly complicated, non-linear one). We illustrate the case $q = 2$.

$$a_1 = 2,$$

$$a_2 + \binom{a_1 + 1}{2} = 4,$$

$$a_3 + a_1 a_2 + \binom{a_1 + 2}{3} = 8,$$

$$a_4 + a_1 a_3 + \binom{a_2 + 1}{2} + \binom{a_1 + 1}{2} a_2 + \binom{a_1 + 3}{4} = 16,$$

from which we obtain successively $a_1 = 2$, $a_2 = 1$, $a_3 = 2$, $a_4 = 3$.

The point of this section is that, by sleight-of-hand with generating functions, we transform this recurrence relation into a very much simpler one, from which (for example) the fact that $a_n > 0$ can be seen directly.

Multiply equation $(*)$ by t^n and sum over n:

$$\frac{1}{1 - qt} = \sum_{n \geq 0} q^n t^n$$

$$= \sum_{n \geq 0} t^n \sum_{m_1 + 2m_2 + \ldots = n} \prod_{i \geq 0} \binom{a_i + m_i - 1}{m_i}$$

$$= \sideset{}{'}\sum_{m_1, m_2, \ldots} \prod_{i \geq 0} \binom{a_i + m_i - 1}{m_i} t^{im_i}.$$

(The last step needs a little explanation. If we sum over all n and then over all choices of $m_1, m_2, \ldots$ satisfying $m_1 + 2m_2 + \ldots = n$, we have simply summed over all sequences $(m_1, m_2, \ldots)$ with only finitely many non-zero terms; this is what is meant by the prime on the summation sign. Furthermore,

$$t^n = \prod_{i \geq 0} t^{im_i},$$

so the power of t can be split up as claimed.)

Now the main technical step: I claim that the above expression is equal to

$$\prod_{i \geq 1} \sum_{m \geq 0} \binom{a_i + m - 1}{m} t^{im}.$$

This is because, to evaluate an infinite product of this form, we choose one term from each factor in all possible ways so that all but finitely many choices are equal to 1, multiply the chosen terms, and add the results; say we choose the m_i^{th} term from the ith factor, where finitely many m_i are non-zero. This gives just the sum previously described.

Now note that
$$\binom{a+m-1}{m} = (-1)^m \binom{-a}{m};$$
so, by the Binomial Theorem, we have
$$\sum_{m\geq 0} \binom{a_i + m - 1}{m} t^{im} = (1 - t^i)^{-a_i}.$$

So we have
$$(1 - qt)^{-1} = \prod_{i\geq 1}(1 - t^i)^{-a_i}.$$

Now comes the trick. We take logarithms of both sides:
$$-\log(1 - qt) = -\sum_{i\geq 1} a_i \log(1 - t^i),$$
whence
$$\sum_{n\geq 1} \frac{(qt)^n}{n} = \sum_{i\geq 1} a_i \sum_{k\geq 1} \frac{t^{ik}}{k}.$$

Now we equate the coefficients of t^n on both sides. On the right, we obtain a term for each pair (i, k) with $ik = n$; in other words, for each divisor i of n.
$$\frac{q^n}{n} = \sum_{i|n} \frac{a_i}{n/i}.$$

Multiplying by n gives
$$q^n = \sum_{i|n} i a_i. \qquad\qquad (**)$$

This is our desired recurrence relation. It is linear, and has many fewer terms than $(*)$. To re-do the case $q = 2$:
$$a_1 = 2,$$
$$a_1 + 2a_2 = 4,$$
$$a_1 + 3a_3 = 8,$$
$$a_1 + 2a_2 + 4a_4 = 16.$$

In Chapter 12, we will discuss *Möbius inversion*, and solve this recurrence relation explicitly. But, in the meantime, observe that q^n is the sum of at most n terms, of which all except na_n are at most $q^{n/2}$ (since they occur in earlier recurrence relations). In general, $q^n > (n-1)q^{n/2}$; so $a_n > 0$. Thus, *there exists an irreducible polynomial of any degree over any finite field*.

With a little more algebra, the recurrence relation $(**)$ can be used to show the uniqueness in Galois' Theorem as well. (In outline: one shows that any element of a field of order q^n satisfies an irreducible polynomial over the subfield of order q whose degree divides n. Now the a_i irreducible polynomials of degree i have at most ia_i roots; and $(**)$ shows that these roots are just sufficient in number to comprise *one* field of order q^n.)

It is instructive to compare the very different proof of Galois' Theorem normally given in algebra text-books. It is possible to use that proof, and the counting of roots as in the preceding paragraph, to give another proof of $(**)$.

THE PERFORMANCE OF QUICKSORT.

A great deal of computer time is spent in sorting lists — arranging the elements in order, if they were originally arranged haphazardly. It is important to be able to do this efficiently, and to estimate how complex a task it is.

Many important algorithms are recursive: they solve a given problem instance by reducing it to smaller instances of the same problem. Thus, the average (or longest) time taken to solve a problem of size n can be expressed in terms of the time for smaller problems, giving rise to a recurrence relation. As an example, I will calculate the average number of comparisons taken by Hoare's QUICKSORT algorithm to sort a randomly ordered list of n items.

The algorithm is defined as follows.

(4.7.2) QUICKSORT
to sort a list L

Let a be the first item of the list.
- Partition the remainder of the list into sublists L^-, L^+ consisting of the elements less than, greater than a respectively.
- Sort L^- and L^+.
- Return $(L^-$ (sorted), a, L^+ (sorted)).

We will calculate the average number of comparisons of individual elements which have to be made, assuming that the algorithm is presented with a list in random order (that is, all orderings equally likely). But first, what answer do we expect? There are $n!$ possible orderings; since each comparison can at best narrow down the number of possibilities to half the previous value (on average), we would expect to need at least $\log_2 n!$ comparisons.[8] By (3.6.1),[9]

$$\log_2 n! = n \log n / \log 2 + O(n) = 1.4427\ldots n \log n + O(n).$$

We will show that the average number of comparisons required by QUICKSORT is only a constant factor worse than this lower bound, namely $2n \log n + O(n)$.

The crucial observation is that, if the list L is in random order, then
- the first element a is equally likely to be the first, second, $\ldots$, n^{th} smallest element;
- the sublists L^- and L^+ are randomly ordered (i.e. all orderings equally likely).

Let q_n be the average number of comparisons required to sort a list of length n. Thus we have

$$q_n = n - 1 + \frac{1}{n} \sum_{k=1}^{n} (q_{k-1} + q_{n-k}).$$

(The first step requires $n-1$ comparisons; if a is the kth smallest element, the second step requires an average of $q_{k-1} + q_{n-k}$ comparisons, and this number has to be averaged over the possible values of k.) We can simplify this to

$$q_n = n - 1 + \frac{2}{n} \sum_{k=0}^{n-1} q_k,$$

since each of $q_0, \ldots, q_{n-1}$ occurs twice in the sum.

The initial value is clearly $q_0 = 0$.

To solve this recurrence relation, we find a differential equation for its generating function. Let

$$Q(t) = \sum_{n \geq 0} q_n t^n$$

be the generating function. Multiplying the recurrence relation by nt^n and summing gives

$$\sum_{n \geq 0} n q_n t^n = \sum_{n \geq 0} n(n-1)t^n + 2 \sum_{n \geq 0} \left(\sum_{i=0}^{n-1} q_i \right) t^n.$$

We analyse the three terms. The second is just the Taylor series of $2t^2/(1-t)^3$. (We have

$$\sum_{n \geq 0} n(n-1)t^{n-2} = 2/(1-t)^3,$$

---

[8] This might be called the 'Twenty Questions' principle. For a proof, see Exercise 23.

[9] The notation $O(f(n))$ means 'a function whose absolute value is bounded by $cf(n)$ for some constant c, as $n \to \infty$'.

most easily by differentiating twice the series for $1/(1-t)$, or alternatively by the Binomial Theorem.)
The first term is $tQ'(t)$, since

$$Q'(t) = \sum_{n \geq 0} nq_n t^{n-1}.$$

The last term is the most difficult; I claim that it is $2tQ(t)/(1-t)$. This is because

$$(t/(1-t))Q(t) = (t + t^2 + t^3 + \ldots) \cdot (q_0 + q_1 t + q_2 t^2 + q_3 t^3 + \ldots),$$

and the t^n term is obtained by multiplying t^{n-i} from the first factor and $q_i t^i$ from the second, and summing over i.

Thus, we have

$$tQ'(t) = \frac{2t^2}{(1-t)^3} + \frac{2t}{(1-t)} Q(t).$$

This is a first-order linear differential equation, for which there is a standard method for solution. Without going through the general case, we have

$$\left((1-t)^2 Q(t)\right)' = (1-t)^2 Q'(t) - 2(1-t)Q(t) = \frac{2t}{(1-t)},$$

so

$$(1-t)^2 Q(t) = -2(t + \log(1-t))$$

(using the fact that $Q(0) = 0$). Hence

$$Q(t) = \frac{-2(t + \log(1-t))}{(1-t)^2}.$$

It still seems a tall order to find the coefficients in this power series explicitly; but it can be done. We have

$$Q(t) = 2 \left(\frac{t^2}{2} + \frac{t^3}{3} + \ldots \right) \cdot (1 + 2t + 3t^2 + \ldots),$$

so

$$q_n = 2 \sum_{i=2}^{n} \left(\frac{1}{i} \right) (n - i + 1)$$

$$= 2(n+1) \sum_{i=1}^{n} \left(\frac{1}{i} \right) - 4n.$$

This is an exact formula, though it involves a sum of n terms. We can produce an approximation by using the fact that[10]

$$\sum_{i=1}^{n} \left(\frac{1}{i} \right) = \log n + O(1),$$

whence

$$q_n = 2n \log n + O(n),$$

as we promised.

---

[10] The sum is an approximation to the area under the curve $y = 1/x$ from $x = 1$ to $x = n$.

4.8. Exercises

FIBONACCI NUMBERS. In these exercises, F_n denotes the nth Fibonacci number.

1. (a) There are n seating positions arranged in a line. Prove that the number of ways of choosing a subset of these positions, with no two chosen positions consecutive, is F_{n+1}.

(b) If the n positions are arranged around a circle, show that the number of choices is $F_n + F_{n-2}$ for $n \geq 2$.

2. Prove the following identities:

(a) $F_n^2 - F_{n+1}F_{n-1} = (-1)^n$ for $n \geq 1$.

(b) $\sum_{i=0}^{n} F_i = F_{n+2} - 1$.

(c) $F_{n-1}^2 + F_n^2 = F_{2n}, \qquad F_{n-1}F_n + F_n F_{n+1} = F_{2n+1}$.

(d) $F_n = \sum_{i=0}^{\lfloor n/2 \rfloor} \binom{n-i}{i}$.

3. Show that F_n is composite for all *odd* $n > 3$.

4. Show that

$$\sum_{i=0}^{\lfloor (n-1)/2 \rfloor} F_{n-2i} = F_{n+1} - 1$$

for $n \geq 1$.

5. Prove that every non-negative integer x less than F_{n+1} can be expressed in a unique way in the form

$$F_{i_1} + F_{i_2} + \ldots + F_{i_r}, \qquad\qquad (*)$$

where $i_1, i_2, \ldots, i_r \in \{1, \ldots, n\}$, $i_1 > i_2 + 1$, $i_2 > i_3 + 1$, ... (in other words, $i_1, \ldots, i_r$ are all distinct and no two are consecutive). Deduce Exercise 1(a).

[HINT: By Exercise 4, the largest expression of the form $(*)$ that can be made using Fibonacci numbers below F_n is $F_n - 1$. So, if $F_n \leq x < F_{n+1}$, then F_n must be included in the sum; and $x - F_n < F_{n-1}$, so F_{n-1} cannot be included.]

6. Fibonacci numbers are traditionally associated with the breeding of rabbits.[11] Assume that a pair of rabbits does not breed in its first month, and that it produces a pair of offspring in each subsequent month. Assume also that rabbits live forever. Show that, starting with one newborn pair of rabbits, the number of pairs alive in the nth month is F_n.

7. Prove that the number of additions required to compute the Fibonacci number F_n according to the 'inefficient' algorithm described in the text is $F_n - 1$.

8. (a) Prove that $F_{m+n} = F_m F_n + F_{m-1} F_{n-1}$ for $m, n \geq 0$ (with the convention that $F_{-1} = 0$).

(b) Use this to derive an algorithm for calculating F_n using only $c \log n$ arithmetic operations. [HINT: see Russian peasant multiplication (Exercise 12 of Chapter 2).]

(c) Given that multiplication is slower than addition, is this algorithm really better than one involving $n - 1$ additions?

---

[11] This example is due to Fibonacci (Leonardo of Pisa) himself.

MISCELLANEOUS RECURRENCES AND GENERATING FUNCTIONS.

9. (a) Solve the following recurrence relations.
  (i) $f(n+1) = f(n)^2$, $f(0) = 2$.
  (ii) $f(n+1) = f(n) + f(n-1) + f(n-2)$, $f(0) = f(1) = f(2) = 1$.
  (iii) $f(n+1) = 1 + \sum_{i=0}^{n-1} f(i)$, $f(0) = 1$.

  (b) Show that the number of ways of writing n as a sum of positive integers, where the order of the summands is significant, is 2^{n-1} for $n \geq 1$.

10. The number $f(n)$ of steps required to solve the 'Chinese rings puzzle' with n rings satisfies $f(1) = 1$ and

$$f(n+1) = \begin{cases} 2f(n), & n \text{ odd}, \\ 2f(n) + 1, & n \text{ even}. \end{cases}$$

Prove that $f(n+2) = f(n+1) + 2f(n) + 1$. Hence or otherwise find a formula for $f(n)$.[12]

11. (a) Let $s(n)$ be the number of sequences $(x_1, \ldots, x_k)$ of integers satisfying $1 \leq x_i \leq n$ for all i and $x_{i+1} \geq 2x_i$ for $i = 1, \ldots, k-1$. (The length of the sequence is not specified; in particular, the empty sequence is included.) Prove the recurrence

$$s(n) = s(n-1) + s(\lfloor n/2 \rfloor)$$

for $n \geq 1$, with $s(0) = 1$. Calculate a few values of s. Show that the generating function $S(t)$ satisfies $(1-t)S(t) = (1+t)S(t^2)$.

  (b) Let $u(n)$ be the number of sequences $(x_1, \ldots, x_k)$ of integers satisfying $1 \leq x_i \leq n$ for all i and $x_{i+1} > \sum_{j=1}^{i} x_j$ for $i = 1, \ldots, k-1$. Calculate a few values of u. Can you discover a relationship between s and u? Can you prove it?

12. Let $F(t)$ be a formal power series with constant term 1. By finding a recurrence relation for its coefficients, show that there is a multiplicative inverse $G(t)$ of $F(t)$. Moreover, if the coefficients of F are integers, so are those of G.

13. A permutation π of the set $\{1, \ldots, n\}$ is called *connected* if there does not exist a number k with $1 \leq k < n$ such that π maps the subset $\{1, 2, \ldots, k\}$ into itself. Let c_n be the number of connected permutations. Prove that

$$\sum_{i=1}^{n} c_i(n-i)! = n!$$

Deduce that, if $F(t) = \sum_{n \geq 1} n! t^n$ and $G(t) = \sum_{n \geq 1} c_n t^n$ are the generating functions of the sequences $(n!)$ and (c_n) respectively, then $1 - G(t) = (1 + F(t))^{-1}$. (Note that $F(t)$ and $G(t)$ diverge for all $t \neq 0$.)

---

[12] The formula, and an algorithm for solution, were given in 1872 in *Théorie du Baguenodier*, by 'Un Clerc de Notaire Lyonnais' (now identified as Louis Gros). See S. N. Afriat, *The Ring of Linked Rings* (1982).

14. Let
$$\prod_{n\geq 1}(1+t^n) = \sum_{n\geq 0} a_n t^n.$$

Prove that a_n is the number of ways of writing n as the sum of *distinct* positive integers. (For example, $a_6 = 4$, since $6 = 5+1 = 4+2 = 3+2+1$.)

15. (a) In an election, there are two candidates, A and B; the number of votes cast is $2n$. Each candidate receives exactly n votes; but, at every intermediate point during the count, A has received more votes than B. Show that the number of ways this can happen is the Catalan number C_n. [HINT: A leads by just one vote after the first vote is counted. Suppose that this next occurs after $2i + 1$ votes have been counted. Then there are $f(i)$ choices for the count between these points, and $f(n-i)$ choices for the rest of the count, where $f(n)$ is the required number; so we obtain the Catalan recurrence.]
HARDER PROBLEM. Can you construct a bijection between the bracketed expressions and the voting patterns in (a)?

(b) In the above election, assume only that, at any intermediate stage, A has received at least as many votes as B. Prove that the number of possibilities is now C_{n+1}. [HINT: Give A an extra vote at the beginning of the count, and B an extra vote at the end.]

16. A clown stands on the edge of a swimming pool, holding a bag containing n red and n blue balls. He draws the balls out one at a time and discards them. If he draws a blue ball, he takes one step back; if a red ball, one step forward. (All steps have the same size.) Show that the probability that the clown remains dry is $1/(n+1)$.

17. Prove that
$$\lim_{n\to\infty} \left(\frac{B_n}{n!}\right)^{1/n} = 0.$$

[HINT: See the footnote on p. 56.]

18. (a) Prove that the exponential generating function for the number $s(n)$ of involutions on $\{1,\ldots,n\}$ (Section 4.4) is $\exp(t + \frac{1}{2}t^2)$.
(b) Prove that the exponential generating function for the number $d(n)$ of derangements of $\{1,\ldots,n\}$ is $1/((1-t)\exp(t))$.

19. The *Bernoulli numbers* $B(n)$ (not to be confused with the Bell numbers B_n) are defined by the recurrence $B(0) = 1$ and
$$\sum_{k=0}^{n} \binom{n+1}{k} B(k) = 0$$

for $n \geq 1$. Prove that the exponential generating function
$$f(t) = \sum_{n\geq 0} \frac{B(n)t^n}{n!}$$

is given by $f(t) = t/(\exp(t)-1)$.
    Show that $f(t) + \frac{1}{2}t$ is an even function of t, and deduce that $B(n) = 0$ for all odd $n \geq 3$.

REMARK. The Bernoulli numbers play an important and unexpected rôle in topics as diverse as numerical analysis, Fermat's last theorem and p-adic integration.

What is the solution of the similar-looking recurrence $b(0) = 1$ and

$$\sum_{k=0}^{n} \binom{n}{k} b(k) = 0$$

for $n \geq 1$?

20. For even n, let e_n be the number of permutations of $\{1, \ldots, n\}$ with all cycles even; o_n, the number of permutations with all cycles odd; and $p_n = n!$ the total number of permutations. Let $E(t)$, $O(t)$ and $P(t)$ be the exponential generating functions of these sequences. Show that
(a) $P(t) = (1 - t^2)^{-1}$;
(b) $E(t) = (1 - t^2)^{-1/2}$; [HINT: Exercise 15 of Chapter 3]
(c) $E(t).O(t) = P(t)$;
(d) $e_n = o_n$ for all even n.
HARD EXERCISE. Find a 'bijective' proof of the last equality. (See R. P. Lewis and S. P. Norton, *Discrete Mathematics* **138** (1995), 315–318, for a solution.)
QUESTIONS ON QUICKSORT AND BINARY TREES.

21. Show that QUICKSORT sometimes requires all $\binom{n}{2}$ comparisons to sort a list. For how many orderings does this occur? One such ordering is the case when the list is already sorted — is this a serious defect of QUICKSORT?

22. Let m_n be the minimum number of comparisons required by QUICKSORT to sort a list of length n. Prove that, for each integer $k > 1$, m_n is a linear function of n on the interval from $2^{k-1} - 1$ to $2^k - 1$, with

$$m_{2^k-1} = (k - 2)2^k + 2.$$

If $n = 2^k - 1$, what can you say about the number of orderings requiring m_n comparisons?

23. This exercise justifies the 'Twenty Questions' principle. We are given N objects and required to distinguish them by asking questions, each of which has two possible answers. The aim of this exercise is to show that, no matter what scheme of questioning is adopted, on average the number of questions required is at least $\log_2 N$. (For some schemes, the average may be much larger. If we ask 'Is it a_1?', 'Is it a_2?', etc., then on average $(N + 1)/2$ questions are needed!)
   A *binary tree* is a graph (see Chapter 2) with the following properties:
   • there is a vertex (the *root*) lying on just two edges;
   • every other vertex lies on one or three edges (and is called a *leaf* or an *internal vertex* accordingly);
   • there are no circuits (closed paths of distinct vertices), and every vertex can be reached by a path from the root.
   It is convenient to arrange the vertices of the tree on successive *levels*, with the root on level 0. Then any non-leaf is joined to two *successors* on the next level, and every vertex except the root has one *predecessor*. The *height* of a vertex is the number of the level on which it lies.
   In our situation, a vertex is any set of objects which can be distinguished by some sequence of questions. The root corresponds to the whole set (before any questions are asked), and leaves are singleton sets. The two successors of a vertex are the sets distinguished by the two possible answers to the next question. The height of a leaf is the number of questions required to identify that object uniquely.
   STEP 1. Show that there are two leaves of maximal height (h, say) with the same predecessor. Deduce that, if there is a leaf of height less than $h - 1$, we can find another binary tree with N leaves having smaller average height. Hence conclude that, in a tree with minimum average height, every leaf has height m or $m + 1$, for some m.

STEP 2. Since there are no leaves at height less than m, there are altogether 2^m vertices on level m.

STEP 3. If there are p internal vertices on level m, show that there are $2p$ leaves of height $m+1$, and $N - 2p = 2^m - p$ of height m; so $N = 2^m + p$, where $0 \leq p < 2^m$.

STEP 4. Prove that $\log_2(2^m + p) \leq m + 2p/(2^m + p)$, and deduce that the average height of leaves is at least $\log_2 N$.

REMARK. Sorting a list is equivalent to finding the permutation which takes the given order of the list into the 'correct' order; thus it involves identifying one of $n!$ possibilities. So any sorting method which compares elements of the list will require, on average, at least $\log_2 n! = n \log n / \log 2 + O(n)$ comparisons, as claimed in the text. Figure 4.1 shows the binary tree for QUICKSORT with $n = 3$.

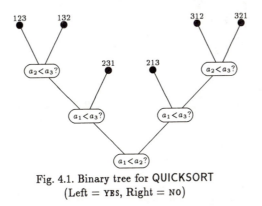

Fig. 4.1. Binary tree for QUICKSORT
(Left = YES, Right = NO)

24. Suppose that the two successors of each non-leaf node in a binary tree are distinguished as 'left' and 'right'. Show that, with this convention, the number of binary trees with n leaves is the Catalan number C_n. [HINT: Removing the root gives two binary trees, a 'left' and a 'right' tree. Use this to verify the recurrence relation.]

5. The Principle of Inclusion and Exclusion

To every thing there is a season, and a time to every purpose under the heaven:
A time to be born, and a time to die; a time to plant, and a time to pluck up that which is planted;
A time to kill, and a time to heal; a time to break down, and a time to build up;
A time to weep, and a time to laugh; a time to mourn, and a time to dance;
A time to cast away stones, and a time to gather stones together; a time to embrace, and a time to refrain from embracing;
A time to get, and a time to lose; a time to keep, and a time to cast away;
A time to rend, and a time to sew; a time to keep silence, and a time to speak;
A time to love, and a time to hate; a time of war, and a time of peace.

Ecclesiastes, Chapter 3

TOPICS: Principle of Inclusion and Exclusion; Stirling numbers; even and odd permutations

TECHNIQUES: Generating function tricks; matrix inverses

ALGORITHMS:

CROSS-REFERENCES: set-partitions, cycles of permutations, inverse of Pascal's triangle (Chapter 3); derangements, exponential generating function, [Bernoulli numbers] (Chapter 4); Möbius inversion (Chapter 12)

Suppose we are given a family of sets, and told the number of elements which lie simultaneously in every set of each possible subfamily. Then we have enough information to work out how many elements lie in none of the sets, or indeed, how many lie in each region of the Venn diagram of the family. The *Principle of Inclusion and Exclusion*, known as PIE for short, is a formula for calculating this. It gives rise to another proof of the theorem about inverting Pascal's triangle, as well as a formula for the number of partitions of an n-set into k parts. This last number is a so-called *Stirling number of the second kind*. We spend the second half of the chapter investigating these numbers and their relatives, and their surprising properties.

5.1. PIE

In a class of 100 pupils, a survey establishes that 45 play cricket, 53 play tiddlywinks, and 55 play Space Invaders. Furthermore, 28 play cricket and tiddlywinks; 32 play cricket and Space Invaders; 35 play tiddlywinks and Space Invaders; and 20 play all three sports. How many pupils don't play any sport?

This problem can be answered by drawing a Venn diagram to represent the three sets. Then the numbers in each region can be worked out in turn, until finally the number in none of the regions is found. For example, 8 pupils play cricket and tiddlywinks but not Space Invaders.

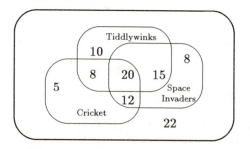

Fig. 5.1. A Venn diagram

The *Principle of Inclusion and Exclusion* gives a formula for this calculation, not relying on our ability to draw meaningful Venn diagrams with arbitrarily many sets.

First, some notation. Let X be our 'universe' (corresponding to the whole class in the example), and let $A_1, A_2, \ldots, A_n$ be subsets of X. (It is not forbidden that some set occurs more than once in the sequence.) If I is a subset of the index set $\{1, \ldots, n\}$, we set

$$A_I = \bigcap_{i \in I} A_i,$$

with the convention that $A_\emptyset = X$. (Intersecting more sets gives a smaller result; so intersecting no sets at all should give the largest possible set.)

(5.1.1) Principle of Inclusion and Exclusion
Let $(A_1, \ldots, A_n)$ be a *family of subsets of X. Then the number of elements of X which lie in none of the subsets A_i is*

$$\sum_{I \subseteq \{1,\ldots,n\}} (-1)^{|I|} |A_I|.$$

PROOF. The sum on the right is a linear combination of cardinalities of sets A_I with coefficients $+1$ or -1. We calculate, for each point of X, its 'contribution' to the sum, that is, the sum of the coefficients of the sets A_I which contain it.

Suppose first that $x \in X$ lies in none of the sets A_i. Then the only term in the sum to which x contributes is that with $I = \emptyset$; and its contribution is 1.

Otherwise, the set $J = \{i \in \{1, \ldots, n\} : x \in A_i\}$ is non-empty; and $x \in A_I$ precisely when $I \subseteq J$. Thus, the contribution of x is

$$\sum_{I \subseteq J} (-1)^{|I|} = \sum_{i=0}^{j} \binom{j}{i} (-1)^i$$
$$= (1-1)^j = 0.$$

by the Binomial Theorem, where $j = |J|$.

Thus, points lying in no set A_i contribute 1 to the sum, while points in some A_i contribute 0; so the overall sum is the number of points lying in none of the sets, as claimed.

PIE has a natural interpretation for small n. For $n = 2$, we take the number of points in X, and subtract the sum of the numbers in A_1 and A_2; the points in $A_1 \cap A_2$ have been subtracted twice, and must be added in again.[1] For $n = 3$, after the pairwise intersections have been added, we find that the points lying in all three sets have been included once too often, and must be removed again.

We proceed to a couple of applications of PIE.

(5.1.2) Corollary. *The number of surjective mappings from an n-set to a k-set is given by*

$$\sum_{i=0}^{k} (-1)^i \binom{k}{i} (k - i)^n.$$

In particular, we have

$$n! = \sum_{i=0}^{n} (-1)^i \binom{n}{i} (n - i)^n.$$

PROOF. We take X to be the set of all mappings from $\{1, \ldots, n\}$ to $\{1, \ldots, k\}$, so that $|X| = k^n$. For $i = 1, \ldots, k$, we let A_i be the set of mappings f for which the point i does *not* lie in the range of f. Then each $f(x)$ can be any of the $k - 1$ points different from i, and so $|A_i| = (k - 1)^n$. More generally, A_I consists of all mappings whose range contains no point of I, and $|A_I| = (k - |I|)^n$.

A mapping is a surjection if and only if it lies in none of the sets A_i. So, by PIE, the number of surjections is equal to

$$\sum_{I \subseteq \{1, \ldots, k\}} (-1)^{|I|} (k - |I|)^n.$$

Put $i = |I|$. There are $\binom{k}{i}$ sets I of cardinality i, where i runs from 1 to k; this gives the result.

If $k = n$, then the permutations of $\{1, \ldots, n\}$ are precisely the surjective mappings from this set to itself.

For a second application, we give a second proof of the formula for the number of derangements.

(5.1.3) Theorem. *The number of derangements of $\{1, \ldots, n\}$ is equal to*

$$n! \sum_{i=0}^{n} \frac{(-1)^i}{i!}.$$

PROOF. This time, we take X to be the set of permutations, and A_i the set of permutations fixing the point i; so $|A_i| = (n - 1)!$, and more generally, $|A_I| =$

---

[1] This gives the familiar identity $|A_1 \cup A_2| + |A_1 \cap A_2| = |A_1| + |A_2|$ (see Section 2.7).

$(n - |I|)!$, since permutations in A_I fix every point in I and permute the remaining points arbitrarily. A permutation is a derangement if and only if it lies in none of the sets A_i; so the number of derangements is

$$\sum_{I \subseteq \{1,\dots,n\}} (-1)^{|I|}(n - |I|)! = \sum_{i=0}^{n}(-1)^i \binom{n}{i}(n - i)!$$

on putting $i = |I|$. The result follows on noting that $\binom{n}{i}(n - i)! = n!/i!$.

5.2. A generalisation

In the introductory example, it is clear that there is enough information to find, not only the number of pupils who play none of the sports, but (for example) the number who play cricket only. This can be formulated in general, as we will do in this section. As a consequence, we give a different proof of Exercise 23 of Chapter 3, about the inverse of the matrix of binomial coefficients.

(5.2.1) Proposition. *Let $(A_1, \dots, A_n)$ be a family of sets, and I a subset of the index set $\{1, \dots, n\}$. Then the number of elements which belong to A_i for all $i \in I$ and for no other values is*

$$\sum_{J \supseteq I}(-1)^{|J \setminus I|}|A_J|.$$

PROOF. We define a new family of sets indexed by $N \setminus I$, where $N = \{1, \dots, n\}$, by setting $B_k = A_{I \cup \{k\}}$ for $k \in N \setminus I$. The Proposition asks us to calculate the number of elements of A_I lying in none of the sets B_k. By PIE, this number is

$$\sum_{K \subseteq N \setminus I} (-1)^{|K|}|B_K|,$$

where $B_\emptyset = A_I$. Now the correspondence $K \leftrightarrow J = I \cup K$ between subsets of $N \setminus I$ and subsets of N containing I is a bijection; and $B_K = A_J$ if K and J correspond. So the result is true.

Next, we turn this result into more abstract form, referring to arbitrary set functions rather than cardinalities of sets.

(5.2.2) Proposition. *Let $N = \{1, \dots, n\}$, and let f and g be functions from $\mathcal{P}(N)$ to the rational (or real) numbers. Then the following are equivalent:*
(a) $g(I) = \sum_{J \supseteq I} f(J)$;
(b) $f(I) = \sum_{J \supseteq I}(-1)^{|J \setminus I|}g(J)$.

PROOF. We argue that it suffices to prove the result when the values of f are non-negative integers. For either of (a) and (b) can be regarded as a system of 2^n linear equations in 2^n unknowns (the values of f or g); this means that the corresponding homogeneous system has only the zero solution in integers. But any rational solution would give rise to an integer solution, on multiplying the solution by a suitable integer (the least common multiple of the denominators). So the linear equations have a unique rational solution. This means that the determinant of the coefficients is non-zero, and this fact doesn't change on passing from the rationals to the reals.

But now, given any non-negative integer values of the function f, we can construct a family $(A_1, \ldots, A_n)$ of sets with the property that the number of points lying in A_i for $i \in I$ but for no other values of A_i is exactly $f(I)$. (Imagine a Venn diagram for n sets; put $f(I)$ elements in the region corresponding to this condition.) Then $g(I) = \sum_{J \supseteq I} f(J)$ is the total number of elements in A_I; and the result follows from (5.2.1).

The same result with the set inclusions reversed is also true:

(5.2.3) Proposition. *Let $N = \{1, \ldots, n\}$, and let f and g be functions from $\mathcal{P}(N)$ to the rational (or real) numbers. Then the following are equivalent:*
(a) $g(I) = \sum_{J \subseteq I} f(J)$;
(b) $f(I) = \sum_{J \subseteq I} (-1)^{|I \setminus J|} g(J)$.

To see this, we define new set functions f' and g' by the rules that $f'(I) = f(N \setminus I)$ and $g'(I) = g(N \setminus I)$, and apply (5.2.2) to these functions. If I' and J' denote $N \setminus I$ and $N \setminus J$ respectively, condition (a) becomes

$$g(I) = g'(I') = \sum_{J' \supseteq I'} f'(J') = \sum_{J \subseteq I} f(J).$$

Similarly, condition (b) translates correctly, because $|J' \setminus I'| = |I \setminus J|$.

(5.2.4) Corollary. *Let f and g be real-valued functions on $\{0, \ldots, n\}$. Then the following are equivalent:*

(a) $g(i) = \sum_{j \leq i} \binom{i}{j} f(j)$;

(b) $f(i) = \sum_{j \leq i} (-1)^{i-j} \binom{i}{j} g(j)$.

PROOF. We define set functions F and G on $\mathcal{P}(N)$ by letting $F(I) = f(i)$ and $G(I) = g(i)$ whenever $|I| = i$. Now, if $|I| = i$, then I has $\binom{i}{j}$ subsets of size j, and the result follows immediately from (5.2.3).

This result gives an alternative proof of the result of Chapter 3, Exercise 24, about inverting Pascal's Triangle. We repeat the result for reference.

(5.2.5) Theorem. *Let n be given, and let A and B be the $(n+1) \times (n+1)$ matrices (with rows and columns indexed from 0 to n) having (i, j) entries*

$$A_{ij} = \binom{i}{j}, \qquad B_{ij} = (-1)^{i+j} \binom{i}{j}.$$

Then $B = A^{-1}$.

PROOF. Let V be the real vector space of functions from $\{0, \ldots, n\}$ to $\mathbb{R}$; each vector f is represented by the $(n+1)$-tuple $(f(0), \ldots, f(n))$. Then the matrices A and B represent linear transformations of V mapping the function f to the function g and back again (in the notation of (5.2.4)); so one is the inverse of the other.

5.3. Stirling numbers

In this section, we look at two 2-parameter families of numbers. They are related to the factorials and Bell numbers in much the same way that the binomial coefficients are related to the powers of 2. (In a sense, they complete the pattern 'subsets, permutations, partitions' of Chapter 3.) The reasons for discussing them here are a bit tenuous: their surprising relationship to each other ((5.3.4) below) parallels that of the binomial coefficients to their signed versions, proved using PIE in the last section; and there is a formula for the Stirling numbers of the second kind, which is an application of PIE, from which some of their most important properties are derived.

Let n and k be positive integers with $k \leq n$.

The *Stirling number of the first kind*, $s(n, k)$, is defined by the rule that $(-1)^{n-k}s(n, k)$ is the number of permutations of $\{1, \ldots, n\}$ with k cycles. (Note the sign. Sometimes a different convention is used, according to which the Stirling numbers are the absolute values of those defined here.)

The *Stirling number of the second kind*, $S(n, k)$, is the number of partitions of $\{1, \ldots, n\}$ with k (non-empty) parts.

The definitions can be extended to all n and k by defining the Stirling numbers to be 0 unless $1 \leq k \leq n$.

(5.3.1) Proposition. *(a)* $\displaystyle\sum_{k=1}^{n}(-1)^{n-k}s(n, k) = \sum_{k=1}^{n}|s(n, k)| = n!;$

(b) $\displaystyle\sum_{k=1}^{n} S(n, k) = B_n,$ *where* B_n *is the* nth *Bell number.*

This is clear from the definition.

Both arrays satisfy recurrence relations, similar to that for Pascal's triangle. Recall that $s(n, 0) = S(n, 0) = 0$ for all n.

(5.3.2) Proposition. *(a)* $s(n, n) = S(n, n) = 1;$
(b) $s(n+1, k) = -ns(n, k) + s(n, k-1);$
(c) $S(n+1, k) = kS(n, k) + S(n, k-1).$

PROOF. (a) is clear; the proofs of (b) and (c) are similar. Consider first partitions of $\{1, \ldots, n+1\}$ with k parts. Either $n+1$ is a singleton part (in which case $\{1, \ldots, n\}$ is partitioned into $k-1$ parts), or $n+1$ is adjoined to one of the k parts into which $\{1, \ldots, n\}$ is partitioned.

The case of permutations requires a little more care. Given a cycle of length l, there are l places at which a new point can be interpolated, giving l different cycles. So, given a permutation of $\{1, \ldots, n\}$ with k cycles, there are n ways of interpolating the point $n+1$ so as to have k cycles resulting (since the cycle lengths sum to n). In addition, we could add the one-point cycle $(n+1)$ to a permutation of $\{1, \ldots, n\}$ with $k-1$ cycles. Thus

$$|s(n+1, k)| = n|s(n, k)| + |s(n, k-1)|,$$

and on putting the signs in correctly we obtain the result.

Using this recurrence, we prove a remarkable 'generating function' form. Recall than $(t)_n = t(t-1) \ldots (t-n+1)$.

(5.3.3) Proposition. *(a)* $(t)_n = \sum\limits_{k=1}^{n} s(n, k) t^k$;

(b) $t^n = \sum\limits_{k=1}^{n} S(n, k)(t)_k$.

PROOF. The proofs are by induction on n. Since $t^1 = (t)_1 = t$ and $s(1,1) = S(1,1) = 1$, the inductions begin at $n = 1$.

PROOF OF (a). Assume that $(t)_n = \sum_{k=1}^{n} s(n, k) t^k$. Then we have

$$(t)_{n+1} = (t)_n(t-n) = \left(\sum_{k=1}^{n} s(n, k) t^k \right) (t-n),$$

and the coefficient of t^k on the right is $-ns(n, k) + s(n, k-1) = s(n+1, k)$.

PROOF OF (b). Assume that $t^n = \sum_{k=1}^{n} S(n, k)(t)_k$. Then

$$t^{n+1} = t^n \cdot t = \sum_{k=1}^{n} (t)_k ((t-k) + k) S(n, k).$$

Since $(t)_k(t-k) = (t)_{k+1}$, we have

$$t^{n+1} = \sum_{k=1}^{n} S(n, k)(t)_{k+1} + \sum_{k=1}^{n} k S(n, k)(t)_k$$

$$= \sum_{k=1}^{n+1} (S(n, k-1) + k S(n, k)) (t)_k$$

$$= \sum_{k=1}^{n+1} S(n+1, k)(t)_k,$$

since $S(n, 0) = S(n, n+1) = 0$.

There are direct combinatorial proofs of this result. Such a proof for (b) is outlined in Exercise 4; but the argument for (a) involves the concept of group action and the Orbit-Counting Lemma, and is deferred until Part 2.

(5.3.4) Corollary. *Let A and B be the $n \times n$ matrices whose (i, j) entries are given by the Stirling numbers $s(i, j)$ and $S(i, j)$ respectively. Then $B = A^{-1}$.*

PROOF. A and B are the transition matrices between two different bases for the space of polynomials of degree n with constant term zero:
- First basis: $t, t^2, \ldots, t^n$;
- Second basis: $(t)_1, (t)_2, \ldots, (t)_n$.

We conclude with a formula for the Stirling numbers of the second kind.

(5.3.5) Proposition. $S(n, k) = \dfrac{1}{k!} \displaystyle\sum_{j=1}^{k} (-1)^{k-j} \binom{k}{j} j^n.$

PROOF. We saw in Section 5.1 that this expression, without the factor $\frac{1}{k!}$, is the number of surjections from $\{1, \ldots, n\}$ to $\{1, \ldots, k\}$. (I have also replaced the dummy variable i by $j = k - i$, and dropped the term with $j = 0$.) So it suffices to prove that the number of surjections is $k!\, S(n, k)$.

Each surjection f defines a partition of $\{1, \ldots, n\}$ with k non-empty parts, viz., $f^{-1}(1), \ldots, f^{-1}(k)$. But every partition arises from exactly $k!$ surjections, since we may assign the numbers $1, \ldots, k$ to the parts in any order. The result is proved.

5.4. Project: Stirling numbers and exponentials

In this section, we explore a different way of looking at the inverse relationship between the two kinds of Stirling numbers: they correspond to substitution of exponential or logarithmic functions into a power series.

We begin with the Stirling numbers of the second kind. First, we obtain an exponential generating function for $S(n, k)$ for fixed k, as n varies.

(5.4.1) Proposition. $\displaystyle\sum_{n \geq 0} \frac{S(n, k)t^n}{n!} = \frac{(\exp(t) - 1)^k}{k!}.$

The proof uses the formula for $S(n, k)$ derived using PIE. We have

$$\sum_{n \geq 0} \frac{S(n, k)t^n}{n!} = \sum_{n \geq 0} \frac{1}{k!} \sum_{j=0}^{k} \binom{k}{j} (-1)^{k-j} j^n \frac{t^n}{n!}$$

$$= \frac{1}{k!} \sum_{j=0}^{k} \binom{k}{j} (-1)^{k-j} \sum_{n \geq 0} \frac{(jt)^n}{n!}$$

$$= \frac{1}{k!} \sum_{j=0}^{k} \binom{k}{j} (-1)^{k-j} \exp(jt)$$

$$= \frac{(\exp(t) - 1)^k}{k!}$$

(The added term $j = 0$ in the inner sum in the first line is zero.)

Note that this gives the e.g.f. of the Bell numbers as a corollary, since

$$\sum_{n \geq 0} \frac{B_n t^n}{n!} = \sum_{n \geq 0} \sum_{k=0}^{n} \frac{S(n, k)t^n}{n!}$$

$$= \sum_{k \geq 0} \frac{(\exp(t) - 1)^k}{k!}$$

$$= \exp(\exp(t) - 1),$$

on reversing the order of summation.

This leads to the following result:

(5.4.2) Theorem. *Let (f_n) and (g_n) be sequences with e.g.f.s $F(t)$ and $G(t)$ respectively. Then the following assertions are equivalent:*

(a) $g_0 = f_0$ *and* $g_n = \sum_{k=1}^{n} S(n,k)f_k$ *for* $n \geq 1$;

(b) $G(t) = F(\exp(t) - 1)$.

PROOF. If (a) holds, then

$$G(t) = f_0 + \sum_{n \geq 1}\sum_{k=1}^{n} S(n,k)f_k \frac{t^n}{n!}$$

$$= \sum_{k \geq 0} \frac{f_k(\exp(t) - 1)^k}{k!}$$

$$= F(\exp(t) - 1).$$

Using the inverse relation between the Stirling numbers, we immediately deduce the following:

(5.4.3) Theorem. *Let (f_n) and (g_n) be sequences with e.g.f.s $F(t)$ and $G(t)$ respectively. Then the following assertions are equivalent:*

(a) $f_0 = g_0$ *and* $f_n = \sum_{k=1}^{n} s(n,k)g_k$ *for* $n \geq 1$;

(b) $F(t) = G(\log(1+t))$.

We can use this result to derive the e.g.f. of the Stirling numbers of the first kind. Let $g_k = 1$ and $g_n = 0$ for $n \neq k$. Then, if f and g are related as in the theorem, we have $f_n = s(n,k)$. Thus, we obtain

(5.4.4) Proposition. $\displaystyle\sum_{n \geq 0} \frac{s(n,k)t^n}{n!} = \frac{(\log(1+t))^k}{k!}.$

5.5. Even and odd permutations

Let π be a permutation of $\{1,\ldots,n\}$, and denote by $c(\pi)$ the number of disjoint cycles of π. The *sign* of π is defined to be $\mathrm{sign}(\pi) = (-1)^{n-c(\pi)}$; and π is said to be *even* or *odd* according as its sign is $+1$ or -1. We observe first:

(5.5.1) Proposition. *For $n \geq 2$, there are equally many even and odd permutations of an n-set.*

PROOF. We use the formula

$$t(t-1)\ldots(t-n+1) = \sum_{k=1}^{n} s(n,k)t^k.$$

Putting $t = 1$ and using the fact that $n \geq 2$, we see that $\sum_{k=1}^{n} s(n,k) = 0$. But $s(n,k)$ is defined to be $(-1)^{n-k}$ times the number of permutations with k cycles; so $\sum_{k=1}^{n} s(n,k)$ is the sum of the signs of the permutations in S_n, and so there are equally many with either sign.

To analyse the sign further, we relate it to the composition of permutations. Recall the convention that composition works from left to right.

(5.5.2) Proposition. *Let π be a permutation of $\{1,\dots,n\}$, and τ a transposition. Then*

$$c(\pi\tau) = c(\pi) \pm 1.$$

PROOF. We examine the effect of composition with a transposition $(i\ j)$. If i and j lie in different cycles of π, then these cycles are 'stitched together' in $\pi\tau$, which has one fewer cycle than π. (For suppose that the cycles are $(a_1\ \dots\ a_k)$ and $(b_1\ \dots\ b_l)$, where $i = a_1$, $j = b_1$. Check that $\pi\tau$ has the cycle $(a_1\ \dots\ a_k\ b_1\ \dots\ b_l)$.) Conversely, if i and j lie in the same cycle of π, then this cycle splits into two in $\pi\tau$.

We see that $\pi\tau$ has the opposite sign to π. Hence, if a permutation π is a product of m transpositions, then its sign is $(-1)^m$; and, in particular, however π is expressed as a product of transpositions, the parity of the number of transpositions is always the same.

(5.5.3) Theorem. *(a) Any permutation is a product of transpositions.*
(b) the map sign is a homomorphism from the symmetric group to the multiplicative group $\{\pm 1\}$ of order 2.

PROOF. (a) It is intuitively clear that, however the numbers $1,\dots,n$ are ordered, it is possible to sort them into the usual order by a sequence of swaps. Formally, if two points i and j lie in the same cycle of π, then composing π with the transposition $(i\ j)$ increases by 1 the number of cycles; so the result follows by induction on $n - c(\pi)$.

(b) We have to show that $\mathrm{sign}(\pi_1\pi_2) = \mathrm{sign}(\pi_1)\mathrm{sign}(\pi_2)$. To show this, express π_2 as a product of (say m) transpositions; composing π_1 with each transposition changes its sign, so the overall effect is to multiply by $(-1)^m$.

It follows that the set of all even permutations in S_n is a normal subgroup. This subgroup is called the *alternating group* A_n. We now have two proofs that $|A_n| = n!/2$ if $n \geq 2$. First, this is immediate from (5.5.1); second, A_n is the kernel of a homomorphism onto a group of order 2.

5.6. Exercises

1. An opinion poll reports that the percentage of voters who would be satisfied with each of three candidates A, B, C for President is 65%, 57%, 58% respectively. Further, 28% would accept A or B, 30% A or C, 27% B or C, and 12% would be content with any of the three. What do you conclude?

2. Make tables of the two kinds of Stirling numbers for small values of n and k.

3. Prove directly that $S(n,1) = 1$, $S(n,2) = 2^{n-1} - 1$, and $S(n, n-1) = \binom{n}{2}$. Find a formula for $S(n, n-2)$.

4. Prove that $|s(n,1)| = (n-1)!$ using the recurrence relation, and show directly that the number of cyclic permutations of an n-set is $(n-1)!$.

5. This exercise outlines a proof that $t^n = \sum_{k=1}^{n} S(n,k)(t)_k$.

(a) Let t be a positive integer, $T = \{1, \ldots, t\}$, and $N = \{1, \ldots, n\}$. The number of functions $f : N \to T$ is t^n. Given such a function f, define an equivalence relation $\equiv$ on N by the rule

$$i \equiv j \quad \text{if and only if} \quad f(i) = f(j).$$

The classes of this equivalence relation can be numbered $C_1, \ldots, C_k$ (say), ordered by the smallest points in the classes. (So C_1 contains 1; C_2 contains the smallest number not in C_1; and so on.) Then the values $f(C_1), \ldots, f(C_k)$ are k distinct elements of T, and so can be chosen in $(t)_k$ ways; the partition can be chosen in $S(n, k)$ ways. Summing over k proves the identity *for the particular value of t*.

(b) Prove that if a polynomial equation $F(t) = G(t)$ is valid for all positive integer values of the argument t, then it is the polynomials F and G are equal.

6. For this exercise, recall the Bernoulli numbers $B(n)$ from Exercise 19 of Chapter 4, especially the fact that their e.g.f. is $t/(\exp(t) - 1)$. Derive the formula

$$B(n) = \sum_{k=1}^{n} \frac{(-1)^k k! S(n, k)}{(k + 1)}$$

for the nth Bernoulli number.

7. Let (f_n) and (g_n) be sequences, with e.g.f.s $F(t)$ and $G(t)$ respectively. Show the equivalence of the following assertions:

(a) $g_n = \sum_{k=0}^{n} \binom{n}{k} f_k$;

(b) $G(t) = F(t) \exp(t)$.

8. Show that a permutation which is a cycle of length m can be written as a product of $m - 1$ transpositions. Deduce that it is an even permutation if and only if its length is odd. Hence show that an arbitrary permutation is even if and only if it has an even number of cycles of even length (with no restriction on cycles of odd length).

9. This exercise outlines the way in which the sign of permutations is normally treated by algebraists. Let $x_1, \ldots, x_n$ be indeterminates, and consider the polynomial

$$F(x_1, \ldots, x_n) = \prod_{i<j} (x_j - x_i).$$

Note that every pair of indeterminates occur together once in a bracket. If π is a permutation, then $F(x_{1\pi}, \ldots, x_{n\pi})$ is also the product of all possible differences (but some have had their signs changed). So

$$F(x_{1\pi}, \ldots, x_{n\pi}) = \text{sign}(\pi) F(x_1, \ldots, x_n),$$

where $\text{sign}(\pi) = \pm 1$ is the number of pairs $\{i, j\}$ whose order is reversed by π. Prove that

- sign is a homomorphism;
- if τ is a transposition, then $\text{sign}(\tau) = -1$.

10. Recall from Section 3.8 that a *preorder* is a reflexive and transitive relation which satisfies trichotomy. Prove that the exponential generating function for the number of preorders on an n-set is $1/(2 - \exp(t))$. [HINT: the e.g.f. for the number of orders is $1/(1 - t)$.]

11. (a) Show that the smallest number of transpositions of $\{1, \ldots, n\}$ whose product is an n-cycle is $n - 1$.

(b) Prove that any n-cycle can be expressed in n^{n-2} different ways as a product of $n-1$ transpositions. [HINT: The product of the transpositions $(x_i \; y_i)$ is an n-cycle if and only if the pairs $\{x_i, y_i\}$ are the edges of a tree (Section 3.10). Double-count (tree, cycle) pairs, using Cayley's Theorem (3.10.1) and the fact that all cycles have the same number of expressions as products of transpositions.]

6. Latin squares and SDRs

> TOPICS: Latin squares, SDRs, Hall's Theorem, orthogonal Latin squares, quasigroups, groups, permanents
>
> TECHNIQUES:
>
> ALGORITHMS:
>
> CROSS-REFERENCES: Network flows (Chapter 11), affine planes and nets (Chapter 9), groups (Chapter 14)

In this chapter, we examine Latin squares, showing that there are many of them (by means of a digression through Hall's theorem on SDRs), and then consider orthogonal Latin squares.

6.1. Latin squares

Latin squares arise in Euler's 'thirty-six officers' problem, but with one level of detail removed. The definition is as follows.

A *Latin square* of order n is an $n \times n$ array or matrix with entries taken from the set $\{1, 2, \ldots, n\}$, with the property that each entry occurs exactly once in each row or column.[1] So, in a solution to Euler's problem, if the officers' ranks are numbered from 1 to 6, they are arranged in a Latin square; and similarly for the regiments.

REMARK. Sometimes it is convenient to regard the entries as coming not from the set $\{1, \ldots, n\}$ but from an arbitrary given set of n elements.

---

[1] Why are they called Latin squares? Wait and see!

The *existence* of Latin squares is not in doubt. The array

$$
\begin{array}{ccccc}
1 & 2 & 3 & 4 & 5 \\
5 & 1 & 2 & 3 & 4 \\
4 & 5 & 1 & 2 & 3 \\
3 & 4 & 5 & 1 & 2 \\
2 & 3 & 4 & 5 & 1
\end{array}
$$

is a Latin square of order 5; the construction obviously generalises. So our goal is to refine this observation, and come up with some estimate of how many different Latin squares there are.

We can interpret a Latin square as follows.[2] Given a class of n boys and n girls, arrange a sequence of n dances so that each boy and girl dance together exactly once. (The (i, j) entry of the Latin square gives the number of the dance at which the ith boy and the jth girl dance together.)

Latin squares were first used in statistical design. Very roughly, suppose that n varieties of a crop have to be tested. A field is laid out in a $n \times n$ array of plots. We assume that there may be some unknown but systematic variation in fertility, or susceptibility to insect attack, moving across or down the field; so we arrange that each variety is planted in one plot in each row or column, to offset this effect.

6.2. Systems of distinct representatives

We have to make quite a long detour to reach our goal. We prove a result known as *Hall's Marriage Theorem*; this was originally shown by Philip Hall, and a refinement (which we need) was shown by Marshall Hall Jr.[3] but there are now many different proofs. This result is closely connected with the theory of flows in networks, and you may meet it in an Operations Research course. Our objective here is different. (We return to networks in Chapter 11.)

Let $A_1, \ldots, A_n$ be subsets of a set X. (We allow the possibility that some of them are equal.) A *system of distinct representatives* (SDR) for these sets is an n-tuple $(x_1, \ldots x_n)$ of elements with the properties

(a) $x_i \in A_i$ for $i = 1, \ldots, n$ (i.e., representatives);

(b) $x_i \neq x_j$ for $i \neq j$ (i.e., distinct).

For any set $J \subseteq \{1, \ldots, n\}$ of indices, we define

$$
A(J) = \bigcup_{j \in J} A_j.
$$

(Don't confuse this with the similar A_J which occurred in PIE, where we had intersection in place of union. Here, $A(\emptyset) = \emptyset$.)

If the sets $A_1, \ldots, A_n$ have a system of distinct representatives, then necessarily $|A(J)| \geq |J|$ for any set $J \subseteq \{1, \ldots, n\}$, since $A(J)$ contains the representative x_j of each set A_j for $j \in J$, and these representatives are all distinct. Hall's Theorem says that this necessary condition is also sufficient:

---

[2] This is in the spirit of Kirkman's Schoolgirls Problem (Chapters 1, 8).

[3] No relation.

(6.2.1) Hall's Theorem. *The family* $(A_1, \ldots, A_n)$ *of finite sets has a system of distinct representatives if and only if the following condition holds:*

(6.2.2) Hall's Condition

(HC) $\qquad\qquad |A(J)| \geq |J|$ for all $J \subseteq \{1, \ldots, n\}$

PROOF. We use induction on the number n of sets. The induction obviously starts: (HC) guarantees a representative for a single set! We call a set J of indices *critical* if $|A(J)| = |J|$. The motivation is that, if J is critical, then every element of $A(J)$ must be used as a representative of one of these sets. We divide the proof into two cases:

CASE 1. No set J is critical except for $J = \emptyset$ and possibly $J = \{1, \ldots, n\}$. Let x_n be any point of A_n (note $A_n \neq \emptyset$ by (HC)) and, for $j = 1, \ldots, n-1$, define $A'_j = A_j \setminus \{x_n\}$. We *claim* that the family $(A'_1, \ldots, A'_{n-1})$ satisfies (HC). Take $J \subseteq \{1, \ldots, n-1\}$, and suppose that $J \neq \emptyset$. Then

$$|A'(J)| \geq |A(J)| - 1$$
$$> |J| - 1,$$

the first inequality true since at worst x_n is omitted, the second since J is not critical by assumption. So $|A'(J)| \geq |J|$, proving the claim.

By induction, $(A'_1, \ldots, A'_{n-1})$ has a SDR $(x_1, \ldots, x_{n-1})$. Then $(x_1, \ldots, x_n)$ is a SDR for the original family, since clearly x_n is distinct from all the other x_j.

CASE 2. Some set $J \neq \emptyset, \{1, \ldots, n\}$ is critical. We may suppose that J is minimal subject to this. Then the family $(A_j : j \in J)$ has a SDR $(x_j : j \in J)$, by induction. For $i \notin J$, set $A^*_i = A_i \setminus A(J)$. We *claim* that the family $(A^*_i : i \notin J)$ satisfies (HC). Take K to be a set of indices disjoint from J. Then

$$|A^*(K)| = |A(J \cup K)| - |A(J)|$$
$$\geq |J \cup K| - |J|$$
$$= |K|,$$

the first equality since in fact $A^*(K) = A(J \cup K) \setminus A(J)$, and the inequality since $|A(J \cup K)| \geq |J \cup K|$ by (HC) but $|A(J)| = |J|$ since J is critical.

So there is a SDR $(x_i : i \notin J)$ for the sets $(A^*_i : i \notin J)$. Combining this with the SDR for the sets $(A_j : j \in J)$ gives the required result.

This theorem is sometimes called *Hall's Marriage Theorem*, because of the following interpretation. Given a set of boys and a set of girls, each girl knowing a specified set of boys, it is possible for all the girls to marry boys that they know if and only if any set of k girls know altogether at least k boys.

(6.2.3) Hall's Theorem Variant. *Suppose that* $(A_1, \ldots, A_n)$ *are sets satisfying (HC), and suppose that* $|A_i| \geq r$ *for* $i = 1, \ldots, n$. *Then the number of different SDRs for the family is at least*

$$\begin{cases} r! & \text{if } r \leq n, \\ r(r-1) \ldots (r-n+1) & \text{if } r > n. \end{cases}$$

NOTE. Two SDRs may use the same elements and still be different, if they assign different elements to the sets. For example, (1, 2) and (2, 1) are different SDRs for the sets $(\{1, 2, 3\}, \{1, 2, 4\})$.

PROOF. This is just a variant on the proof of Hall's Theorem. We use induction on n; if $n = 1$, then a single set of size at least r has at least r SDRs! So assume true for families with fewer than n sets.

In Case 2 of the proof above, we have $r \leq |J| \leq n$, and the family $(A_j : j \in J)$ has at least $r!$ SDRs, each of which can be extended to the whole family.

In Case 1, there are at least r choices for the representative x_n. For each choice, the family $(A_i' : 1 \leq i \leq n - 1)$ consists of $n - 1$ sets each of size at least $r - 1$ satisfying (HC), so by induction it has a least $(r - 1)!$ SDRs if $r \leq n$, or at least $(r-1) \ldots ((r-1) - (n-1) + 1)$ if $r > n$. Multiplying gives the result.

We need the following consequence of Hall's Theorem:

(6.2.4) Theorem. *Let* $(A_1, \ldots, A_n)$ *be a family of subsets of* $\{1, \ldots, n\}$, *and let* r *be a positive integer such that*
(a) $|A_i| = r$ *for* $i = 1, \ldots, n$;
(b) each element of $\{1, \ldots, n\}$ *is contained in exactly* r *of the sets* $A_1, \ldots, A_n$.
Then the family $(A_1, \ldots, A_n)$ *satisfies (HC), and so has an SDR.*

PROOF. Let J be a set of indices. We count choices of (j, x), where $j \in J$ and $x \in A_j$. There are $|J|$ choices for j, and for each j there are r choices for $x \in A_j$, or $r|J|$ altogether. On the other hand, $x \in A(J)$, so there are $|A(J)|$ choices for x; and x lies in r sets, not all of which might have index in J, so there are *at most* r choices for j. Thus

$$r|J| \leq |A(J)|r,$$

and since $r > 0$ we get (HC).

(6.2.5) Corollary. *Under the hypotheses of the last theorem, the family of sets has at least* $r!$ *SDRs.*

This just combines Theorem (6.2.4) with the Hall Variant (6.2.3).

6.3. How many Latin squares?

Now we return to Latin squares. We want to construct Latin squares 'row by row', and so we want to be sure that if we have fewer than n rows, there are many ways to add another row. So we define a $k \times n$ *Latin rectangle*, for $k \leq n$, to be a $k \times n$ array with entries from $\{1, \ldots, n\}$, having the property that each entry occurs exactly once in each row and at most once in each column.

(6.3.1) Proposition. *Given a $k \times n$ Latin rectangle with $k < n$, there are at least $(n - k)!$ ways to add a row to form a $(k + 1) \times n$ rectangle.*

PROOF. The elements of the new row must all be distinct, and each must *not* be among those already used in its column. So we let A_i be the set of entries not occurring in the ith column of the rectangle, and we have:

$(x_1, \ldots, x_n)$ *is a possible* $(k + 1)^{\text{st}}$ *row for the rectangle if and only if it is a SDR for the family* $(A_1, \ldots, A_n)$.

Now clearly each set A_i has size $n - k$, since k of the n entries have already been used. Consider a particular entry, say x. This occurs k times in the rectangle (one in each of the rows), in k distinct columns; so there are $n - k$ columns where it does not occur. So the hypotheses of Corollary (6.2.5) are satisfied, with $r = n - k$.

(6.3.2) Theorem. *The number of Latin squares of order n is at least*

$$\prod_{k=1}^{n} k!.$$

PROOF. Add rows one at a time: there are at least $n!$ choices for the first row, at least $(n - 1)!$ for the second, and so on.

This problem incorporates two counting problems we met earlier. The first row of a Latin square of order n is simply a permutation of $\{1, \ldots, n\}$, and there are exactly $n!$ choices for it. Given the first row, we may (by re-labelling) assume that it is $(1\ 2\ \ldots\ n)$; then a legitimate second row is precisely a permutation satisfying $i\pi \neq i$ for $i = 1, \ldots, n$, that is, a *derangement*. We know that the number of derangements is the nearest integer to $n!/e$ for $n \geq 4$, this is better than the lower bound of $(n - 1)!$ which we used, so the estimate for the number of Latin squares can be improved a bit. However, the number of choices of the third row depends on the way the first two rows were chosen, so we cannot get the exact answer simply by multiplying n numbers together.

EXAMPLE. There are 2 Latin squares of order 2, and $3! \cdot 2! = 12$ of order 3. However, for order 4, there are $24 \cdot 3$ choices of the first two rows which can be extended in 4 different ways, and $24 \cdot 6$ which have just 2 extensions; so the number of Latin squares is

$$24 \cdot 3 \cdot 4 + 24 \cdot 6 \cdot 2 = 576.$$

(See Exercise 1.)

REMARK. Let $L(n)$ be the number of Latin squares of order n. We have shown that $L(n) \geq n!(n - 1)! \ldots 1!$. This bound was improved, about fifteen years ago, to

$$L(n) \geq (n!)^{2n}/n^{n^2}.$$

We explore this in Section 6.5. On the other hand, we have

$$L(n) \leq n^{n^2},$$

since there are n^{n^2} ways of filling in the n^2 positions of the array with entries from $\{1, \ldots, n\}$. We can improve this to

$$L(n) \le (n!)^n$$

by observing that each entire row is chosen from the set of permutations of $\{1, \ldots, n\}$, and there are $n!$ permutations. A further improvement is made by noticing that all the rows after the first are derangements of the first row, so roughly $L(n) \le (n!)^n / e^{n-1}$.

To compare these bounds, it is helpful to estimate $\log L(n)$ rather than $L(n)$ itself. The simplest possible upper bound, namely $L(n) \le n^{n^2}$, gives

$$\log L(n) \le n^2 \log n.$$

On the other hand, we have

$$\log L(n) \ge \sum_{k=1}^{n} \log k!$$

$$\ge \sum_{k=1}^{n} (k \log k - k)$$

$$\ge \tfrac{1}{2} n^2 \log n + O(n^2),$$

where we used the simple bound $k! \ge (k/e)^k$ from Chapter 2, Exercise 3, and the fact that $\sum_{k=1}^{n} k \log k = \int_1^n x \log x \, dx + O(n \log n)$. So roughly the upper and lower bounds for $\log L(n)$ differ by a factor of 2. The improved lower bound mentioned above removes this factor, giving

$$\log L(n) = n^2 \log n + O(n^2).$$

6.4. Quasigroups

There is another way of looking at Latin squares. Let $G = \{g_1, \ldots, g_n\}$ be a set of size n. If $A = (a_{ij})$ is any $n \times n$ matrix with entries from the set $\{1, \ldots, n\}$, we can define a binary operation, or 'multiplication', on G by the rule

$$g_i \circ g_j = g_k \quad \text{if and only if} \quad a_{ij} = k.$$

Conversely, any binary operation on G gives rise to such a matrix, once we have numbered the elements of G as $g_1, \ldots, g_n$.[4]

A binary structure like G above is called a *quasigroup* if the following axioms hold:

(*left division*) for all $g_j, g_k \in G$, there is a unique $g_i \in G$ with $g_i g_j = g_k$;
(*right division*) for all $g_i, g_k \in G$, there is a unique $g_j \in G$ with $g_i g_j = g_k$.
Now the following result follows from the definitions:

(6.4.1) Proposition. *A binary structure G is a quasigroup if and only if the corresponding matrix A is a Latin square.*

---

[4] The matrix is the *multiplication table* of the binary structure.

PROOF. The left divisibility condition just says that each column of the matrix contains each entry exactly once; and similarly for right division.

There are various advantages to turning Latin squares into algebraic objects like quasigroups. For one thing, we can obtain a kind of measure of the strength of various algebraic axioms by seeing how many Latin squares correspond to structures satisfying these axioms. For example, there are very many quasigroups; but there are many fewer groups (see next section), so the group axioms are very powerful! Another is that algebraic constructions can be transferred to Latin squares. One example of this is the *direct product*.

Let G and H be binary structures (the binary operation in each of them will be denoted by $\circ$). The *direct product* $G \times H$ is defined, just as for groups, as follows: it is the set of ordered pairs (g, h), for $g \in G, h \in H$, with operation

$$(g_1, h_1) \circ (g_2, h_2) = (g_1 \circ g_2, h_1 \circ h_2).$$

Now it is easily established that the direct product of quasigroups is a quasigroup. (For left divisibility, suppose that in the above equation g_2, h_2, g_3, h_3 are given. Then g_1 is determined by left divisibility in G, and similarly h_1 in H.)

The direct product can be translated into a direct product operation on Latin squares, which we write with the same notation, i.e. the direct product of A and B is $A \times B$. If B has order n, we take n copies of A with disjoint sets of entries, and replace each entry of B by the corresponding copy of A. For example,

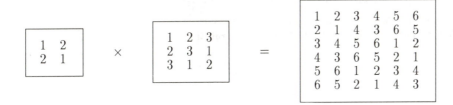

6.5. Project: Quasigroups and groups

The best-known examples of quasigroups are groups: these are quasigroups with an identity element whose composition is associative. In this section, we describe a refinement of the estimate for the number of quasigroups, using the proof of the van der Waerden permanent conjecture; and we show that two of the most basic theorems about groups (Lagrange's Theorem and Cayley's Theorem)[5] can be used to put an upper bound on the number of groups. We see that groups are very rare among quasigroups; in other words, the associative law is a very powerful condition.

QUASIGROUPS: PERMANENTS AND SDRS.
First we need a couple of definitions, whose relevance will not be immediately apparent. A matrix is said to be *stochastic* if its entries are non-negative real numbers and its row sums are equal to 1. The term suggests a connection with probability. A system is initially in one of m states $S_1, \ldots, S_m$,

---

[5] These theorems and their historical context are described in Chapter 14.

and can make a transition to one of n states $T_1, \ldots, T_n$. If the probability of jumping from S_i to T_j is p_{ij}, then the $m \times n$ matrix with (i, j) entry p_{ij} is stochastic. A stochastic matrix is called *doubly stochastic* if, in addition, its column sums are all equal to 1. (This implies in particular that the matrix is square.) This condition doesn't have an obvious probabilistic interpretation, though it is connected to stationary states and reversibility of Markov chains.

Let A be an $n \times n$ matrix with (i, j) entry a_{ij}. Then there is a well-known formula for the determinant of A:

$$\det(A) = \sum_{\pi \in S_n} \prod_{i=1}^{n} \text{sign}(\pi) \, a_{i\,i\pi}.$$

(Recall our convention that permutations act on the right, and the definition of the sign of a permutation in Chapter 5.)

If we leave out the sign factor in this expression, we obtain the *permanent* of A:

$$\text{per}(A) = \sum_{\pi \in S_n} \prod_{i=1}^{n} a_{i\,i\pi}.$$

Though the formula is simpler, the permanent is much harder to manipulate or evaluate than the determinant! It is clear that the matrix with every entry $1/n$ has permanent $n!/n^n$ (the sum has $n!$ terms, each the product of n factors $1/n$).

The *van der Waerden permanent conjecture* asserted:

> The permanent of an $n \times n$ doubly stochastic matrix A is at least $n!/n^n$, with
> equality if and only if every entry of A is equal to $1/n$.

This conjecture was proved in 1979–1980, independently, by Egorychev and Falikman. Earlier, Bang and Friedland had shown the slightly weaker result that the permanent of a doubly stochastic matrix is at least e^{-n}. (Note that $e^{-n} < n!/n^n$, by Exercise 3 of Chapter 2.) If you want to see how it was done, look it up in Marshall Hall's *Combinatorial Theory* (1989).

What is the relevance to this chapter? Given a family $(A_1, \ldots, A_n)$ of n subsets of $\{1, \ldots, n\}$, we define the *incidence matrix* A of the family by the rule that the (i, j) entry of A is given by

$$A_{ij} = \begin{cases} 1 & \text{if } i \in A_j, \\ 0 & \text{if } i \notin A_j. \end{cases}$$

Then we have:

(6.5.1) Proposition. *With the above notation,* $\text{per}(A)$ *is equal to the number of SDRs of the family of sets.*

PROOF. In the evaluation of the permanent, the product corresponding to a permutation π is zero unless $i\pi \in A_i$ for all i, when it is one. In this case, $(1\pi, \ldots, n\pi)$ is a SDR for $(A_i, \ldots, A_n)$. Conversely, any SDR arises from such a permutation. Hence the permanent is equal to the number of SDRs.

(6.5.2) Proposition. *Let* $(A_1, \ldots, A_n)$ *be a family of subsets of* $\{1, \ldots, n\}$. *Suppose that*
- *each set* A_j *has cardinality* r;
- *each point* i *lies in* r *of the sets* $A_1, \ldots, A_n$.

Then the number of SDRs of the family is at least $n!(r/n)^n$.

REMARK. You should stop and compare this with the lower bound $r!$ proved in Section 6.3.

PROOF. The incidence matrix A has all row and column sums r. So $(1/r)A$ is doubly stochastic, whence $\text{per}((1/r)A) \geq n!/n^n$, from which the result follows since $\text{per}((1/r)A) = (1/r)^n \text{per}(A)$.

(6.5.3) Proposition. *The number $L(n)$ of $n \times n$ Latin squares satisfies*

$$L(n) \geq \frac{n!^{2n}}{n^{n^2}}.$$

PROOF. Just as in Section 6.3, we have

$$L(n) \geq \prod_{r=1}^{n} n!(r/n)^n.$$

What about the number of quasigroups? Given a quasigroup, if we number its elements $1, \ldots, n$ in any order, its multiplication table is a Latin square. So each quasigroup gives at most $n!$ Latin squares; this is insignificant compared with $L(n)$, and the estimate $n^2 \log n + O(n^2)$ holds for the logarithm of the number of quasigroups too.

GROUPS: LAGRANGE AND CAYLEY.

We will now show that the number of groups on the set $\{g_1, \ldots, g_n\}$ is very small compared with the number of quasigroups. If $G(n)$ is this number, we prove that $G(n) \leq n^{n(\log_2 n + 1)}$. In other words, $\log G(n) = O(n(\log n)^2)$, much smaller than $\log L(n)$.

The proof, not surprisingly, requires some algebra. In fact, little is needed; just two of the basic theorems proved in the nineteenth century. (Using more powerful tools, better estimates can be derived.) The results we need are:

- *Lagrange's Theorem:* The order of a subgroup of a group G divides the order of G.
- *Cayley's Theorem:* Any group of order n is isomorphic to a subgroup of the symmetric group S_n.

We also need the concept of the subgroup H *generated by* a set $\{g_1, \ldots, g_k\}$ of elements of G. This is the smallest subgroup of G containing $g_1, \ldots, g_k$, and consists of all elements of G which can be written as products of these elements and their inverses. (See Chapter 14 for further discussion.) By convention, the empty product is 1 (the identity of G), which lies in every subgroup.

(6.5.4) Lemma. *Any group G of order n can be generated by at most $\log_2 n$ elements.*

PROOF. We prove by induction that if $g_1, g_2, \ldots$ are chosen so that, for all k, g_{k+1} does not lie in the subgroup G_k generated by $g_1, \ldots, g_k$ (and $g_1 \neq 1$), then the order of G_k is at least 2^k. For the inductive step, $|G_{k+1}| > |G_k|$ (since $g_k \in G_{k+1} \setminus G_k$), and $|G_k|$ divides $|G_{k+1}|$ by Lagrange's Theorem; so we have $|G_{k+1}| \geq 2|G_k|$, and the induction goes through.

By Cayley's Theorem, a copy of any group of order n can be found inside S_n, and gives rise to at most $n!$ operations on a set $\{g_1, \ldots, g_n\}$ (this being the number of labellings of the group elements). So $G(n)$ is no greater than $n!$ times the number of subgroups of order n of S_n. By the Lemma, the latter number does not exceed the number of choices of $\log_2 n$ elements of S_n; so

$$G(n) \leq (n!)^{\log_2 n + 1} \leq n^{n(\log_2 n + 1)}.$$

6.6. Orthogonal Latin squares

Two Latin squares $A = (a_{ij})$ and $B = (b_{ij})$ are said to be *orthogonal* if, for any pair (k, l) of elements from $\{1, \ldots, n\}$, there are unique values of i and j such that $a_{ij} = k$, $b_{ij} = l$; in other words, there is a unique position where A has entry k and B has entry l. A set $\{A_1, \ldots, A_r\}$ of Latin squares is called a *set of mutually orthogonal Latin squares*, or *set of MOLS* for short, if any two squares in the set are orthogonal. (Sometimes the terms *pairwise orthogonal* and *POLS* are used instead.)

Sometimes a pair of orthogonal Latin squares is called a *Graeco-Latin square*. The reason comes from a different representation sometimes used. Instead of numbers, the entries can be taken from any set of size n; the first n letters of the alphabet are commonly used. Now if we use letters of different alphabets, say the Latin alphabet for A and the Greek for B, then the two squares can be combined into one unambiguously; and A and B are orthogonal if and only if each combination of a Latin and a Greek letter occurs exactly once in the square.[6] For example, here are two orthogonal Latin squares of order 3 and the corresponding Graeco-Latin square.

$$
\begin{array}{ccc}
1 & 2 & 3 \\
2 & 3 & 1 \\
3 & 1 & 2
\end{array}
\qquad , \qquad
\begin{array}{ccc}
1 & 2 & 3 \\
3 & 1 & 2 \\
2 & 3 & 1
\end{array}
\qquad \longrightarrow \qquad
\begin{array}{ccc}
a\alpha & b\beta & c\gamma \\
b\gamma & c\alpha & a\beta \\
c\beta & a\gamma & b\alpha
\end{array}
$$

A problem which has had much attention is:

What is the maximum size $f(n)$ of a set of MOLS of order n?

This question is closely connected with the existence question for projective and affine planes, as we will see in Chapter 9.

We observe first that $f(n) \leq n - 1$ for all n. For let $A_1, \ldots, A_r$ be mutually orthogonal Latin squares; without loss of generality, we may assume that each square has $(1,1)$ entry 1. Now each square has $n - 1$ further entries 1, none occurring in the first row or column; and, by orthogonality, these 1s cannot occur in the same position in two different squares. Since there are only $(n-1)^2$ available positions, there cannot be more than $n - 1$ squares.

(6.6.1) Proposition. *If n is a prime power, then $f(n) = n - 1$.*

This result uses Galois' Theorem on the existence of finite fields (see Section 4.7). We use the fact that there is a field F of any given prime power order n. Now we take the elements of F to index the rows and columns of all the squares. For each non-zero element $m \in F$, we define a matrix A_m whose (i,j) entry is $(A_m)_{i,j} = im + j$.

Now each A_m is a Latin square. For, if $im + j_1 = im + j_2$, then $j_1 = j_2$; and, if $i_1 m + j = i_2 m + j$, then $i_1 m = i_2 m$, and so $i_1 = i_2$ (since m is non-zero and so has an inverse).

Moreover, these squares are orthogonal. For, given elements $a, b \in F$, and $m_1 \neq m_2$, the equations

$$
im_1 + j = a,
$$
$$
im_2 + j = b,
$$

have a unique solution (i, j).

This doesn't appear to help evaluate $f(n)$ in general. But it gives us a lower bound. To show this, we use the direct product construction for Latin squares, and make the following observation:

---

[6] The Latin letters alone form a Latin square. (Hence the name.)

(6.6.2) Proposition. *If A_1 and A_2 are orthogonal Latin squares of order n, and B_1 and B_2 orthogonal Latin squares of order m, then the Latin squares $A_1 \times B_1$ and $A_2 \times B_2$ of order nm are orthogonal.*

PROOF. As we saw in the last section, direct products are easier to define for quasigroups. So we re-formulate orthogonality for quasigroups. For convenience, we take the same set $G = \{g_1, \ldots, g_n\}$ of symbols for both quasigroups, but distinguish the binary operations. Let $(G, \circ)$ and $(G, *)$ be quasigroups. These quasigroups are said to be *orthogonal* if the following holds:

(*orthogonality*) for all $g_k, g_l \in G$, there exist unique elements $g_i, g_j \in G$ such that

$$g_i \circ g_j = g_k \text{ and } g_i * g_j = g_l.$$

This is equivalent to orthogonality of the corresponding squares. Now it is a simple exercise to prove that, if $(G, \circ)$ and $(G, *)$ are orthogonal quasigroups, and $(H, \circ)$ and $(H, *)$ are another pair of orthogonal quasigroups (possibly of different order), then $(G \times H, \circ)$ and $(G \times H, *)$ are also orthogonal.

(6.6.3) Proposition. *Let $n = p_1^{a_1} \ldots p_r^{a_r}$, where $p_1, \ldots, p_r$ are distinct primes and $a_1, \ldots, a_r > 0$, and let q be the minimum of $p_1^{a_1}, \ldots, p_r^{a_r}$. Then $f(n) \geq q - 1$.*

PROOF. Let $q_i = p_i^{a_i}$. Then we can find $q_1 - 1$ MOLS of order q_1, $q_2 - 1$ of order q_2, and so on. Since a subset of a set of MOLS is again a set of MOLS, if q is the minimum of $q_1, q_2, \ldots$, we can find $q - 1$ MOLS of each of these orders; taking their products gives a set of $q - 1$ MOLS of order n.

REMARK. More generally, we have

$$f(n_1 n_2) \geq \min\{f(n_1), f(n_2)\}.$$

(6.6.4) Corollary. *If $n \not\equiv 2 \pmod{4}$, then there exist two orthogonal Latin squares of order n.*

PROOF. If $q = 2$ in the Proposition, then n is divisible by 2 but not by 4, so that $n \equiv 2 \pmod 4$.

Euler conjectured that the converse is also true; in other words, that if $n \equiv 2 \pmod 4$, then orthogonal Latin squares do not exist. For $n = 6$, this is his 'thirty-six officers' problem posed in the first chapter. It turned out that Euler was right about the 36 officers (no solution exists), but wrong for all larger values of n. More generally, it is known that $f(n) \to \infty$ as $n \to \infty$. (This means that, for any given r, there exist r MOLS of order n for all but finitely many values of r. For example, two MOLS of order n exist for all n except $n = 1, 2$ and 6.)

6.7. Exercises

1. (a) Show that the number of $n \times n$ Latin squares is 1, 2, 12, 576 for $n = 1, 2, 3, 4$ respectively.

(b) Prove that, up to permutations of the rows, columns, and symbols in a Latin square, there are unique squares of orders 1, 2, 3, and two different squares of order 4.

(c) Show that one of the two types of Latin square of order 4 has an orthogonal 'mate' and the other does not.

3. A Latin square $A = (a_{ij})$ of order n is said to be *row-complete* if every ordered pair (x, y) of distinct symbols occurs exactly once in consecutive positions in the same row (i.e., as $(a_{ij}, a_{i\,j+1})$ for some i, j). (Note that there are $n(n-1)$ ordered pairs of distinct symbols, and each of the n rows contains $n-1$ consecutive pairs of symbols.)

(a) Prove that there is no row-complete Latin square of order 3 or 5, and construct one of order 4.

(b) Define analogously a *column-complete* Latin square.

(c) Suppose that the elements of $\mathbb{Z}/(n)$ are written in a sequence $(x_1, x_2, \ldots, x_n)$ with the property that every non-zero element of $\mathbb{Z}/(n)$ can be written uniquely in the form $x_{i+1} - x_i$ for some $i = 1, \ldots, n-1$. Let A be the Latin square (with rows, columns and entries indexed by $0, \ldots, n-1$ instead of $1, \ldots, n$) whose (i, j) entry is $a_{ij} = x_i + x_j$. (This is the addition table of $\mathbb{Z}/(n)$, written in a strange order.) Prove that A is both row-complete and column-complete.

(d) If n is even, show that the sequence

$$(0, 1, n - 1, 2, n - 2, \ldots, \tfrac{1}{2}n - 1, \tfrac{1}{2}n + 1, \tfrac{1}{2}n)$$

has the property described in (c).[7]

REMARK. Row- and column-complete Latin squares are useful for experimental design where treatments may affect adjacent plots.

4. (a) Find a family of three subsets of a 3-set having exactly three SDRs.

(b) How many SDRs does the family

$$\{\{1, 2, 3\}, \{1, 4, 5\}, \{1, 6, 7\}, \{2, 4, 6\}, \{2, 5, 7\}, \{3, 4, 7\}, \{3, 5, 6\}\}$$

have?[8]

5. Let $(A_1, \ldots, A_n)$ be a family of subsets of $\{1, \ldots, n\}$. Suppose that the incidence matrix of the family is invertible. Prove that the family possesses a SDR.

6. Use the truth of the van der Waerden permanent conjecture to prove that the number $d(n)$ of derangements of $\{1, \ldots, n\}$ satisfies

$$d(n) \geq n! \left(1 - \frac{1}{n}\right)^n.$$

How does this estimate compare with the truth?

7. Prove the following generalisation of Hall's Theorem:

If a family $(A_1, \ldots, A_n)$ of subsets of X satisfies $|A(J)| \geq |J| - r$ for all $J \subseteq \{1, \ldots, n\}$, then there is a subfamily of size $n - r$ which has a SDR.

[HINT: add r 'dummy' elements which belong to all the sets.]

---

[7] I am grateful to R. A. Bailey for this exercise.

[8] This family is the set of triples of the Steiner triple system of order 7; see Chapter 8.

7. Extremal set theory

TOPICS: Intersecting families; Sperner families; de Bruijn–Erdős Theorem; [regular families]

TECHNIQUES: LYM method

ALGORITHMS:

CROSS-REFERENCES: Hall's Theorem (Chapter 6); Steiner triple systems (Chapter 8), projective planes (Chapter 9), designs (Chapter 16)

In mathematical usage, a family resembles a set except that an element is allowed to be included more than once. However, at risk of confusion, the term 'family of sets' is used for what should be called a 'set of sets'; that is, no repetitions are allowed. Whether, and how much, such repetitions matter depends on the problem. Three situations may occur:

- In the Principle of Inclusion and Exclusion (Section 5.1) and Hall's Marriage Theorem (Section 6.2), repetitions have no effect at all.
- Problems considered in this chapter typically ask for the largest family of sets with some property; often they would be meaningless if repetitions are allowed.
- In design theory (Chapter 16), the issue is more complicated: see Section 16.2 for discussion.

So there is no 'right' answer to whether to allow repetitions.

Extremal set theory considers families of subsets of a set satisfying some restriction (perhaps in terms of inclusion or intersection of its members). It then asks the questions:

- What is the maximal size of such a family?
- Can one describe all families which meet this bound?

Like many topics, it is best introduced by example. In this chapter, we'll consider three example results in extremal set theory. In the first, the proof of the bound is trivial, but there are far too many families meeting it to allow any decent description. The second is just the opposite: the proof of the bound is quite ingenious, but not much more work is needed to give a precise description of families meeting it. The last case is somewhere between; it is included because it ties in with another of our topics, finite geometry.

Let $X = \{1, 2, \ldots, n\}$. The set of all subsets of X is called the *power set* of X, and denoted $\mathcal{P}(X)$, or sometimes 2^X. (The latter notation relates to the fact that $|\mathcal{P}(X)| = 2^{|X|}$, with a natural bijection between these sets, as we saw in Chapters 2 and 3.) By a *family of sets* is meant a subset $\mathcal{F}$ of $\mathcal{P}(X)$. The conditions we will impose on a family all relate to pairs of sets in the family; they are as follows:
(a) any two sets have non-empty intersection;
(b) no set contains another;
(c) two sets have exactly one common point.

7.1. Intersecting families

A family $\mathcal{F}$ of subsets of X is *intersecting* if $A, B \in \mathcal{F} \Rightarrow A \cap B \neq \emptyset$.

(7.1.1) Proposition. *An intersecting family of subsets of* $\{1, \ldots, n\}$ *satisfies* $|\mathcal{F}| \leq 2^{n-1}$. *Moreover, there are intersecting families of size* 2^{n-1}.

PROOF. Let $X = \{1, \ldots, n\}$. The 2^n subsets of X can be divided into 2^{n-1} complementary pairs $\{A, X \setminus A\}$; clearly an intersecting family contains at most one set from each pair. This proves the bound. But the family of all sets containing a particular element (say 1) of X has cardinality 2^{n-1} and is intersecting.

There are far too many intersecting families of size 2^{n-1} for there to be any hope of classifying them. Here are a couple of examples in addition to the ones in the proof of the Proposition.

EXAMPLE 1. If n is odd, the set of all subsets A containing more than half the points of X is intersecting, and has size 2^{n-1} (since, as required by the proof, it does contain one of each pair of complementary sets). If n is even, we modify the construction as follows: take all sets with strictly more than $n/2$ points; then divide the sets of size $n/2$ into complementary pairs, and take one of each pair in any manner whatever. This gives lots of different examples. (Note that if $|A| \geq n/2$ and $|B| > n/2$, then

$$|A \cap B| = |A| + |B| - |A \cup B| > n/2 + n/2 - n = 0,$$

so the families constructed really are intersecting.)

EXAMPLE 2. Let $X = \{1, \ldots, 7\}$, and let $\mathcal{B}$ consist of the seven subsets

$$\{\{1, 2, 3\}, \ \{1, 4, 5\} \ \{1, 6, 7\} \ \{2, 4, 6\} \ \{2, 5, 7\} \ \{3, 4, 7\} \ \{3, 5, 6\}\}.$$

(Then $(X, \mathcal{B})$ is a *Steiner triple system* of order 7 — see the next chapter for definitions.) Let $\mathcal{F}$ be the set of all those subsets of X which contain a member of $\mathcal{B}$. Then $\mathcal{F}$ is intersecting, and $|\mathcal{F}| = 64 = 2^{7-1}$ (see Exercise 1).

If we further restrict the sets to all have the same size k, what can be said? If $n < 2k$, then any family of k-subsets of an n-set is intersecting, and there is no restriction; so we should assume that $n \geq 2k$ to get meaningful results. If $n = 2k$, then an intersecting family contains at most one of each pair of disjoint sets, and so contains at most $\frac{1}{2}\binom{n}{k} = \binom{n-1}{k-1}$ sets. In general, there is always an intersecting family of size $\binom{n-1}{k-1}$, consisting of all those k-sets containing some fixed point of X; and, for large enough n, this is best possible. More generally, there is a t-*intersecting* family $\mathcal{F}$ of k-sets (i.e., satisfying $|F_1 \cap F_2| \geq t$ for all $F_1, F_2 \in \mathcal{F}$) of size $\binom{n-t}{k-t}$ (consisting of all k-sets containing a fixed t-set), and this is also best possible for large enough n:[1]

(7.1.2) Erdős–Ko–Rado Theorem. *Given k and t, there exist n_1, n_2 such that*
(a) if $n \geq n_1$, a t-intersecting family of k-subsets of an n-set has size at most $\binom{n-t}{k-t}$;
(b) if $n \geq n_2$, a t-intersecting family of k-subsets of an n-set which has size $\binom{n-t}{k-t}$ consists of all k-subsets containing some t-subset of the n-set.

A special case of this theorem is given as Exercises 2 and 3.

7.2. Sperner families

The family $\mathcal{F}$ of sets is called a *Sperner family* if no member of $\mathcal{F}$ properly contains any other, that is,

$$A, B \in \mathcal{F} \Rightarrow A \not\subset B \text{ and } B \not\subset A.$$

For any fixed k, the set of all subsets of X of size k forms a Sperner family containing $\binom{n}{k}$ sets. Since the binomial coefficients increase to the midpoint and then decrease (see Exercise 3 of Chapter 3), the largest Sperner families *of this type* occur when $k = n/2$ (if n is even) and when $k = (n-1)/2$ or $(n+1)/2$ (if n is odd). It turns out that these are the largest Sperner families without restriction.

(7.2.1) Sperner's Theorem. *Let $\mathcal{F}$ be a Sperner family of subsets of the n-element set X. Then $|\mathcal{F}| \leq \binom{n}{\lfloor n/2 \rfloor}$. Moreover, if equality holds, then $\mathcal{F}$ consists either of all subsets of X of size $\lfloor n/2 \rfloor$, or all subsets of size $\lceil n/2 \rceil$ (these are the same if n is even).*

PROOF. The ingenious proof uses the concept of a *chain* of subsets, a sequence

$$\emptyset = A_0 \subset A_1 \subset \ldots \subset A_n = X.$$

How many chains are there? If π is any permutation, then we get a chain by setting $A_i = \{1\pi, \ldots, i\pi\}$ for $i = 0, \ldots, n$. Conversely, in a chain, the points are added one at a time, so we can uniquely recover the permutation. Thus there are as many chains as permutations, viz. $n!$.

Next we ask: How many chains contain a fixed set A? If $|A| = k$, then it must occur that $A = A_k$, and the chain is obtained by welding together a chain for A and a chain for $X \setminus A$. So A lies in $k!(n-k)!$ chains, a proportion $1/\binom{n}{k}$ of the total. We could also see this by observing that each of the $\binom{n}{k}$ sets of size k lies in equally many chains, by symmetry.

---

[1] Unusually for the twentieth century, this theorem was proved in 1947, but was not published until 1963.

Now let $\mathcal{F}$ be a Sperner family. By assumption, any chain contains at most one member of $\mathcal{F}$. So the number of chains which do contain a member of $\mathcal{F}$ is

$$\sum_{A \in \mathcal{F}} |A|!(n - |A|!) = n! \left(\sum_{A \in \mathcal{F}} \frac{1}{\binom{n}{|A|}} \right).$$

Since there are only $n!$ chains altogether, we see that

$$\sum_{A \in \mathcal{F}} \frac{1}{\binom{n}{|A|}} \leq 1.$$

Now, as we already observed, the middle binomial coefficients are the largest; so their reciprocals are the smallest, and if we set $m = \lfloor n/2 \rfloor$, we have

$$\sum_{A \in \mathcal{F}} \frac{1}{\binom{n}{m}} \leq 1,$$

whence $|\mathcal{F}| \leq \binom{n}{m}$, the required bound.

When is this bound met? Examining the argument, we see that attaining the bound forces that $\binom{n}{|A|} = \binom{n}{m}$ for every $A \in \mathcal{F}$, in other words, every set in $\mathcal{F}$ has size either $m = \lfloor n/2 \rfloor$, or $n - m = \lceil n/2 \rceil$. If n is even, then these two numbers are equal, and $\mathcal{F}$ must consist of all the sets of size m. But if n is odd, there is further work required. In that case $n = 2m + 1$ and all the sets have size m or $m + 1$, but we have to show that either they all have size m, or they all have size $m + 1$.

Looking at the proof again we see that, if the bound is met, then every chain contains one member of $\mathcal{F}$; so, if A is an m-set and B a $(m + 1)$-set with $A \subset B$, then $A \in \mathcal{F}$ if and only if $B \notin \mathcal{F}$. Now suppose that A is a m-set in $\mathcal{F}$, and A' any other m-set. It is possible to find a sequence of sets beginning at A and ending at A', every term being of size m or $m + 1$, and each two consecutive terms related by inclusion:

$$A \subset B_0 \supset A_1 \subset \ldots$$

We see, following this sequence, that all of its m-sets belong to $\mathcal{F}$, while none of its $(m + 1)$-sets do. So $A' \in \mathcal{F}$. Since A' was arbitrary, $\mathcal{F}$ consists of all m-sets. Similarly, if there is a $(m + 1)$-set in $\mathcal{F}$, then it consists of all $(m + 1)$-sets.

The technique used here is called the *LYM technique*[2]. Roughly speaking, it depends on the fact that a Sperner family and a chain have at most one set in common, and the number of chains containing a set takes only a few values. A simpler example along the same lines is given in Exercise 2.

7.3. The de Bruijn–Erdős Theorem

The third result is a specialisation of the first. Instead of assuming that two sets meet in at least one point, we assume that they meet in *exactly* one.

The proof of this theorem is a bit harder than what we've had before; if you have trouble following it, concentrate on understanding the result. The proof uses Hall's 'Marriage Theorem' (6.2.1) on the existence of systems of distinct representatives.

---

[2] This is an acronym for its inventors, Lubell, Yamamoto and Meshalkin, who found it independently.

(7.3.1) De Bruijn–Erdős Theorem. *Let $\mathcal{F}$ be a family of subsets of the n-set X. Suppose that any two sets of $\mathcal{F}$ have exactly one point in common. Then $|\mathcal{F}| \leq n$. If equality holds, then one of the following situations occurs:*

(a) *up to re-numbering the points and sets, we have $\mathcal{F} = \{A_1, \ldots, A_n\}$, where $A_i = \{i, n\}$ for $i = 1, \ldots, n$ (so $A_n = \{n\}$);*

(b) *up to re-numbering the points and sets, we have $\mathcal{F} = \{A_1, \ldots, A_n\}$, where $A_n = \{1, 2, \ldots, n-1\}$, and $A_i = \{i, n\}$ for $1 \leq i \leq n-1$;*

(c) *for some positive integer q, we have $n = q^2 + q + 1$, each set in $\mathcal{F}$ has size $q+1$, and each point lies in $q+1$ members of $\mathcal{F}$.*

REMARK. Case (b) is illustrated in Fig. 7.1. The last two cases of equality overlap: when $n = 3$ (and $q = 1$), both describe a 'triangle'. For $q > 1$, the structure described in (c) is called a *projective plane* of order q. These planes will be considered further in Section 9.5. The first example (with $q = 2$) is the Steiner triple system of order 7, to be discussed in the next chapter.

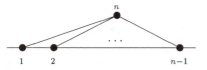

Fig. 7.1. An extremal family

PROOF. First, we can suppose that no point of X lies in every set of $\mathcal{F}$. For, if x were such a point, then the sets of $\mathcal{F}$ would partition $X \setminus \{x\}$, and we would be in case (a) of the theorem. Also, we may assume that $X \notin \mathcal{F}$; for if it were, there could be at most one further set in $\mathcal{F}$, a singleton.

Let $\mathcal{F} = \{A_1, \ldots, A_b\}$. Moreover, for $j = 1, \ldots, b$, let $B_j = X \setminus A_j$, and $k_j = |A_j|$; and for $i = 1, \ldots, n$, let r_i be the number of sets in $\mathcal{F}$ which contain i. (We call r_i the *replication number* of the point i.)

We claim that, if $i \notin A_j$, then $r_i \leq k_j$. This is because each member of $\mathcal{F}$ containing i meets A_j in a unique point, and these points are all distinct.

Now we do some counting. First count pairs (i, A_j) with $i \in A_j$. Each point i lies in r_i sets A_j, and each set A_j contains k_j points i. So we have

$$\sum_{i=1}^{n} r_i = \sum_{j=1}^{b} k_j = N, \text{ say.} \tag{1}$$

Since any two members of $\mathcal{F}$ meet in just one point, counting triples (i, A_j, A_k) with $i \in A_j \cap A_k$ and $j \neq k$ gives

$$\sum_{i=1}^{n} r_i(r_i - 1) = b(b-1). \tag{2}$$

Since pairs of blocks intersect in one point, any pair of points lies in at most one block, counting (i, i', A_j) with $i \neq i'$ and $\{i, i'\} \subseteq A_j$ gives

$$\sum_{j=1}^{b} k_j(k_j - 1) \leq n(n - 1). \tag{3}$$

Either $b < n$ and we are done, or there are distinct blocks $A_1, \ldots, A_n$ different from X. Suppose that the latter holds.

We claim that $B_1, \ldots, B_n$ satisfy Hall's condition. Let J be a subset of $\{1, \ldots, n\}$; then $B(J)$ is the set of points not contained in A_j for any $j \in J$. If $J = \{j\}$, then $B(J) = B_j = X \setminus A_j \neq \emptyset$, by assumption; so (HC) holds in this case. If $2 \leq |J| \leq n-1$, then $|B(J)| \geq n-1$ (for, if $i, j \in J$, then every point except perhaps $A_i \cap A_j$ lies in $B(J)$). If $|J| = n$, the conclusion is clear.

Thus there is a SDR for the family $(B_j : j = 1, \ldots, n)$. If we choose the numbering so that i is a representative of B_i, we have the conclusion that $i \notin A_i$ for $i = 1, \ldots, n$. From our earlier observation, this means that $r_i \leq k_i$ for $i = 1, \ldots n$. Hence we have

$$b(b - 1) = \sum_{i=1}^{n} r_i^2 - N \leq \sum_{j=1}^{b} k_j^2 - N \leq n(n - 1), \tag{4}$$

whence $b \leq n$. Equality forces not only $r_i = k_i$ for $1 \leq i \leq n$, but also (from (3)) every two points lie in a unique block. From this, the argument earlier in the proof shows that, if $i \notin A_j$, then $r_i = k_j$.

Suppose that there are points x, y with $r_x \neq r_y$. Then each set of $\mathcal{F}$ contains at least one of x and y; for, if $x, y \notin A_j$, then $r_x = k_j = r_y$. If z is any further point, then we may suppose that $r_x \neq r_z$ (interchanging x and y if neccessary), and so any set contains at least one of x and z. But only one set, say A, contains both y and z. So every set except A contains x. This forces the structure defined under (b).

Thus, we may suppose that r_x is constant, say $r_x = q + 1$. Now $|A| = q + 1$ for all $A \in \mathcal{F}$, since for every set A there is a point $x \notin A$, and $|A| = r_x$. Take a point x. Then $q + 1$ sets of $\mathcal{F}$ contain x, and each contains q further points of X; and there are no overlaps among these points. Thus $n = 1 + (q + 1)q = q^2 + q + 1$, as claimed.

(This argument is an improvement, due to R. A. Bailey, of the one I used originally.)

7.4. Project: Regular families

A family $\mathcal{F}$ of subsets of X is *regular* if every point lies in a constant number r of elements of $\mathcal{F}$. It is interesting to ask questions of extremal set theory restricted to regular families. This section considers regular intersecting families. First, however, we show that regular families do exist!

(7.4.1) Theorem. *Let b, k, n, r be positive integers satisfying*

$$bk = nr, \qquad k < n, \qquad b \leq \binom{n}{k}.$$

Then there is a regular family $\mathcal{F}$ of k-subsets of an n-set with $|\mathcal{F}| = b$.

PROOF.[3] There is a simple way to make a family 'more regular'. Let r_x be the *replication number* of

---

[3] This argument is due to David Billington.

x, the number of sets of the family which contain x. If $r_x > r_y$, then there must exist a $(k-1)$-set U, containing neither x nor y, such that $\{x\} \cup U \in \mathcal{F}$, $\{y\} \cup U \notin \mathcal{F}$. Now form a new family $\mathcal{F}'$ by removing $\{x\} \cup U$ from $\mathcal{F}$ and including $\{y\} \cup U$ in its place. In the new family, $r'_x = r_x - 1$, $r'_y = r_y + 1$, and all other replication numbers are unaltered. Starting with any family of k-sets, we reach by this process a family in which all the replication numbers differ by at most 1 (an *almost regular* family), containing the same number of sets as the original family.

But, by double counting, the average replication number is $bk/n = r$; and an almost regular family whose average replication number is an integer must be regular.

This idea can be modified to prove a theorem of Brace and Daykin:

(7.4.2) Theorem. *If k is not a power of 2, and $n = 2k$, there exists a regular intersecting family of size $\frac{1}{2}\binom{n}{k} = \binom{n-1}{k-1}$ of k-subsets of an n-set.*

We begin with two remarks:

REMARK 1. As we already saw in Section 7.1, an intersecting family of k-subsets of a $2k$-set has size at most $\binom{2k-1}{k-1}$, with equality if and only if it contains one of each complementary pair of k-sets.

REMARK 2. The replication number of a regular family as in the theorem is $r = \frac{1}{2}\binom{2k-1}{k-1}$. This is an integer if and only if k is not a power of 2 (Exercise 4). So the condition on k is necessary.

We need a slightly more complicated version of the replacement procedure, in order to preserve the intersecting property. Let x and y be points with $r_x \geq r_y + 2$. Then there are two disjoint $(k-1)$-subsets U and V of $X \setminus \{x, y\}$ such that $\{x\} \cup U, \{x\} \cup V \in \mathcal{F}$. The complements of these sets are $\{y\} \cup V$, $\{y\} \cup U$ respectively, and are not in $\mathcal{F}$. If we replace both of these sets by their complements, we obtain an intersecting family $\mathcal{F}'$ in which $r'_x = r_x - 2$, $r'_y = r_y + 2$, and the other replication numbers are unaltered. Applying a sequence of such operations to an arbitrary intersecting family, we obtain a family in which the new replication differ by at most 2, and are congruent mod 2 to their initial values.

Let $\mathcal{F}$ be any intersecting family of size $\binom{2k-1}{k-1}$, in which all the replication numbers are congruent to $\frac{1}{2}\binom{2k-1}{k-1}$ mod 2. [Let $\mathcal{F}_0$ be the family of all sets containing the point x. Its replication numbers are $r_x = \binom{2k-1}{k-1}$, $r_y = \binom{2k-2}{k-1}$ for $y \neq x$, which are all even. If a family in which all replication numbers are odd is required, replace a single set by its complement.] Now apply the above process. If a collection of numbers differ by at most 2, and all have the same parity as their (integral) average, then all the numbers are equal.

7.5. Exercises

1. Verify the claim in Example 2 of Section 7.1.

2. If $n = 2k$, an intersecting family of k-subsets of an n-set has size at most $\frac{1}{2}\binom{n}{k} = \binom{n-1}{k-1}$, because it contains at most one of each complementary pair of k-sets. We proceed to generalise this result and argument. What follows could be regarded as a very simple version of the LYM technique. PROVE:

> Suppose that k divides n. Then an intersecting family $\mathcal{F}$ of k-subsets of an n-set X has size at most $\binom{n-1}{k-1}$.

[HINT: Let $\mathcal{C}$ be the set of all partitions of X into n/k subsets of size k. We don't need to know $|\mathcal{C}|$ (though this could be counted), merely the fact that each k-set lies in $|\mathcal{C}|/\binom{n-1}{k-1}$ members of $\mathcal{C}$. Prove this by double-counting pairs (B, C), where B is a k-set and $C \in \mathcal{C}$ with B a member of C.]

Now double-count pairs (B, C), with $B \in \mathcal{F}$, $C \in \mathcal{C}$, and $B \in C$, to obtain

$$|\mathcal{F}| \cdot |\mathcal{C}| \Big/ \binom{n-1}{k-1} \leq |\mathcal{C}| \cdot 1,$$

using the fact that, since the parts of a partition are disjoint, at most one of them lies in any intersecting family.

3. (HARDER PROBLEM). Prove that, if k divides n and $n \geq 3k$, then any intersecting family of size $\binom{n-1}{k-1}$ of k-subsets of the n-set X consists of all k-sets containing some point of X. [HINT: it follows from the argument of Exercise 2 that, if $|\mathcal{F}| = \binom{n-1}{k-1}$, then given *any* partition of X into disjoint k-sets, exactly one of these k-sets belongs to $\mathcal{F}$. Exploit this fact.]

4. Show that $\binom{2k-1}{k-1}$ is even if and only if k is not a power of 2.

5. (a) If n is not a power of 2, construct a regular intersecting family of subsets of an n-set, having size 2^{n-1}.
    (b) If $n = 2, 4$ or 8, show that there is no such family.

REMARK. Families with these properties do exist for all powers of 2 greater than 8; this was shown by Aaron Meyerowitz.

6. Prove that, in any intersecting family of size $\binom{2k-1}{k-1}$ of k-subsets of a $2k$-set, the replication numbers all have the same parity.

7. Let $\mathcal{F}$ be any intersecting family of subsets of the n-set X. Show that there is an intersecting family $\mathcal{F}' \supseteq \mathcal{F}$ with $|\mathcal{F}'| = 2^{n-1}$. [HINT: A *blocking set* for $\mathcal{F}$ is a set Y which meets every member of $\mathcal{F}$ but contains none. Adjoin to $\mathcal{F}$ all sets which contain a member of $\mathcal{F}$, all blocking sets of size greater than $\frac{1}{2}n$, and (if n is even) one of each complementary pair of blocking sets of size $\frac{1}{2}n$.]

    By proving that the Steiner triple system of order 7 has no blocking sets, give another proof of Exercise 1.

8. Let $\mathcal{F}$ be a Sperner family of subsets of the n-set X. Define $b(\mathcal{F})$ to be the family of all subsets Y of X such that
    (i) $Y \cap F \neq \emptyset$ for all $F \in \mathcal{F}$;
    (ii) Y is minimal subject to (i) (i.e., no proper subset of Y satisfies (i)).
(a) Prove that $b(\mathcal{F})$ is a Sperner family.
(b) Show that, for any $F \in \mathcal{F}$ and any $y \in F$, there exists $Y \in b(\mathcal{F})$ with $Y \cap F = \{y\}$.
(c) Deduce that $b(b(\mathcal{F})) = \mathcal{F}$.
(d) Let $\mathcal{F}_k$ denote the Sperner family of all k-subsets of X. Prove that $b(\mathcal{F}_k) = \mathcal{F}_{n+1-k}$ for $k > 0$. What is $b(\mathcal{F}_0)$?

8. Steiner triple systems

... how did the *Cambridge and Dublin Mathematical Journal*, Vol. II, p. 191 [1846] manage to steal so much from ... *Crelle's Journal*, Vol. LVI, p. 326 [1859], on exactly the same problem in combinations?

T. P. Kirkman (1887)

TOPICS: Steiner triple systems; packings and coverings; [tournament schedules; finite geometries]

TECHNIQUES: Direct and recursive constructions; [use of linear algebra and finite fields for constructions]

ALGORITHMS:

CROSS-REFERENCES: Extremal set theory (Chapter 7); [finite fields (Chapter 4); finite geometry (Chapter 9)]

This chapter is devoted to the proof of existence of Steiner triple systems. The topic is somewhat specialised; but the technique, involving a mixture of direct and recursive constructions (the latter building up large objects of some type from smaller ones) is of very wide applicability.

8.1. Steiner systems

In 1845, the following problem in extremal set theory was posed in an unlikely forum, the *Lady's and Gentleman's Diary*:

> Given integers l, m, n with $l < m < n$, what is the greatest number of m-element subsets of an n-element set with the property that any l-element subset lies in at most one of the chosen sets?

The problem proved too difficult for the journal's readership, and so it was specialised to the case $l = 2$, $m = 3$. This provided the incentive for a 40-year-old Lancashire vicar, T. P. Kirkman, to take up mathematics: his first published paper, the following year, contained a contribution to this case.[1]

Returning to the general problem for a moment, we observe:

---

[1] Kirkman is now remembered almost entirely for his work on this problem, but he also wrote extensively on projective geometry, groups, polyhedra, and knots, and was regarded as one of the leading British mathematicians of his day. An account of his life and work can be found in the article 'T. P. Kirkman: Mathematician' by Norman Biggs in the *Bulletin of the London Mathematical Society* **13** (1981), 97–120.

(8.1.1) Proposition. *Let $\mathcal{B}$ be a family of m-subsets of an n-set, such that any l-set lies in at most one member of $\mathcal{B}$. Then*

$$|\mathcal{B}| \le \binom{n}{l} \Big/ \binom{m}{l}.$$

Equality holds if and only if any l-subset lies in exactly one member of $\mathcal{B}$.

PROOF. We count pairs (L, B), where L is an l-set and $B \in \mathcal{B}$ with $L \subseteq B$. Each $B \in \mathcal{B}$ contains $\binom{m}{l}$ subsets of size l, so there are $|\mathcal{B}| \cdot \binom{m}{l}$ pairs. On the other hand, there are altogether $\binom{n}{l}$ subsets of size l, and each lies in at most one set B, so there are at most $\binom{n}{l}$ pairs. Thus

$$|\mathcal{B}| \cdot \binom{m}{l} \le \binom{n}{l}.$$

Equality is only possible if every l-set lies in a (unique) member of $\mathcal{B}$.

A pair $(X, \mathcal{B})$, where X is an n-set and $\mathcal{B}$ a family of m-subsets satisfying the hypotheses of the proposition and attaining the bound is called a *Steiner system* $S(l, m, n)$.[2] A very important specialisation of the above problem is the following:

For which values of l, m, n does a Steiner system $S(l, m, n)$ exist?

A Steiner system $S(2, 3, n)$ is called a *Steiner triple system*. To reiterate: a Steiner triple system consists of a set X of points and a set $\mathcal{B}$ of 3-element subsets of X (called *triples* or *blocks*), with the property that any two points of X lie in a unique triple. The number n is called the *order* of the Steiner triple system. In this chapter, we settle the existence question for Steiner triple systems. I will usually abbreviate 'Steiner triple system' to STS, and write STS(n) for a Steiner triple system of order n.

First, some examples.

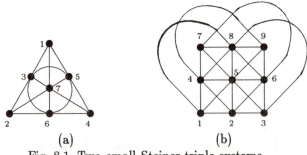

(a)   (b)

Fig. 8.1. Two small Steiner triple systems

---

[2] The name is a double misnomer: the question posed by Steiner was not equivalent to the existence of Steiner systems, though they are the same in the special case $l = 2$, $m = 3$; and this special case was settled by Kirkman seven years before the question was asked by Steiner. However, the terminology is now standard, and the term 'Kirkman system' has a different meaning.

Fig. 8.1(a) shows a STS(7). More formally, $(X, \mathcal{B})$ is a STS(7), where

$$X = \{1, 2, 3, 4, 5, 6, 7\}$$
$$\mathcal{B} = \{123, 145, 167, 246, 257, 347, 356\}.$$

(We write 123 for the set $\{1, 2, 3\}$, and so on.)

Fig. 8.1(b) shows a STS(9). Note that it solves the 'nine schoolgirls' problem posed in Chapter 1: the walking scheme is

Day 1:	123	456	789
Day 2:	147	258	369
Day 3:	159	267	348
Day 4:	357	168	249

Moreover, there are trivial Steiner triple systems of orders 3 (three points forming a triple), 1 (a single point, no triples), and 0 (no points or triples). Before reading further, show that there is no Steiner triple system of order 2, 4, 5 or 6.

The next theorem determines completely the possible orders of Steiner triple systems.

(8.1.2) Theorem. *These exists a Steiner triple system of order n if and only if either*
- *$n = 0$; or*
- *$n \equiv 1$ or 3 (mod 6).*

This theorem asserts that a numerical condition is *necessary* and *sufficient* for the existence of something. So the proof has two parts. First, we must show that the order of a Steiner triple system satisfies the constraint: the argument is given below. Second, given a number n of the correct form, we have to construct a STS(n). This is more difficult, and will take the next two sections.

PROOF OF NECESSITY. Suppose that $(X, \mathcal{B})$ is a STS of order n. Clearly, we may suppose that $n > 0$. We establish two important properties by 'double counting'.

1. *Any point lies in $(n - 1)/2$ triples.*

   Choose a point x, and count pairs (y, B), where y is a point different from x, and B a triple containing x and y. First, there are $n - 1$ choices for y and, for each choice, there is a unique triple containing x and y: altogether $n - 1$ pairs. Second, if x lies in r triples, then (since each triple contains two points other than x) there are $2r$ choices of the pair (y, B). Hence $2r = n - 1$, and r is as claimed.

2. *There are $n(n - 1)/6$ triples altogether.*

   We count pairs (x, B), where x is a point and B a triple containing x. Each of the n points lies in $(n - 1)/2$ triples, so there are $n(n - 1)/2$ pairs. If there are b triples, each containing 3 points, then there are $3b$ choices. So $3b = n(n - 1)/2$, giving the claimed value for b.

   Now the necessity of the condition follows. For, if $n > 0$, then both $(n - 1)/2$ and $n(n - 1)/6$ must be whole numbers. The first condition asserts that n is odd, whence $n \equiv 1, 3$ or 5 (mod 6). Suppose that $n \equiv 5$ (mod 6), say $n = 6k + 5$. Then the number of triples is

$$n(n - 1)/6 = (6k + 5)(3k + 2)/3;$$

but this is not an integer, since neither $6k + 5$ nor $3k + 2$ is a multiple of 3. So n must be congruent to 1 or 3 modulo 6.

Note that, if $n \equiv 1$ or 3 (mod 6), then both $(n-1)/2$ and $n(n-1)/6$ are integers; but this in itself is no guarantee that a STS(n) exists, of course.

8.2. A direct construction

The proof of sufficiency given in the next section involves a *recursive* construction, in which larger Steiner triple systems are built up from smaller ones. In this section, we show that a *direct* construction can be used to prove half of the theorem. Specifically:

(8.2.1) Proposition. If $n \equiv 3 \pmod{6}$, there exists a STS(n).

PROOF. Suppose that $n \equiv 3 \pmod{6}$; then $n = 3m$ where m is odd. The point set is made up of three copies of the integers mod m. Formally,

$$X = \{a_i, b_i, c_i : i \in \mathbb{Z}/(m)\}.$$

Blocks are of two types:
(a) all sets of the form $a_i a_j b_k$, $b_i b_j c_k$, or $c_i c_j a_k$, where $i, j, k \in \mathbb{Z}/(m)$, $i \neq j$, and $i + j = 2k$ (in $\mathbb{Z}/(m)$);
(b) all sets of the form $a_i b_i c_i$, for $i \in \mathbb{Z}/(m)$.

Before verifying that this works, observe that the equation $i + j = 2k$ has a unique solution (in $\mathbb{Z}/(m)$) for any one of the variables, given the other two. This is clear for i and j. For k it depends on the fact that (since m is odd) any element of $\mathbb{Z}/(m)$ can be uniquely divided by 2: depending on the parity of l, either $l/2$ or $(l + m)/2$ is the unique solution of $2x = l$.

First let us count the triples. (This is not necessary for the proof but is a helpful check.) There are $\binom{m}{2} = m(m - 1)/2$ choices of i and j, and for each choice, a unique k and hence three triples of the first type. There are clearly m triples of the second type. This makes altogether

$$3m(m - 1)/2 + m = 3m(3m - 1)/6$$

triples, as required. Now let us verify that they do form a Steiner triple system, by showing that any two points lie in a unique triple.

There are several cases:
(i) Points a_i and a_j, $i \neq j$. A triple containing them must be of type (a); by our remark above, there is a unique such triple.
(ii) Points b_i and b_j, or c_i and c_j: these cases are similar.
(iii) Points a_i and b_i. These lie in a unique triple of type (b); and in no triple of type (a), since if $a_i a_j b_i$ were such a triple, then $i + j = 2i$, whence $j = i$.
(iv) Points b_i and c_i, or c_i and a_i: similarly.
(v) Points a_i and b_k, $k \neq i$: these lie in a unique block of type (a).
(vi) Points b_i and c_k, or c_i and a_k: similarly.

For $n = 9$, we get a different-looking STS of order 9. In fact it turns out to be the same as before, just drawn differently (see Fig. 8.2, in which three triples are not shown).

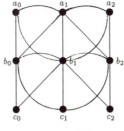

Fig. 8.2. STS(9)

8.3. A recursive construction

Before embarking on the main business, we attend to one important detail: the construction of a STS(13).

For this, it would be sufficient to give a list of 13 points and $13.12/6 = 26$ triples, and leave the verification to the reader. However, the construction is a special case of something more general, so we give it abstractly.

We take X to be the set $\mathbb{Z}/(13)$ of residue classes modulo 13. Consider first the sets

$$B_1 = \{0, 1, 4\}, \qquad B_2 = \{0, 2, 8\}.$$

We *claim* that the following holds.

> For any non-zero $z \in X$, there is a unique way to write $z = u - v$ with u, v chosen from the same set B_i.

This is seen by listing all possibilities:

$1 = 1 - 0$	$2 = 2 - 0$	$3 = 4 - 1$	$4 = 4 - 0$
$5 = 0 - 8$	$6 = 8 - 2$	$7 = 2 - 8$	$8 = 8 - 0$
$9 = 0 - 4$	$10 = 1 - 4$	$11 = 0 - 2$	$12 = 0 - 1$

and noting that each of the $2 \cdot 3 \cdot 2 = 12$ expressions $u - v$ for $u, v \in B_i$, $i = 1, 2$ has been used once.

Now let

$$\mathcal{B} = \{B_1 + z, B_2 + z : z \in X\},$$

where $B_i + z = \{t + z : t \in B_i\}$. (So $\mathcal{B}$ consists of the triples 014, 125, 236, ..., 028, 139, ... ; 26 in all.)

We claim that $(X, \mathcal{B})$ is a STS. Clearly X is a 13-set and $\mathcal{B}$ a set of 3-subsets of X. Let x, y be distinct points of X. If $x, y \in B_i + z$, then $x - z, y - z \in B_i$, and $(x - z) - (y - z) = x - y$. By the claim above, there is a unique choice of i, u, v so that $x - y = u - v$ with $u, v \in B_i$; and then $x - z = u$, so $z = x - u = y - v$ is also determined. So there is a unique triple containing x and y.

This technique works whenever we can find sets $B_1, \ldots$ in $\mathbb{Z}/(n)$ such that any non-zero element of $\mathbb{Z}/(n)$ can be written uniquely in the form $u - v$, with u and v chosen from the same B_i. For example, with $n = 7$, it is possible to use a single set $B_1 = \{0, 1, 3\}$, giving rise to the familiar STS(7) labelled in a different way (Fig. 8.3).

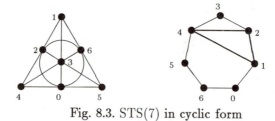

Fig. 8.3. STS(7) in cyclic form

This construction works much more generally. We use it for all primes n (indeed, all prime powers) which are congruent to 1 (mod 6) in a Project (Section 8.5).

Now we come to the main technical result. This is a recursive construction, building larger systems from smaller ones. First, a definition.

A *subsystem* of the Steiner triple system $(X, \mathcal{B})$ is a subset Y of X with the property that any triple in $\mathcal{B}$ which contains two points of Y is contained within Y. If $\mathcal{C}$ is the set of triples contained within the subsystem Y, then $(Y, \mathcal{C})$ is a Steiner triple system in its own right, and we may refer to this as a subsystem without confusion.

(8.3.1) Proposition. *Suppose that there exists a STS of order v containing a subsystem of order u, and also there exists a STS of order w. Then there exists a STS of order $u + w(v - u)$. If $w > 0$, it contains a copy of the STS(v) as a subsystem. Moreover, if $0 < u < v$ and $w > 0$, then it can be assumed to have a subsystem of order 7.*

EXAMPLE. Given this result, we can give two constructions of a STS of order 19 (the smallest value for which we haven't yet constructed a STS). In the proposition, take either

- $u = 1, v = 3, w = 9$ ($19 = 1 + 9(3 - 1)$); or
- $u = 1, v = 7, w = 3$ ($19 = 1 + 3(7 - 1)$).

The idea behind the construction is described like this. Imagine that the STS(v) is drawn on a piece of paper, with the points of the STS(u) on the left-hand side. Make w copies of this page. Now bind them into a book by glueing them together on the left, so that the points of the STS(u) on the different pages become identified (and lie on the spine of the book). We have $u + w(v - u)$ points, as required. Moreover, we have some triples already, all those lying on a single page of the book (possibly using points of the spine). Any further triple uses three points from different pages. We use the STS(w) to help us choose these triples; so we imagine that the pages are numbered by its points.

Formally, then, let the point set of the STS(v) be $\{a_1, \ldots, a_u\} \cup \{b_i : i \in \mathbb{Z}/(m)\}$, where $m = v - u$, and the points $\{a_1, \ldots, a_u\}$ form the STS(u). Let the points of the STS(w) be $\{c_1, \ldots, c_w\}$. Take

$$x = \{a_1, \ldots, a_u\} \cup \{d_{p,i} : p = 1, \ldots, w; i \in \mathbb{Z}/(m)\}.$$

The blocks are of two types:
(a) the blocks of the STS(v), copied onto each 'page' (each set consisting of all the a_i and all the $d_{p,i}$ with fixed p) by the mapping that fixes all a_i and maps b_i to $d_{p,i}$;
(b) all sets of the form $d_{p_1,i_1} d_{p_2,i_2} d_{p_3,i_3}$ for which the 'page numbers' $c_{p_1}, c_{p_2}, c_{p_3}$ form a triple of the STS(w) and $i_1 + i_2 + i_3 = 0$ (in $\mathbb{Z}/(m)$).

Let us check that it works. Take two points. If they lie on the same page, then they lie in a unique triple of the first type, by the defining property of the STS(v). If they lie on different pages, then they have the form d_{p_1,i_1} and d_{p_2,i_2}, where $p_1 \neq p_2$. Then the third point on the triple must have the form d_{p_3,i_3}; p_3 is uniquely determined by the requirement that $c_{p_1} c_{p_2} c_{p_3}$ is a triple of the STS(w), and i_3 by the requirement that $i_1 + i_2 + i_3 = 0$.

It remains to show the last part, about the subsystem of order 7. Suppose that $0 < u < v$ and $w > 1$. Since $u > 0$, we may take a point a of the subsystem. Since $v > u$, we have $m = v - u$ even; choose the numbering of the points b_i so that $a b_0 b_{m/2}$ is a triple (otherwise it is arbitrary). Since $w > 1$, there is a triple in the STS(w), say $c_1 c_2 c_3$. Now it is easily checked that the seven points

$$\{a\} \cup \{d_{p,i} : p = 1, 2, 3; i = 0, m/2\}$$

form a subsystem (see Fig. 8.4).

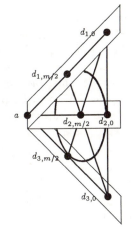

Fig. 8.4. A STS(7) subsystem

Now let A be the set of positive integers n for which there is a Steiner triple system of order n — we have to show that A contains all numbers $n \equiv 1$ or 3 (mod 6). Also, we let B be the set of positive integers n for which there exists a Steiner triple system of order n containing a subsystem of order 7.

We note that B is a subset of A. Also, the following implications hold:

$$n \in A \Rightarrow 3n \in A$$
$$n \in B \Rightarrow 3n \in B$$
$$n \in A, n > 1 \Rightarrow 3n - 2 \in B$$
$$n \in A, n > 3 \Rightarrow 3n - 6 \in B$$
$$n \in B \Rightarrow 3n - 14 \in B$$

These are justified by (8.3.1), with the following values of (u, v, w):

$$(0, n, 3)$$
$$(0, n, 3)$$
$$(1, n, 3)$$
$$(3, n, 3)$$
$$(7, n, 3)$$

We divide the potential members of B into congruence classes modulo 18: any admissible n is congruent to 1, 3, 7, 9, 13 or 15 (mod 18). Now we have

$$6k + 1 \in A \Rightarrow 18k + 1 = 3(6k + 1) - 2 \in B$$
$$6k + 3 \in A \Rightarrow 18k + 3 = 3(6k + 3) - 6 \in B$$
$$6k + 3 \in A \Rightarrow 18k + 7 = 3(6k + 3) - 2 \in B$$
$$6k + 3 \in B \Rightarrow 18k + 9 = 3(6k + 3) \in B$$
$$6k + 9 \in B \Rightarrow 18k + 13 = 3(6k + 9) - 14 \in B$$
$$6k + 7 \in A \Rightarrow 18k + 15 = 3(6k + 7) - 6 \in B$$

We claim that every admissible number $n \geq 15$ lies in B. Suppose not, and take a least counterexample. If $n \equiv 1 \pmod{18}$, then we must have $n < 55$ — for if $18k + 1 \geq 55$, then $6k + 1 \geq 19$, so $6k + 1 \in B$ (since $6k + 1$ is at least 15 and is smaller than the least counterexample), and $18k + 1 \in B$ also. So $n = 19$ or 37. Checking the other congruence classes this way, we find the possible values of n to be 15, 19, 21, 25, 27, 33, 37. So the claim will be proved if we can show that each of these numbers is in B. Suitable values of (u, v, w) in (8.3.1), with the relevant equation, are given in the following table.

$$(1, 3, 7) \qquad 15 = 1 + 7(3 - 1)$$
$$(1, 3, 9) \qquad 19 = 1 + 9(3 - 1)$$
$$(0, 7, 3) \qquad 21 = 0 + 3(7 - 0)$$
$$(1, 9, 3) \qquad 25 = 1 + 3(9 - 1)$$
$$(1, 3, 13) \qquad 27 = 1 + 13(3 - 1)$$
$$(3, 13, 3) \qquad 33 = 3 + 3(13 - 3)$$
$$(1, 13, 3) \qquad 37 = 1 + 3(13 - 1)$$

We use the fact that $7, 9, 13 \in A$, as established earlier.

So B does contain all admissible $n \geq 15$. Since $B \subseteq A$, and since A contains 1, 3, 7, 9 and 13, the theorem is proved.

The proof of the theorem is constructive: given a number $n \equiv 1$ or $3 \pmod 6$, if n is sufficiently large, then we can read off from the proof a number n' such that an STS(n') can be used to construct an STS(n). For example, how to construct a STS(625)? Since $625 = 18 \cdot 34 + 13$, and $6 \cdot 34 + 9 = 213$, we require a STS(213). Then, since $213 = 18 \cdot 11 + 15$, and $6 \cdot 11 + 7 = 73$, we need a STS(73). Then, since $73 = 18 \cdot 4 + 1$, and $6 \cdot 4 + 1 = 25$, we need a STS(25). Then the recursion 'bottoms out', since the proof tells us how to construct this system.

The construction given here is by no means the only one possible. Exercises 2 and 13 yield a completely different STS(625).

This is not the end of the story — one can ask how many different ways there are of forming a Steiner triple system on a set of n points, where $n \equiv 1$ or $3 \pmod 6$. But we now pursue a different question: if n is not of this form, how close can we get to a STS?

8.4. Packing and covering

Steiner triple systems represent special solutions to an extremal set problem — indeed, to two such problems, as we now discuss. This situation, where a structure satisfying a condition containing the words 'exactly one' is an extreme case for both 'at most one' and 'at least one', is very common; the extremal problems are referred to as *packing* and *covering* problems.

Let X be a set with n elements. A (2,3)-*packing* is a set $\mathcal{B}$ of triples such that any two points of X are contained in *at most* one member of $\mathcal{B}$; and a (2,3)-*covering* is a set $\mathcal{B}$ of triples such that any two points are contained in *at least* one member of $\mathcal{B}$. Obviously any subset of a packing is a packing, and any superset of a covering is a covering; so we let $p(n)$ denote the size of the largest (2,3)-packing, and $c(n)$ the size of the smallest (2,3)-covering, of an n-set.

(8.4.1) Proposition. *(a)* $p(n) \leq n(n-1)/6$.
(b) $c(n) \geq n(n-1)/6$.
(c) Equality holds in either bound if and only if there exists a STS(n).

PROOF. The arguments are straight double counting. For packings, each of the $n(n-1)/2$ pairs is contained in at most one triple, and each of the $p(n)$ triples contains exactly three pairs. For coverings, the inequality reverses.

Thus, if $n \equiv 1$ or $3 \pmod 6$, we have $p(n) = c(n) = n(n-1)/6$. For other values, $p(n)$ is smaller than this bound, and $c(n)$ is larger. It is possible to prove a general result improving the inequalities:

(8.4.2) Proposition. *(a)* $p(n) \leq \lfloor \frac{n}{3} \lfloor \frac{n-1}{2} \rfloor \rfloor$.
(b) $c(n) \geq \lceil \frac{n}{3} \lceil \frac{n-1}{2} \rceil \rceil$.

PROOF. We follow the argument for the necessary condition for the existence of a STS(n) (8.1.2). Let $\mathcal{B}$ be a packing. Then, by double counting, any point x lies in at most $\frac{n-1}{2}$ triples of $\mathcal{B}$. However, the number of triples containing x is an integer, so we can round this number down to $\lfloor \frac{n-1}{2} \rfloor$. Then, again by double counting, the number of triples is at most $\frac{n}{3}\lfloor\frac{n-1}{2}\rfloor$; and, again, we can round this number down. The argument for coverings is similar, except that we round up.

These bounds are not always attained. But there is one case where they are met:

(8.4.3) Proposition. *If $n \equiv 0$ or $2 \pmod{6}$, then $p(n) = n(n-2)/6$.*

PROOF. n is even, so $\lfloor\frac{n-1}{2}\rfloor = \frac{n-2}{2}$. Then $\frac{n(n-2)}{6}$ is an integer, so this quantity is our upper bound for $p(n)$.

On the other hand, there exists a Steiner triple system of order $n + 1$, since this number is congruent to 1 or 3 (mod 6). This STS has $\frac{(n+1)n}{6}$ blocks, of which each point lies in $\frac{n}{2}$. So, if we remove one point and all triples containing it, we obtain a packing of size

$$\frac{(n+1)n}{6} - \frac{n}{2} = \frac{n(n-2)}{6}.$$

8.5. Project: Some special Steiner triple systems

This section describes some constructions of Steiner triple systems by algebraic, rather than combinatorial, methods. The resulting systems have a high degree of symmetry.

PROJECTIVE TRIPLE SYSTEMS.

In this subsection and the next, we construct examples of highly symmetric Steiner triple systems, using linear algebra over the fields $\mathbb{Z}/(2)$ and $\mathbb{Z}/(3)$. These systems are instances of more general 'finite geometries', to be treated in Chapter 9.

Let F be the field $\mathbb{Z}/(2)$ of order 2. Let V be a vector space of dimension d over F. Then V can be realised concretely as the set of all d-tuples of elements of F, so that $|V| = 2^d$. We take X to be the set of non-zero vectors in V, and

$$\mathcal{B} = \{\{x, y, z\} : x, y, z \text{ distinct}, x + y + z = 0\}.$$

CLAIM. $(X, \mathcal{B})$ is a Steiner triple system of order $2^d - 1$.

PROOF. It's clear that, if $x + y + z = 0$, then any two of x, y, z determine the third. We have to show that, if x and y are distinct and non-zero, then z is distinct from both and non-zero. So suppose that $0 \neq x \neq y \neq 0$. Then $z = -(x + y) = x + y$ (since $-1 = 1$ in F). Since $y \neq 0$, we have $z \neq x$; since $x \neq 0$, we have $z \neq y$; and since $x \neq y$, we have $z = x + y = x - y \neq 0$.

We denote this system by $P(d - 1)$; it is called a *projective triple system* or *projective geometry* of dimension $d - 1$ over F. (There are geometric reasons for letting the dimension be $d - 1$ rather than d; these will appear later.) Fig. 8.5 shows the familiar STS(7) presented as $P(2)$.

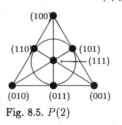

Fig. 8.5. $P(2)$

Projective systems have an important, and characteristic, property. A *triangle* in a STS is a set of three points not forming a triple.

(8.5.1) Theorem. *A STS is projective if and only if any triangle is contained in a subsystem of order 7.*

PROOF. Let $(X, \mathcal{B})$ be a projective triple system, and $\{x, y, z\}$ a triangle. Then $x + y + z \neq 0$; so the seven points x, y, z, $x + y$, $y + z$, $z + x$, $x + y + z$ are all distinct and are easily seen to form a subsystem.

For the converse, let $(X, \mathcal{B})$ be a STS in which every triangle is contained in a 7-point subsystem. We have to construct the algebraic structure of a vector space over $\mathbb{Z}/(2)$. This is an example of the procedure of 'coordinatisation' in geometry.

Let 0 be a symbol not in X, and let $V = X \cup \{0\}$. We define an operation $+$ on V by the rules that, for all $v \in V$,

$$0 + v = v + 0 = v,$$
$$v + v = 0,$$

and, if $x, y \in X$ with $x \neq y$, then $x + y$ is the third point of the triple containing x and y.

This operation is obviously commutative; 0 is the identity, and every element is its own inverse. We show that it is associative. There are several cases, most of which are trivial (for example, $(x + 0) + y = x + y = x + (0 + y)$). The only non-trivial case occurs when $\{x, y, z\}$ is a triangle, in which case the structure of the STS(7) gives the required conclusion (see Fig. 8.6).

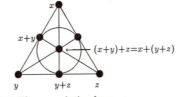

Fig. 8.6. The associative law

We conclude that

$(V, +)$ *is an abelian group.*

Next we define a scalar multiplication on V, by elements of F, by the rules

$$0 \cdot v = 0,$$
$$1 \cdot v = v,$$

for all $v \in V$. We have

$(V, +, \cdot)$ *is a vector space over* $\mathbb{Z}/(2)$.

Again, most of the axioms are trivial. The most interesting is

$$(\alpha + \beta) \cdot v = \alpha \cdot v + \beta \cdot v.$$

In the case $\alpha = \beta = 1$, we have $\alpha + \beta = 0$, and the result follows from the fact that $v + v = 0$.

Now X is the set of non-zero vectors, and $\mathcal{B}$ the set of triples with sum 0, in V; so the system is projective.

AFFINE TRIPLE SYSTEMS.

There is a similar construction involving the field $\mathbb{Z}/(3)$. Let V be a d-dimensional vector space over this field. Let $X = V$, and

$$\mathcal{B} = \{\{x, y, z\} \subseteq X : x, y, z \text{ distinct}, x + y + z = 0\}.$$

CLAIM. $(X, \mathcal{B})$ is a Steiner triple system of order 3^d.

Proof. Again, if $x + y + z = 0$, then any two of x, y, z determine the third. Suppose that $x \neq y$. Then $z \neq x$, since if $x = z = -(x + y)$ then $y = -2x = x$; and similarly $z \neq y$, so all three points are distinct.

This system is called an *affine triple system* or *affine geometry* of dimension d over $\mathbb{Z}/(3)$. (Note the dimension!) It has a property resembling that of projective triple systems:

In an affine triple system, any triangle is contained in a subsystem of order 9.

(See Exercise 5.)

The converse, surprisingly, is false. The first counterexample has order 81, and was constructed by Marshall Hall. As a result, the term *Hall triple system* is used for any Steiner triple system which is not affine but has the property that every triangle is contained in a subsystem of order 9. It is known that the order of a Hall triple system must be a power of 3, and that they exist for all orders which are powers of 3 and at least 81.

Nobody knows any example of a Steiner triple system of order n in which each triangle lies in a unique subsystem of order $k < n$, for any k other than 7 or 9.

NETTO SYSTEMS.

These Steiner triple systems are constructed using the method we saw already for the STS(13).

(8.5.2) Proposition. Let $B_1, \ldots, B_t$ be 3-subsets of $\mathbb{Z}/(n)$. Suppose that, for any non-zero element $u \in \mathbb{Z}/(n)$, there is a unique value of $i \in \{1, \ldots, t\}$ and unique $x, y \in B_i$ such that $u = x - y$. Set

$$\mathcal{B} = \{B_i + z : 1 \leq i \leq t, z \in \mathbb{Z}/(n)\},$$

where $B_i + z = \{b + z : b \in B_i\}$. Then $(\mathbb{Z}/(n), \mathcal{B})$ is a Steiner triple system.

Proof. Take two distinct elements $x, y \in \mathbb{Z}/(n)$; we have to show that a unique triple in $\mathcal{B}$ contains x and y. When do we have $x, y \in B_i + z$? This condition implies that $x - z, y - z \in B_i$; and $(x - z) - (y - z) = x - y \neq 0$. So, given x and y, there is a unique choice of i; and the elements $x - z$ and $y - z$ (and hence z) are also determined.

Note that the number of triples is $tn = n(n - 1)/6$; so $n = 6t + 1$, or $n \equiv 1 \pmod 6$. Note also that the cyclic permutation $x \mapsto x + 1 \pmod n$ preserves the Steiner triple system.

We will see that, for any prime number $p \equiv 1 \pmod 6$, there exist sets $B_1, \ldots, B_t$ satisfying the hypothesis of (8.5.2). For this, we use the following fact:

> If $p \equiv 1 \pmod 6$, then the field $\mathbb{Z}/(p)$ contains a primitive sixth root of unity
> (an element z satisfying $z^6 = 1$, $z^k \neq 1$ for $0 < k < 6$).

The algebraic explanation of this fact is that the multiplicative group of $\mathbb{Z}/(p)$ is a cyclic group of order $p - 1$, and so (if $6|p - 1$) contains a cyclic subgroup of order 6.

Since $0 = z^6 - 1 = (z^3 - 1)(z + 1)(z^2 - z + 1)$, and $z^3 \neq 1$, $z \neq -1$, we have $z^2 - z + 1 = 0$. We note the equations

$$1 = 1 - 0, \quad z = z - 0, \quad z^2 = z - 1, \quad z^3 = 0 - 1, \quad z^4 = 0 - z, \quad z^5 = 1 - z.$$

Now set $t = (p - 1)/6$, and let $s_1, \ldots, s_t$ be coset representatives for the distinct cosets of the subgroup generated by z in the multiplicative group of $\mathbb{Z}/(p)$. Then let $B_i = \{0, s_i, s_i z\}$ for $i = 1, \ldots, t$. Then every non-zero residue mod p is uniquely expressible in the form $s_i z^j$, where $1 \leq i \leq t$ and $0 \leq j \leq 5$. According to the displayed equations, it is uniquely expressible in the form $x - y$ for some $x, y \in B_i$ and some i. This proves the claim.

The STS we have constructed is called a *Netto system* of order p, denoted by $N(p)$.

The construction can be generalised, using finite fields. In Section 4.7, we briefly discussed the theorem of Galois, guaranteeing a unique field $\text{GF}(q)$ of any prime power order q. It is also true that the multiplicative group of $\text{GF}(q)$ is cyclic. So the construction of a Netto system $N(q)$ of prime power order $q \equiv 1 \pmod 6$ works exactly as for prime order p, with $\text{GF}(q)$ replacing $\mathbb{Z}/(p)$ in the construction. See Exercise 2 for an example of this.

8.6. Project: **Tournaments and Kirkman's schoolgirls**

In this section, we construct Kirkman's own solution to his Schoolgirls Problem.

We begin with a detour. The schoolgirls enjoy playing hockey, and the school has a team in a league, playing matches against other school teams at weekends during term. In the course of a season, every team plays against every other team once. If there are n teams in the competition, what is the least number of rounds required to play all the matches?

The number of matches to be played is $\binom{n}{2} = n(n-1)/2$. If n is even, then $n/2$ matches can be played in every round, so we need (at least) $n-1$ rounds. If n is odd, then only $(n-1)/2$ matches can be played in a round, with one team having a bye; so n rounds are required. A *tournament schedule* for n teams is an arrangement of all pairs of teams into the minimum numbers of rounds just calculated (viz. $n-1$ if n is even, n if n is odd).

Of course, we cannot guarantee that tournament schedules exist on the basis of this argument; but there is a simple construction, as follows. First, consider the case where n is odd. Draw a regular n-gon in the plane, and number its vertices $0, \ldots, n-1$ corresponding to the teams (these numbers are regarded as belongong to the integers mod n). For each edge of the n-gon, there are $(n-3)/2$ diagonals parallel to this edge; this parallel class determines the matches in a round, with the team corresponding to the vertex opposite the chosen edge having a bye. Fig. 8.7 shows the case $n = 5$.

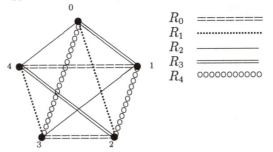

Fig. 8.7. Tournament schedule: five teams

This construction can be presented algebraically: the edge and diagonals in the parallel class not containing the vertex i have the form $\{j, k\}$, where $j + k = 2i$ (in $\mathbb{Z}/(n)$).

For n even, we temporarily remove one team from the competition, and construct a tournament schedule with $n-1$ rounds for the remaining teams as above. Then we decree that, in each round, the extra team will play the team which would otherwise have had a bye in that round.

Now we present Kirkman's marching orders for his schoolgirls. First we construct a STS(15). Divide the 15 schoolgirls into a group X of 7 girls and a group Y of 8. We take $X = \{x_0, \ldots, x_6\}$ to be the point set of a Steiner triple system STS(7). Also, we take Y to 'be' the teams in a tournament with 7 rounds $R_0, \ldots, R_6$. Each R_i consists of four disjoint pairs of girls; we add girl x_i to each of these pairs to form a triple. In this way we get 28 triples which, together with the 7 triples of the STS(7), form 35 triples, the right number for a STS(15).

We check that it really is a STS(15). Any two girls in X belong to a unique triple of the subsystem. Any two girls in Y form a pair belonging to one round R_i of the tournament, and so lie in a triple with x_i. Finally, take a girl in X (say x_i) and a girl $y \in Y$: the unique triple containing them is $\{x_i, y, y'\}$, where $\{y, y'\}$ belongs to round R_i.

Finally, we have to divide the triples into seven sets of five, corresponding to the walking groups on the seven days of the week. For this, we exploit the cyclic structure of both the STS(7) and the tournament schedule. We can take the triples of the STS(7) to be $\{B_0, \ldots, B_6\}$, where $B_i = \{x_{i+1}, x_{i+2}, x_{i+4}\}$. Label the girls in Y as $\{y_0, \ldots, y_6, z\}$, where the ith round R_i of the tournament consists of $\{y_i, z\}$ and all $\{y_j, y_k\}$ with $j + k = 2i$. Then $\{x_0, y_0, z\}$, $\{y_4, y_6, x_5\}$, $\{y_2, y_3, x_6\}$, and $\{y_1, y_5, x_3\}$ are triples (since, for example, $4 + 6 = 2 \times 5$). Together with $\{x_1, x_2, x_4\}$, these make up the groups for day 0: every girl is in one group. Now the groups for day i are obtained by adding i to the subscripts of the x's and y's.

In general, a Steiner triple system, whose triples can be partitioned into classes with the property that each point lies in a unique triple of every class, is called a *Kirkman system*.

8.7. Exercises

1. Kirkman's original (incomplete, but basically correct) proof of the existence of Steiner triple systems went as follows. Kirkman defined two kinds of structure: S_n, what we have called a Steiner triple system of order n; and S'_n, whose exact details don't concern us here. He claimed to show:

(a) S_1 exists;
(b) if S_n exists, then S_{2n+1} exists;
(c) if S_n exists and $n > 1$, then S'_{n-2} exists;
(d) if S'_n exists, then S_{2n-1} exists.

Prove that, from (a)–(d), it follows that S_n exists for all positive integers $n \equiv 1$ or 3 (mod 6). For which values of n does S'_n exist?

2. Construct a Netto system of order 25.

[HINT: As in Section 4.7, we have to find an irreducible quadratic over $\mathbb{Z}/(5)$, use it to construct $GF(25)$, and then find a primitive sixth root of unity in this field. But all this can be simplified. We know that z must satisfy $z^2 - z + 1 = 0$, and this quadratic is irreducible over $\mathbb{Z}/(5)$; so let

$$GF(25) = \{a + bz : a, b \in \mathbb{Z}/(5)\},$$

where $z^2 = z - 1$. All that remains is to find the coset representatives of the subgroup generated by z.]

3. Prove that, given any STS(7), its points can be numbered $1, \ldots, 7$ so that its triples are those listed in Fig. 8.1(a). Prove a similar statement for STS(9).

[HINT: show that any two triples of a STS(7) must meet; while, in a STS(9), there are just two triples disjoint from a given triple, and these are disjoint from one another.]

Formally, an *isomorphism* between Steiner triple systems $(X_1, \mathcal{B}_1)$ and $(X_2, \mathcal{B}_2)$ is a bijective map $f : X_1 \to X_2$ which carries the triples in $\mathcal{B}_1$ to those in $\mathcal{B}_2$. You are asked to prove that Steiner triple systems of orders 7 and 9 are unique up to isomorphism.

HARDER PROBLEM. Prove that there are just two non-isomorphic Steiner triple systems of order 13.

REMARK. After this, things get more difficult. There are exactly 80 non-isomorphic STS of order 15, and millions of non-isomorphic STS(19) (the exact number has never been determined).

4. An *automorphism* of a Steiner triple system is an isomorphism from the system to itself. Prove that a Steiner triple system of order 7 or 9 has 168 or 432 automorphisms respectively.

5. (a) Prove that, in an affine triple system, each triangle lies in a subsystem of order 9.

(b) Prove that an affine triple system is a Kirkman system.

6. Verify the following values of the packing and covering functions for small n.

n	3	4	5	6	7	8	9
$p(n)$	1	1	2	4	7	8	12
$c(n)$	1	3	4	6	7	11	12

EXERCISES ON STEINER QUADRUPLE SYSTEMS.

A *Steiner quadruple system* (SQS) is a pair $(X, \mathcal{B})$, where X is a set, and $\mathcal{B}$ a collection of 4-element subsets of X called *quadruples*, with the property that any three points of X are contained in a unique quadruple. The number $n = |X|$ is called the *order* of the quadruple system.

7. If a SQS of order n exists, with $n \geq 2$, then $n \equiv 2$ or $4 \pmod 6$.
   [This condition is also sufficient, but the proof is more difficult.]

8. If $(X, \mathcal{B})$ is a SQS of order n, then $|\mathcal{B}| = n(n-1)(n-2)/24$.

9. Let X be a vector space over $\mathbb{Z}/(2)$, and let $\mathcal{B}$ be the set of 4-subsets $\{x, y, z, w\}$ of X for which $x + y + z + w = 0$. Show that $(X, \mathcal{B})$ is a SQS.

10. Let $(X, \mathcal{B})$ be a SQS of order $n \geq 2$. Take a disjoint copy $(X', \mathcal{B}')$ of this system. Take a tournament schedule on X with rounds $R_1, \ldots, R_{n-1}$, and one on X' with rounds $R'_1, \ldots, R'_{n-1}$. (This is possible since n is even — see Section 8.6.) Now let $Y = X \cup X'$, and $\mathcal{C} = \mathcal{B} \cup \mathcal{B}' \cup \mathcal{R}$, where $\mathcal{R}$ is the set of 4-sets $\{x, y, z', w'\}$ such that
   • $x, y \in X$, $z', w' \in X'$;
   • for some i $(1 \leq i \leq n-1)$, $\{x, y\} \in R_i$ and $\{z', w'\} \in R'_i$.
   Show that $(Y, \mathcal{C})$ is a SQS of order $2n$.

EXERCISES ON SUBSYSTEMS.

11. Let $(X, \mathcal{B})$ be a STS of order n, and Y a subsystem of order m, where $m < n$. Prove that $n \geq 2m + 1$. Show further that $n = 2m + 1$ if and only if every triple in $\mathcal{B}$ contains either 1 or 3 points of Y.

12. Let $(X, \mathcal{B})$ be a STS of order $n = 2m + 1$, and Y a subsystem of order m; say $Y = \{y_1, \ldots, y_m\}$. For $i = 1, \ldots, m$, let R_i be the set of all pairs $\{z, z'\} \subseteq X \setminus Y$ for which $\{y_i, z, z'\} \in \mathcal{B}$. Show that $\{R_1, \ldots, R_m\}$ is a tournament schedule on $X \setminus Y$.
   Show further that this construction can be reversed: a STS(m) and a tournament schedule of order $m + 1$ can be used to build a STS($2m + 1$).

13. Let $(X, \mathcal{B})$ and $(Y, \mathcal{C})$ be STS, of orders m and n respectively. Let $Z = X \times Y$, and let $\mathcal{D}$ consist of all triples of the following types:
   • $\{(x, y_1), (x, y_2), (x, y_3)\}$ for $x \in X$, $\{y_1, y_2, y_3\} \in \mathcal{C}$;
   • $\{(x_1, y), (x_2, y), (x_3, y)\}$ for $\{x_1, x_2, x_3\} \in \mathcal{B}$, $y \in Y$;
   • $\{(x_1, y_1), (x_2, y_2), (x_3, y_3)\}$ for $\{x_1, x_2, x_3\} \in \mathcal{B}$, $\{y_1, y_2, y_3\} \in \mathcal{C}$.
   (Note that a triple in $\mathcal{B}$ and one in $\mathcal{C}$ give rise to six triples of the third type, corresponding to the six possible bijections from one to the other.)
   Show that $(Z, \mathcal{D})$ is a STS of order mn. Show further that, if $m > 1$ and $n > 1$, then $(Z, \mathcal{D})$ contains a subsystem of order 9.

14. What can you say about the set

$$\{n : \text{ there exists a STS}(n) \text{ with a subsystem of order 9}\}?$$

15. COMPUTING PROJECT. Recall the 'nine schoolgirls problem' posed in Chapter 1: nine schoolgirls are to walk, each day in sets of three, for four days, so that each pair of girls walks together once. We've seen that this problem has a unique solution: there is a unique STS(9) up to isomorphism (Exercise 4), and there is a unique way of partitioning its twelve triples into four sets of three with the required property. Now we add a further twist to the problem:

> Arrange walks for the girls for twenty-eight days (divided into seven groups of four) so that
> • in each group of four days, any two girls walk together once;
> • in the entire month, any three girls walk together once.

In other words, we are asked to partition the $\binom{9}{3} = 84$ triples of girls into seven 12-sets, each of which forms a Steiner triple system.

There are 840 different Steiner triple systems on a given 9-set,[3] and so potentially $\binom{840}{7}$ possibilities to check — rather a large number! We make one simplifying assumption. (This means that, if we fail to find a solution, we have not demonstrated that no solution exists.) We assume that

> the required seven STS(9)s can be obtained by applying all powers of a permutation θ of order 7 to a given one.

We can assume that the starting system is the one of Fig. 8.1(b), with point set $X = \{1, \ldots, 9\}$, and triple set

$$B = \{123, 456, 789, 147, 258, 369, 159, 267, 348, 357, 168, 249\}.$$

We can also assume, without loss, that θ fixes the points 1 and 2, and acts as a 7-cycle on the others. (That no generality is lost here depends on the symmetry of the STS(9): all pairs of points are 'alike'.) Finally, there is a unique power of θ which maps 3 to 4; so we may assume that θ itself does so. Thus, in cycle notation,

$$\theta = (1)(2)(3\ 4\ a\ b\ c\ d\ e\ ),$$

where $a, \ldots, e$ are $5, \ldots, 9$ in some order; in other words, $\left(\begin{smallmatrix} 5 & 6 & 7 & 8 & 9 \\ a & b & c & d & e \end{smallmatrix}\right)$ is a permutation in two-line notation.

Thus our algorithm is as follows:
• set up the system (X, B);
• generate in turn all permutations $\left(\begin{smallmatrix} 5 & 6 & 7 & 8 & 9 \\ a & b & c & d & e \end{smallmatrix}\right)$; for each, let $\theta = (1)(2)(3\ 4\ a\ b\ c\ d\ e)$, and check whether $B, B\theta, \ldots, B\theta^6$ are pairwise disjoint. Report success if so.

Program this calculation. (You should find two permutations which give rise to a solution.)

16. Here is a related problem. Cayley showed that it is impossible to partition the $\binom{7}{3} = 35$ triples from a 7-set into five disjoint Steiner triple systems. In fact, no more than two disjoint STS(7)s can be found. Verify this observation.

---

[3] For a proof of this fact, see Chapter 14.

9. Finite geometry

In Plane Geometry that afternoon, I got into an argument with Mr Shull, the teacher, about parallel lines. I say they have to meet. I'm beginning to think everything comes together somewhere.

William Wharton, *Birdy* (1979)

TOPICS: Finite fields; Gaussian coefficients; projective and affine geometries; projective planes

TECHNIQUES: Linear algebra

ALGORITHMS: Row-reduction

CROSS-REFERENCES: Binomial coefficients (Chapter 3); Orthogonal Latin squares (Chapter 6); de Bruijn–Erdős Theorem (Chapter 7); Steiner triple systems (Chapter 8)

Projective geometry over finite fields is a topic of great importance, for many reasons. It provides a large collection of highly symmetric structures, with interesting groups of collineations; it is a so-called 'q-analog' of the family of subsets of a set, providing an interesting perspective; and it ties in with almost everything else we have met so far.

9.1. Linear algebra over finite fields

We already met in Section 4.7 the basic fact about the existence of finite fields:

(9.1.1) Finite fields
There exists a field with q elements if and only if q is a prime power. If so, then the field is unique up to isomorphism. It is called the Galois field of order q, and denoted by $\mathrm{GF}(q)$.

This fact is proved in any good algebra textbook. I have included an outline of the proof at the end of this chapter (Section 9.9). If you haven't met it before, and have trouble with the algebra involved, you may take it on trust, and keep in mind the case when the order is prime. (The Galois field of prime order p is the field $\mathbb{Z}/(p)$ of integers modulo p.)

In traditional linear algebra, it is usually assumed that the field over which we work is the field of real numbers (or possibly some variant, such as the rational or complex numbers). However, almost everything works the same over finite fields. The definition of linearly independent set, spanning set, basis, subspace; the formula

$$\dim(U \cap W) + \dim(U + W) = \dim(U) + \dim(W),$$

the representation of linear maps by matrices, and the rank and nullity formula, all work as usual.

Row operations and reduced echelon form also work in the same way; but, since we will need these, I will sketch them. The three types of *row operation* on a matrix are:

- multiply a row by a non-zero scalar;
- add a multiple of one row to another;
- interchange two rows.

These operations do not change the linear dependence or independence of the set of rows of the matrix, and also do not change the *row space* (the subspace spanned by the rows).

A matrix $A = (a_{ij})$ is said to be in *reduced echelon form* if the following three conditions hold:

- given any row of A, either it is zero, or the first non-zero entry is a 1 (a so-called 'leading 1');
- for any $i > 1$, if the ith row is non-zero, then so is the $(i-1)^{\text{st}}$, and its leading 1 is to the left of the leading 1 in the ith row;
- if a column contains the leading 1 of some row, then all its other entries are 0.

Now the following result holds:

(9.1.2) Proposition. *Any matrix can be put into reduced echelon form by applying a series of elementary row operations; and the reduced echelon form is unique.*

If a matrix is in reduced echelon form, then its rows are linearly independent if and only if the last row is non-zero — this is a familiar test for linear independence of a set of vectors.

Note that, for linear algebra, the weaker notion of 'echelon form' (where the third condition in the definition is deleted) suffices; but, for us, a crucial fact about reduced echelon form is its uniqueness, and this is not true for the weaker form.

9.2. Gaussian coefficients

We are now going to do some counting in vector spaces over finite fields. Let $V(n, q)$ denote an n-dimensional vector space over $\text{GF}(q)$. First, the number of vectors:

(9.2.1) Proposition. *The number of vectors in $V(n, q)$ is equal to q^n.*

PROOF. As usual, by choosing a basis, we represent the vectors by all n-tuples of elements of $\text{GF}(q)$; and there are q^n of these.

The *Gaussian coefficient* $\begin{bmatrix} n \\ k \end{bmatrix}_q$ is defined to be the number of k-dimensional subspaces of $V(n, q)$.

(9.2.2) Gaussian coefficients

$$\begin{bmatrix} n \\ k \end{bmatrix}_q = \frac{(q^n - 1)(q^{n-1} - 1)\dots(q^{n-k+1} - 1)}{(q^k - 1)(q^{k-1} - 1)\dots(q - 1)}.$$

PROOF. First we show:

The number of linearly independent k-tuples in $V(n, q)$ is equal to

$$(q^n - 1)(q^n - q)\dots(q^n - q^{k-1}).$$

This is proved by examining the number of choices of each vector. A k-tuple of vectors is linearly independent if and only if no vector lies in the subspace spanned by the preceding vectors. Thus, the first vector can be anything except zero ($q^n - 1$ choices); the second must lie outside the 1-dimensional subspace spanned by the first ($q^n - q$ choices); and, in general, the ith must lie outside the $(i - 1)$-dimensional subspace spanned by its predecessors ($q^n - q^{i-1}$ choices). Multiplying these numbers gives the result.

Now a k-dimensional subspace is spanned by k linearly independent vectors, and we have counted these. But a given subspace U will have many different bases. How many? Just the number of linearly independent k-tuples in a k-dimensional subspace, which is found from the same formula by putting k in place of n. We must divide by this number to obtain the number of subspaces. Cancelling powers of q gives the quoted formula.

Now the number of k-dimensional subspaces of $V(n, q)$ is equal to the number of $k \times n$ matrices over $\mathrm{GF}(q)$ which are in reduced echelon form and have no zero rows. This gives another way to calculate $\begin{bmatrix} n \\ k \end{bmatrix}_q$.

EXAMPLE. Let $n = 4$ and $k = 2$. Our formula gives

$$\begin{bmatrix} 4 \\ 2 \end{bmatrix}_q = \frac{(q^4 - 1)(q^3 - 1)}{(q^2 - 1)(q - 1)}$$
$$= (q^2 + 1)(q^2 + q + 1) = q^4 + q^3 + 2q^2 + q + 1.$$

We check by counting matrices. The possible shapes are

$$\begin{pmatrix} 1 & 0 & * & * \\ 0 & 1 & * & * \end{pmatrix}, \quad \begin{pmatrix} 1 & * & 0 & * \\ 0 & 0 & 1 & * \end{pmatrix}, \quad \begin{pmatrix} 1 & * & * & 0 \\ 0 & 0 & 0 & 1 \end{pmatrix},$$

$$\begin{pmatrix} 0 & 1 & 0 & * \\ 0 & 0 & 1 & * \end{pmatrix}, \quad \begin{pmatrix} 0 & 1 & * & 0 \\ 0 & 0 & 0 & 1 \end{pmatrix}, \quad \begin{pmatrix} 0 & 0 & 1 & 0 \\ 0 & 0 & 0 & 1 \end{pmatrix},$$

where $*$ denotes an arbitrary element. So there are $q^4 + q^3 + q^2 + q^2 + q + 1$ matrices.

REMARK 1. If we regard the Gaussian coefficient as a function of the real variable q (where n and k are fixed integers), then we find that

$$\lim_{q \to 1} \begin{bmatrix} n \\ k \end{bmatrix}_q = \binom{n}{k}.$$

For, by l'Hôpital's rule, we have

$$\lim_{q \to 1} \frac{q^a - 1}{q^b - 1} = \lim_{q \to 1} \frac{aq^{a-1}}{bq^{b-1}} = \frac{a}{b}$$

for $a, b \neq 0$; so

$$\lim_{q \to 1} \begin{bmatrix} n \\ k \end{bmatrix}_q = \frac{n(n-1)\dots(n-k+1)}{k(k-1)\dots 1} = \binom{n}{k}.$$

For this reason, the Gaussian coefficients are sometimes called the 'q-analogs' of the binomial coefficients.

REMARK 2. The Gaussian coefficients can be given a combinatorial interpretation for all positive integer values of q greater than 1, not just prime powers. For let Q be any set of size q, containing two distinguished elements called 0 and 1. Then the definition of a matrix in reduced echelon form over Q makes sense, even though the algebraic interpretation is lost. The number of $k \times n$ matrices in reduced echelon form with no zero rows is given by a polynomial in q. But, for infinitely many values (all the prime powers), this polynomial coincides with the Gaussian coefficient (which is also a polynomial); so they are identically equal.

The matrix interpretation enables us to give a recurrence relation for the Gaussian coefficients:

(9.2.3) Theorem.

$$\begin{bmatrix} n+1 \\ k \end{bmatrix}_q = \begin{bmatrix} n \\ k-1 \end{bmatrix}_q + q^k \begin{bmatrix} n \\ k \end{bmatrix}_q.$$

PROOF. Consider $k \times (n+1)$ matrices in reduced echelon form, with no zero rows. Divide them into two classes: those for which the leading 1 in the last row occurs in the last column; and the others. Those of the first type correspond to $(k-1) \times n$ matrices in reduced echelon with no zero rows, since the last row and column are zero apart from the bottom-right entry. Those of the second type consist of a $k \times n$ matrix in reduced echelon with no zero rows, with a column containing arbitrary elements adjoined on the right. Since there are q^k choices for this column, the recurrence relation follows.

Note that this relation reduces to the binomial recurrence when $q = 1$. However, unlike the binomial recurrence, it is not 'symmetric'. (For a symmetric form, see Exercise 3.) In fact, we have:

(9.2.4) Proposition.

$$\begin{bmatrix} n \\ k \end{bmatrix}_q = \begin{bmatrix} n \\ n-k \end{bmatrix}_q.$$

PROOF. This follows from the bijection between k-dimensional subspaces of $V = V(n, q)$ and $(n - k)$ dimensional subspaces of its dual space V^* (where a subspace of V corresponds to its annihilator in V^*).

Thus, we obtain another recurrence:

$$\begin{bmatrix} n+1 \\ k \end{bmatrix}_q = \begin{bmatrix} n \\ k \end{bmatrix}_q + q^{n+1-k} \begin{bmatrix} n \\ k-1 \end{bmatrix}_q$$

(see Exercise 4).

From the formula for the Gaussian coefficients, we can deduce another result analogous to a binomial coefficient identity:

$$(q^k - 1) \begin{bmatrix} n \\ k \end{bmatrix}_q = (q^n - 1) \begin{bmatrix} n-1 \\ k-1 \end{bmatrix}_q.$$

In fact, quite a lot of the combinatorics of binomial coefficients can be extended to their q-analogs; but we have enough for our needs now.

We can use the recurrence relation above to prove a pretty analogue of the Binomial Theorem (3.3.1):

(9.2.5) q-binomial Theorem

For $n \geq 1$,

$$\prod_{i=0}^{n-1} (1 + q^i t) = \sum_{k=0}^{n} q^{k(k-1)/2} \begin{bmatrix} n \\ k \end{bmatrix}_q t^k.$$

PROOF. The proof is a straightforward induction. For $n = 1$, both sides are $1 + t$. Suppose that the result is true for n. Then

$$\prod_{i=0}^{n} (1 + q^i t) = \left(\sum_{k=0}^{n} q^{k(k-1)/2} \begin{bmatrix} n \\ k \end{bmatrix}_q t^k \right) \cdot (1 + q^n t).$$

The coefficient of t^k on the right is

$$q^{k(k-1)/2} \begin{bmatrix} n \\ k \end{bmatrix}_q + q^{(k-1)(k-2)/2} \begin{bmatrix} n \\ k-1 \end{bmatrix}_q q^n$$

$$= q^{k(k-1)/2} \left(\begin{bmatrix} n \\ k \end{bmatrix}_q + q^{n-k+1} \begin{bmatrix} n \\ k-1 \end{bmatrix}_q \right)$$

$$= q^{k(k-1)/2} \begin{bmatrix} n+1 \\ k \end{bmatrix}_q,$$

as required.

Letting $q \to 1$, we obtain the usual Binomial Theorem.

It's now easy to count the non-singular matrices.

(9.2.6) Theorem. *The number of non-singular $n \times n$ matrices over* $\mathrm{GF}(q)$ *is*

$$(q^n - 1)(q^n - q) \ldots (q^n - q^{n-1}).$$

PROOF. A square matrix is non-singular if and only if its rows are linearly independent. We counted linearly independent k-tuples above.

Note that the non-singular $n \times n$ matrices form a group, the so-called *general
linear* group $\mathrm{GL}(n, q)$. The theorem above computes the order of this group.

9.3. Projective geometry

The definition of projective geometry seems strange at first meeting. We'll make a
short detour to see where it came from.

One of the goals of painting is to create a 2-dimensional picture whose effect on
a viewer approximates that of the 3-dimensional scene it depicts. In the European
renaissance, painters began to approach this problem mathematically. Let us idealise
the situation, and assume that the painter's eye is a point, and take this point to be
the origin of a coordinate system for space. He sees an object by means of a ray of
light from the object to his eye. Another object seen by a ray in the same direction
will appear in the same place. (In practice, of course, the nearer object will hide
the further one). Thus, the points of the painter's perceptual space can be identified
with semi-infinite rays through the origin.

Fig. 9.1. Perspective

The painter wants to represent his perceptual space in a plane. He sets up a
'picture plane' Π, not passing through his eye. A typical ray will meet Π in a single
point, which can be taken to represent that ray (and hence, to represent objects
for which that ray is the line of sight). Assuming that Π is a mathematical plane,
extending infinitely in all directions, and let Π' be the plane through the painter's
eye parallel to Π; then the rays represented are all those on one side of Π'.

Mathematically, it is simpler to replace rays by lines through the origin, extending in both directions. (The painter doesn't have eyes in the back of his head, and so he will not actually picture objects behind him.) With this convention, every line through the origin is represented by a unique point in the picture plane Π, except for the lines in Π' (that is, the lines parallel to Π). This led to the convention of adjoining mathematical 'ideal points' to Π to represent these lines, forming the *real projective plane.*

Thus, the real projective plane can be regarded in either of two ways: the picture plane Π with 'ideal points' added, or the set of all lines through the origin (1-dimensional subspaces) of 3-dimensional space $\mathbb{R}^3$. The second representation has the disadvantage that points of the plane 'are' lines rather than points, but the (more than compensating) advantage that all points are alike.

What about lines? Given a line L of $\mathbb{R}^3$, not containing the origin, the set of lines joining its points to the origin sweep out a plane (minus one line, the 'point at infinity'), which intersects Π in a line. This is the line which the painter draws to represent L. In other words, in the second (3-space) model, a line of the projective plane is a 2-dimensional subspace of $\mathbb{R}^3$. Note that any two lines of the projective plane meet. For example, if L, L' are lines in 3-space which are parallel but not in Π', then their representations in Π meet at the point where the line through the origin parallel to L intersects Π.

This gives us the clue for the general definition. The n-dimensional *projective space* over a field F, denoted $\mathrm{PG}(n, F)$, is defined by means of an $(n+1)$-dimensional vector space $V = V(n+1, F)$. The points of projective space are the 1-dimensional subspaces of V; the lines are the 2-dimensional subspaces; planes are 3-dimensional subspaces; and so on. Note that a line, normally regarded as 1-dimensional, is represented by a 2-dimensional vector space. We saw the motivation for this already; but, in an attempt to reduce confusion, we use the term k-*flat* for the object in projective geometry represented by a $(k+1)$-dimensional vector subspace.

Now some familiar geometric properties hold. For example:

(a) Two points lie in a unique line.

(b) Two intersecting lines lie in a unique plane.

These properties follow from elementary linear algebra. For (a), the two points are 1-dimensional subspaces, and their span is 2-dimensional. For (b), the two lines are 2-dimensional subspaces U_1 and U_2; the fact that they intersect in a point means that $\dim(U_1 \cap U_2) = 1$, and so $\dim(U_1 + U_2) = 3$, whence the two lines span a plane.

Slightly less familiarly, the converse of (b) holds:

(c) Two coplanar lines intersect.

(This follows by reversing the argument, noting that $\dim(U_1 + U_2) = 3$ implies $\dim(U_1 \cap U_2) = 1$.) In other words, there are no parallel lines!

If F is a finite field $\mathrm{GF}(q)$, then we denote the projective space by $\mathrm{PG}(n, q)$. Now we can count objects in $\mathrm{PG}(n, q)$ in terms of Gaussian coefficients. For example:

(9.3.1) Proposition. $\mathrm{PG}(n, q)$ *has* $\begin{bmatrix} n+1 \\ 1 \end{bmatrix}_q = (q^{n+1} - 1)/(q - 1)$ *points. It has* $\begin{bmatrix} n+1 \\ k+1 \end{bmatrix}_q$ k-*flats, each of which contains* $(q^{k+1} - 1)/(q - 1)$ *points.*

In particular, the projective plane $PG(2, q)$ has $q^2 + q + 1$ points and $q^2 + q + 1$ lines; each line contains $q + 1$ points and each point lies in $q + 1$ lines; two points lie in a unique line, and two lines intersect in a unique point. Thus, it is an example of a family of sets satisfying the hypotheses and the final conclusion of the de Bruijn–Erdős Theorem (see Section 7.3).

9.4. Axioms for projective geometry

How do we recognise a projective space? Let us assume that we are given the points and the lines only. (In fact, all the flats can be recovered from these data: a set of points is a flat if and only if it contains the unique line through any two of its points. See Exercise 6.) Now, as just remarked, two points lie on a unique line. But this alone is not enough to force the structure to be a projective space. For example, any Steiner triple system (Chapter 8) has this property, if we take the lines to be the triples; and certainly not every Steiner triple system is a projective space $PG(n, q)$. (Three points per line forces $q = 2$, so that the total number of points would be $2^{n+1} - 1$. But there are Steiner triple systems where the number of points is not of this form.)

In Section 8.5, we defined a class of Steiner systems which were referred to as projective. If you read that section, you will be reassured to know that those systems are precisely the projective spaces $PG(n, 2)$. As defined there, the points are the non-zero vectors of $V(n + 1, 2)$, and the lines are the triples of vectors with sum zero. But, over $GF(2)$, a 1-dimensional space contains the zero vector and a unique non-zero vector, so there is a one-to-one correspondence between the non-zero vectors and the subspaces they span. Moreover, a 2-dimensional subspace contains the zero vector and three non-zero vectors; it is not hard to see that the sum of these three vectors is zero, and conversely that any three vectors with sum zero, together with the zero vector, form a 2-dimensional subspace.

The characterisation was given by Veblen and Young, and can be stated as follows. A *triangle* consists of three non-concurrent lines.

(9.4.1) Veblen–Young Theorem. *Let $\mathcal{L}$ be a family of subsets (called lines) of the set X. Suppose that the following conditions hold:*
(a) *every line contains at least three points;*
(b) *two points of X lie in a unique line;*
(c) *there exist two disjoint lines;*
(d) *if a line meets two sides of a triangle, not at their intersection, then it meets the third side also.*
Then X and $\mathcal{L}$ can be identified with the points and lines of the projective space $PG(n, q)$ for some $n \geq 3$ and some prime power q.

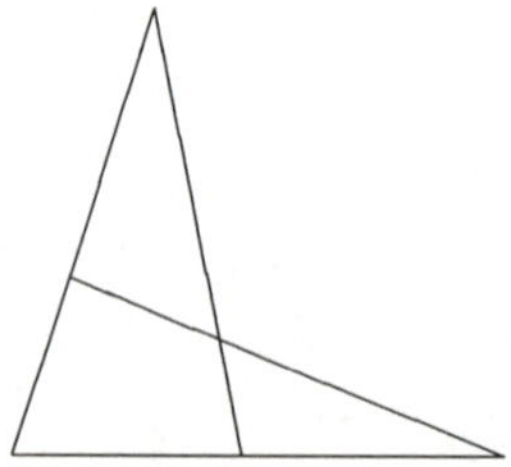

Fig. 9.2. Veblen–Young axiom

This theorem will not be proved here. (But see below for the case $q = 2$.) Nevertheless, we make some remarks. The purpose and necessity of Axioms (a) and (b) is I hope obvious. Axiom (d), known as the *Veblen–Young Axiom*, is the crucial condition. The puropse of axiom (c) is to exclude degenerate cases (where X is the empty set or a singleton, or where there is a unique line), and also to exclude projective planes. (In a projective plane, any two lines intersect, so the Veblen–Young Axiom certainly holds. But there are projective planes which are not of the form $\mathrm{PG}(2, q)$. We consider projective planes further in the next section.)

We conclude this section with a proof of the Veblen–Young Theorem in the case where each line has three points. Axioms (a) and (b) assert that we have a Steiner triple system. Axiom (c) is not really needed, since the Steiner triple systems with 0, 1, 3 and 7 points are projective spaces; so we assume only Axioms (a), (b), (d). The proof resembles the arguments in Section 8.5. We have to identify the points with the non-zero vectors of a vector space over $\mathrm{GF}(2)$; so let $V = X \cup \{0\}$, where 0 is a new symbol. Define addition by the rules

- $v + 0 = 0 + v = v, \quad v + v = 0 \quad$ for all $v \in V$;
- $u + v = w \quad$ if $\{u, v, w\} \in \mathcal{L}$.

Now show that $(V, +)$ is an abelian group. (The Veblen–Young Axiom is needed to establish the associative law.) Then define scalar multiplication by

- $0 \cdot v = 0, 1 \cdot v = v \quad$ for all $v \in V$;

and show that V is a vector space over $\mathrm{GF}(2)$. The points are the non-zero vectors; check that the lines are the triples with sum zero.

9.5. Projective planes

We first met projective planes in the section on extremal set theory; we noticed there that the Steiner triple system of order 7 is an example. We repeat the definition.

A *projective plane* of order q consists of a set X of $q^2 + q + 1$ elements called *points*, and a set $\mathcal{B}$ of $(q + 1)$-element subsets of $\mathcal{B}$ called *lines*, having the property that any two points lie on a unique line.

(This is slightly different from the definition we gave before, where this property was derived from the property that any two lines meet in a unique point; but the two definitions are equivalent.)

The only possible projective plane of order 1 is a triangle. From now on, we assume that the order is greater than 1. The geometry $\mathrm{PG}(2, q)$ is a projective plane for any prime power q.

We list some basic properties of projective planes.

(9.5.1) Proposition. *In a projective plane of order q, the following hold:*

- *any point lies on $q + 1$ lines;*
- *two lines meet in a unique point;*
- *there are $q^2 + q + 1$ lines.*

PROOF. Take a point p. There are $q(q + 1)$ points different from p; each line through p contains q further points, and there are no overlaps between these lines (apart from p). So there must be $q + 1$ lines through p. Now let L_1 and L_2 be lines, and p a point of L_1. Then the $q + 1$ points of L_2 are all joined to p by different lines; since there are only $q + 1$ lines through p, they all meet L_2 in a point; in particular, L_1 meets L_2. Finally, counting pairs (p, L) with $p \in L$, we obtain

$$|\mathcal{B}| \cdot (q+1) = (q^2 + q + 1) \cdot (q+1),$$

so $|\mathcal{B}| = q^2 + q + 1$.

This shows that there is a 'duality principle' for projective planes. Let $(X, \mathcal{B})$ be a projective plane. Let $X' = \mathcal{B}$ and $\mathcal{B}' = \{\beta_x : x \in X\}$, where

$$\beta_x = \{L \in \mathcal{B} : x \in L\};$$

then $(X', \mathcal{B}')$ is also a projective plane of order q. Its points and lines correspond to the lines and points of the original plane.

For which numbers q do projective planes of order q exist? We have seen that they exist for all prime powers. The main non-existence theorem is the celebrated *Bruck–Ryser Theorem*:

(9.5.2) Bruck–Ryser Theorem

If a projective plane of order n exists, where $n \equiv 1$ or 2 (mod 4), then n is the sum of two squares of integers.

The proof is given in Section 9.8. The theorem shows, for example, that there is no projective plane of order 6, a fact connected with Euler's officers, as we will see. However, since $10 = 1^2 + 3^2$, the question of whether or not a projective plane of order 10 exists is not resolved by our results so far. This question was finally settled in the negative by Lam, Swiercz and Thiel in 1989, after several large computations taking a number of years. The existence question for a plane of order 12 is unresolved at present.

How do we recognise the special planes $\mathrm{PG}(2, q)$? It turns out that they are precisely the (finite) projective planes in which the classical theorems of Desargues[1] and Pappus[2] are valid.

(9.5.3) Desargues' Theorem for Π

Let $a_1 b_1 c_1$ and $a_2 b_2 c_2$ be triangles in the projective plane Π such that the lines $a_1 a_2$, b_1, b_2 and $c_1 c_2$ are concurrent. Let $p = b_1 c_1 \cap b_2 c_2$, $q = c_1 a_1 \cap c_2 a_2$, and $r = a_1 b_1 \cap a_2 b_2$. Then p, q, r are collinear.

---

[1] Desargues was a contemporary of Descartes; their advocacy of geometric and algebraic methods respectively created a rivalry between them.

[2] Pappus was one of the last of the classical Greek geometers. His work, the *Collection*, was important in the preservation of their heritage.

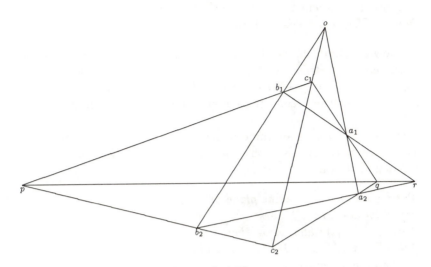

Fig. 9.3. Desargues' Theorem

(9.5.4) Pappus' Theorem for Π

Let a, b, c, d, e, f be points of the projective plane Π, such that a, c, e are collinear and b, d, f are collinear. Let $p = ab \cap de$, $q = bc \cap ef$, $r = cd \cap fa$. Then p, q, r are collinear.

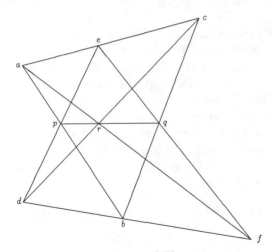

Fig. 9.4. Pappus' Theorem

(9.5.5) Theorem. *The following conditions are equivalent for a finite projective plane* Π:

- Π *is isomorphic to* $\mathrm{PG}(2,q)$ *for some prime power* q;
- *Desargues' Theorem holds in* Π;
- *Pappus' Theorem holds in* Π.

We now develop a connection with the theory of Latin squares. First, we define a related geometric structure. An *affine plane* of order q consists of a set X of q^2 points, and a set $\mathcal{B}$ of q-element subsets of X called *lines*, such that two points lie on a unique line. The Steiner triple system on 9 points is an example of an affine plane.

Two distinct lines of an affine plane clearly have at most one common point. Unlike a projective plane, lines may be disjoint. We call two lines *parallel* if they are either equal or disjoint.

(9.5.6) Proposition. *In an affine plane of order* q,
(i) any point lies on $q+1$ *lines;*
(ii) there are $q(q+1)$ *lines altogether;*
(iii) (Euclid's parallel postulate) if p *is a point and* L *a line, there is a unique line* L' *through* p *parallel to* L;
(iv) parallelism is an equivalence relation; each parallel class contains q *lines which partition the point set.*

PROOF. We begin as before. If p is a point, the $q^2 - 1$ points different from p have the property that each lies on a unique line through p, and each line through p contains $q-1$ further points; so there are $(q^2 - 1)/(q - 1) = q + 1$ lines through p. Now double counting shows that there are $q^2 \cdot (q + 1)/q = q(q + 1)$ lines altogether.

Let p be a point and L a line. If $p \in L$, then clearly L is the unique line through p parallel to itself, since any two such lines intersect in p. Suppose that $p \notin L$. Then p lies on $q + 1$ lines, of which q join it to the points of L; so exactly one is disjoint from L.

The relation of parallelism is, by its definition, reflexive and symmetric, and we have to show that it is transitive. In other words, two lines L, L' parallel to the same line L' are parallel to one another. This is clear if two of the three lines are equal, so suppose not. If L and L' have a point p in common, then they both pass through p and are disjoint from L', which is impossible. So L and L' are disjoint.

Clearly each parallel class contains exactly one line through any point. Thus, the $q + 1$ lines through a point p contain representatives of all the parallel classes. To see the same thing another way, observe that each parallel class contains $q^2/q = q$ lines, since these lines are pairwise disjoint and cover the point set; so there are $q(q + 1)/q = q + 1$ parallel classes.

(9.5.7) Theorem. *A projective plane of order* q *exists if and only if an affine plane of order* q *exists.*

PROOF. We have to construct each type of plane from the other. Suppose that $(X, \mathcal{B})$ is a projective plane. Let L be a line, and set $X_0 = X \setminus L$ and

$$\mathcal{B}_0 = \{L' \setminus L : L' \in \mathcal{B}, L' \neq L\}.$$

(In other words, we remove a line and all of its points.) There are $(q^2+q+1)-(q+1) = q^2$ points in X_0; each line has $(q+1)-1 = q$ points, since any line meets L in a unique point; and two points lie in a unique line. So $(X_0, \mathcal{B}_0)$ is an affine plane.

Conversely, suppose that $(X_0, \mathcal{B}_0)$ is an affine plane. Let Y be the set of parallel classes of lines in this plane. We take the point set X to be $X_0 \cup Y$; then $|X| = q^2 + q + 1$. There are two types of new lines. For each line $L \in \mathcal{B}_0$, set $L^* = L \cup \{C\}$, where C is the parallel class containing L; also take Y as a new line. Thus the new structure is $(X, \mathcal{B})$, where

$$\mathcal{B} = \{L^* : L \in \mathcal{B}_0\} \cup \{Y\}.$$

Any line has $q + 1$ points, since one new point is added to each old line L, and there are $q+1$ parallel classes. We have to show that two points of X lie in a unique line. There are several cases:

- two points x, y in X_0 lie in a unique old line, hence a unique new line of the first kind;
- given a point $x \in X_0$ and a parallel class $C \in Y$, there is a unique line containing x in the parallel class C, hence a unique new line of the first kind containing both;
- two parallel classes lie in a unique new line of the second type, namely Y.

So $(X, \mathcal{B})$ is a projective plane.

The process used above to extend an affine plane to a projective plane is called 'adding a line at infinity'. The line Y is the line at infinity, and its points are the points at infinity, the points where parallel lines of the affine plane meet. This is exactly the procedure which turns the Euclidean 'picture plane' into the real projective plane.

We now make the connection with orthogonal Latin squares, and exhibit affine planes as the solution of a different kind of extremal problem. Recall the definition of a Latin square of order n (from Chapter 6): it is an $n \times n$ matrix with entries $1, 2, \ldots, n$, having the property that each entry occurs exactly once in each row or column. Also, two Latin squares $A = (a_{ij})$ and $B = (b_{ij})$ are orthogonal if, for any pair (k, l) of elements from $\{1, \ldots, n\}$, there are unique values of i and j such that $a_{ij} = k$, $b_{ij} = l$. A set $\{A_1, \ldots, A_r\}$ of Latin squares is called a *set of mutually orthogonal Latin squares* (MOLS) if any two squares in the set are orthogonal. We saw that there cannot be more than $n - 1$ MOLS of order n.

(9.5.8) Theorem. *There exist $n - 1$ MOLS of order n if and only if there is an affine plane of order n.*

PROOF. Given a set $\{A_1, \ldots, A_r\}$ of MOLS, we build a geometry of points and lines resembling a 'partial affine plane'. We take the points to be the cells of an $n \times n$ array:

$$X = \{(i, j) : i, j = 1, \ldots, n\}.$$

There are three types of lines:
(a) *horizontal lines*, of the form $\{(x, j) : x = 1, \ldots, n\}$, where j is fixed ($j = 1, \ldots, n$);
(b) *vertical lines*, of the form $\{(i, y) : y = 1, \ldots, n\}$, where i is fixed ($i = 1, \ldots, n$);
(c) for each square A_m ($m = 1, \ldots, r$), and for each entry k ($k = 1, \ldots, n$), the set $\{(i, j) : (A_m)_{ij} = k\}$.

Clearly there are n^2 points, and any line contains n points.

We claim that two points lie on at most one line. This is clear for horizontal or vertical lines; and the definition of a Latin square guarantees that two points of a type (c) line lie in different rows and columns. Furthermore, lines of type (c) coming from the same square A_m are disjoint. So consider two lines of type (c), defined by square A_{m_1} and entry k_1 and by square A_{m_2} and entry k_2 respectively. Could they have two points (i_1, j_1) and (i_2, j_2) in common? If so, then in both these positions the square A_{m_1} has entry k_1 and A_{m_2} has entry k_2, contradicting orthogonality.

Now any point p lies on $r + 2$ lines: one horizontal, one vertical, and one for each of the squares. These lines contain $(r + 2)(n - 1)$ points other than p. So $1 + (r + 2)(n - 1) \le n^2$, whence $r \le n - 1$, giving another proof (more-or-less the same as the earlier one) of the upper bound. Equality holds if and only if any two points lie on a line, that is, the geometry is an affine plane.

Conversely, suppose that an affine plane of order n occurs. It has n^2 points and $n + 1$ parallel classes of lines. We select two parallel classes $\{H_1, \ldots, H_n\}$ and $\{V_1, \ldots, V_n\}$ of lines (to be the horizontal and vertical lines). Now any point lies on a unique horizontal line H_j and a unique vertical line V_i; we can give this point the coordinates (i, j).

Now let $\{L_1, \ldots, L_n\}$ be any further parallel class, and define a matrix A by the rule that $A_{ij} = k$ if and only if $(i, j) \in L_k$. It is easily checked that this matrix is a Latin square. Furthermore, the matrices obtained from different parallel classes are orthogonal. So we obtain a set of $n - 1$ MOLS from our affine plane.

REMARK. Given any set of r MOLS of order n, a 'geometry' can be constructed as in the above proof. It has n^2 points and $n(r + 2)$ lines, with each line having n points, two points in at most one line, and the lines falling into $r + 2$ parallel classes. Such a geometry is called a *net*.

9.6. Other kinds of geometry

Finite geometers have produced a bewildering variety of new types of geometries, usually defined by lists of axioms: affine spaces, polar spaces (and affine polar spaces), partial and semi-partial geometries, generalised polygons, near-polygons, buildings, etc. In this section, I will say a little about two of these types, which are closely related to projective spaces.

We have already seen the relation between projective and affine planes. Not surprisingly, the same can be done in any dimension. We define the n-dimensional *affine geometry* AG(n, q) over the field GF(q) to be obtained from the projective geometry PG(n, q) by designating a hyperplane H (a subspace of codimension 1) as being 'at infinity' and deleting it, together with all the subspaces it contains.

Just as in the plane case, there is a cartesian representation. If the underlying vector space $V(n + 1, q)$ consists of vectors with coordinates $(x_1, \ldots, x_{n+1})$, we can take the hyperplane at infinity to have equation $x_{n+1} = 0$; then any non-infinite point has a unique representative with $x_{n+1} = 1$, say $(x_1, \ldots, x_n, 1)$, and we can represent it uniquely by the n-tuple $(x_1, \ldots, x_n)$. We can regard this as a vector of

$V(n, q)$. Now the whole geometry can be represented in $V = V(n, q)$, as follows: k-flats turn out to be all *cosets* $W = v$ of k-dimensional vector subspaces W of V. (This works even for points: the only 0-dimensional subspace is $\{0\}$, and its cosets are all the singleton sets $\{v\}$, which can be identified with individual vectors $v \in V$.) Now it is clear that a flat of dimension k contains q^k points. The number of such flats is $q^{n-k} \begin{bmatrix} n \\ k \end{bmatrix}_q$: for there are $\begin{bmatrix} n \\ k \end{bmatrix}_q$ choices of the vector subspace W, and q^n choices of the coset representative v, but q^k of these give rise to the same coset. Summarising:

(9.6.1) Proposition. $AG(n, q)$ has q^n points. It has $q^{n-k} \begin{bmatrix} n \\ k \end{bmatrix}_q$ flats of dimension k, each of which contains q^k points.

There are theorems about recognition of affine spaces, like the Veblen–Young Theorem but more complicated. We won't pursue this any further (but see the discussion of affine Steiner triple systems in Section 8.5).

Now we examine briefly a class of geometries which axiomatise (among other things) the nets, which arose in connection with orthogonal Latin squares and affine planes in Section 9.5.

Let s, t, α be positive integers. A *partial geometry* with parameters s, t, α is a geometry of points and lines for which the following axioms hold:
- every line is incident with $s + 1$ points, and any point with $t + 1$ lines;
- two points are incident with at most one line (and two lines with at most one point — but this is equivalent to the preceding!);
- if the point p is not incident with the line L, then there are exactly α points of L collinear with p (or, equivalently, exactly α lines through p concurrent with L).

The comments in parentheses demonstrate that the *dual* of a partial geometry with parameters s, t, α is a partial geometry with parameters t, s, α. (The dual is defined in the same way as for projective planes in Section 9.5.) Note that $1 \leq \alpha \leq \min(s, t) + 1$.

Part of the motivation for studying partial geometries is that they include many other types of structure as special cases. Let us just notice two cases.

A partial geometry with $\alpha = s + 1$ has the property that any two points lie on a unique line. (For let p and q be points, and L a line containing q. If L also contains p, we're done; else, by the third axiom and the fact that $\alpha = |L|$, every point of L (and in particular q) is collinear with p. Conversely, a structure in which two points lie on a unique line and every line has a constant number of points, is a partial geometry with $\alpha = s + 1$. These include projective and affine planes, projective and affine spaces of arbitrary dimension (where lines are 1-flats), Steiner triple systems, and *complete graphs* (with two points per line).

A net (obtained from a family of r MOLS, as in Section 9.5) is a partial geometry with $s = n - 1$, $t = r - 1$, $\alpha = r - 1$. (The parameters s and t are clear. Now, if p is not on the line L, then every line through p meets L except for the unique line parallel to L.

Conversely, let $\mathcal{G}$ be a partial geometry with $\alpha = t$. Let $n = s + 1$ and $r = t + 1$. Calling two lines *parallel* if they are equal or disjoint, we see that, given any point

p and line L, there is a unique line L' through p parallel to L. Hence parallelism is an equivalence relation, and each parallel class covers all the points of $\mathcal{G}$ once. Now every line has n points. It follows that every parallel class has n lines (since a line not in that parallel class meets each line in the class once), and so there are n^2 points altogether. Thus $\mathcal{G}$ is a net.

We conclude that nets are the same as partial geometries with $\alpha = t$. In particular, $\alpha = t = 1$ defines a square grid.

A very important kind of partial geometry consists of *generalised quadrangles*, defined by the condition that $\alpha = 1$. We see that square grids are generalised quadrangles; but there are many others. Exercise 12 gives a simple construction of one.

9.7. Project: Coordinates and configurations

As you might expect, the projective planes $\mathrm{PG}(2,q)$ have many special properties not shared by arbitrary planes. The proofs of these properties must involve the algebraic structure: in other words, we work with coordinates rather than with the geometric configurations they represent. In this section, we will see how to set up coordinates, and then use them to prove one of the most famous theorems of finite geometry, *Segre's Theorem.*

Let $F = \mathrm{GF}(q)$. The points of $\mathrm{PG}(2,q)$ are 1-dimensional subspaces of the vector space $V = V(3, F)$. Each point is spanned by a non-zero vector (x, y, z); but, of course, any non-zero multiple (cx, cy, cz) would span the same point. We use the notation $[x, y, z]$ for the point spanned by (x, y, z), so that $[x, y, z] = [cx, cy, cz]$ for any $c \neq 0$. Then x, y, z are called *homogeneous coordinates* for the point.

(An alternative procedure would be to call two non-zero vectors *equivalent* if one is a constant multiple of the other, and then define points to be equivalence classes of vectors.)

Any line can be represented by a linear equation $ax + by + cz = 0$, where a, b, c are not all zero. We see that multiplying a, b, c by a constant doesn't change the set of points on the line; so we can also represent lines by equivalence classes (or 1-dimensional subspaces) $[a, b, c]$. (In algebraic terms, lines, or 2-dimensional subspaces of V, are represented by 1-dimensional subspaces of the dual space V^*.)

We can find unique representatives of the points and lines at the cost of distinguishing cases. For this purpose, we take the line $z = 0$ (represented by $[0, 0, 1]$) to be the line at infinity. Now any point not on this line (i.e., in the affine plane) has $z \neq 0$, and so has a unique representative $[x, y, 1]$ (obtained by multiplying through by the inverse of the third coordinate): this corresponds to the usual Cartesian coordinates (x, y) in the affine plane. There are q^2 points of this form. Similarly, points on the line $z = 0$ either have $x \neq 0$ (in which case there is a unique representative $[1, m, 0]$), or have $x = 0$ as well (there is a unique such point, namely $[0, 1, 0]$). This gives the $q + 1$ points on the line at infinity, making $q^2 + q + 1$ lines altogether.

Now we consider the lines. One of them is the line at infinity, $[0, 0, 1]$. For most other lines, as usual in coordinate geometry, we can take the equation to be $y = mx + c$: this line has slope m and y-intercept c in the standard way. (Its affine points are those $[x, y, 1]$ for which x and y satisfy this equation.) In terms of homogeneous coordinates, the equation is $y = mx + cz$, or $[m, -1, c]$; it contains the point $[1, m, 0]$ of the line at infinity. The remaining lines (those with 'infinite slope') have equation $x = c$, which in homogeneous coordinates is $x = cz$ or $[-1, 0, c]$; they pass through the point $[0, 1, 0]$.

We are going to find all the ovals in the planes $\mathrm{PG}(2, q)$ with q odd. First we have to define ovals, and prove a few of their properties.

An *oval* in a projective plane is a set $\mathcal{O}$ of points with the properties that no three of its points are collinear, and it has a unique tangent at each of its points (a line meeting it in no further point). It's clear that this definition is abstracted from the intuitive notion of an oval in the real plane (exemplified by any smooth convex curve); but intuition doesn't always serve us well in finite geometry.

Given an oval $\mathcal{O}$, any line of the plane meets $\mathcal{O}$ in at most two points; we call a line L a *secant*, *tangent* or *passant* according as $|L \cap \mathcal{O}| = 2$, 1 or 0. If p is a point of $\mathcal{O}$, then p lies on $q + 1$ lines (where q is the order of the plane), of which one is a tangent and the other q are secants, each containing one further point of $\mathcal{O}$; so $|\mathcal{O}| = q + 1$.

In $\mathrm{PG}(2, q)$, there is an important special class of ovals, called *conics*. A conic $\mathcal{C}$ is the set of points satisfying a non-singular quadratic equation: thus

$$\mathcal{C} = \{[x, y, z] : ax^2 + by^2 + cz^2 + fyz + gzx + hxy = 0\},$$

where the quadratic form is non-singular (this means that it cannot be transformed into a form in less than three variables by any non-singular linear substitution of the variables x, y, z). Note that, because every term in the quadratic form has degree 2, if (x, y, z) satisfies the equation, so does (cx, cy, cz); so our definition does make sense.

Any conic is an oval. To see this, take a line L, which (by choice of coordinates, i.e., a linear substitution) we can assume is the line $z = 0$. The points of $\mathcal{C} \cap L$ are those $[x, y, 0]$ for which $ax^2 + by^2 + hxy = 0$. Now we cannot have $a = b = h = 0$; for then the quadratic would be $z(gx + fy + cz) = 0$, and a linear substitution would change it to $xz = 0$, involving only two variables. If $a \neq 0$, then the point $[1, 0, 0]$ doesn't satisfy the equation; any other point has a representative $[x, 1, 0]$, and lies on the conic if and only if $ax^2 + hx + b = 0$, and this quadratic equation has at most two solutions. The argument is similar if $b \neq 0$. Finally, if $a = b = 0$, the equation is $hxy = 0$, and there are two points which satisfy it, namely $[1, 0, 0]$ and $[0, 1, 0]$.

In the affine plane, there are three familiar types of conic: the ellipse, parabola, and hyperbola. But the three are equivalent in the projective plane. If we take a conic $\mathcal{C}$ in $\mathrm{PG}(2, q)$, and choose a line L to be the line at infinity, then the conic becomes a hyperbola, parabola or ellipse in the usual fashion if L is a secant, tangent or passant respectively. For example, consider the conic with equation $xy = z^2$. If we choose $z = 0$ to be the line at infinity, the affine form of the equation is $xy = 1$, a hyperbola (put $z = 1$); if $y = 0$ is the line at infinity, the affine form is $x = z^2$, a parabola.

(9.7.1) Segre's Theorem. *If q is an odd prime power, then any oval in $\mathrm{PG}(2, q)$ is a conic.*

PROOF. Let $\mathcal{O}$ be an oval. We begin with some combinatorial analysis which applies in any plane of odd order; then we introduce coordinates.

STEP 1. Any point not on $\mathcal{O}$ lies on 0 or 2 tangents.

Let p be a point not on $\mathcal{O}$. Since $|\mathcal{O}| = q + 1$ is even, and an even number of points lie on secants through p, an even number must lie on tangents also. Let x_i be the number of points outside $\mathcal{O}$ which lie on i tangents. Now we have

$$\sum x_i = q^2,$$
$$\sum i x_i = (q + 1)q,$$
$$\sum i(i - 1)x_i = (q + 1)q.$$

(These are all obtained by double counting. The first holds because there are q^2 points outside $\mathcal{O}$; the second because there are $q+1$ tangents (one at each point of $\mathcal{O}$), each containing q points not on $\mathcal{O}$; and the third because any two tangents intersect at a unique point outside $\mathcal{O}$.)

From these equations, we see that $\sum i(i-2)x_i = 0$. But the term $i=1$ in the sum vanishes (any point lies on an even number of tangents); the terms $i=0$ and $i=2$ clearly vanish, and $i(i-2)>0$ for any other value of i. So $x_i = 0$ for all $i \neq 0$ or 2, proving the assertion.

REMARK. Points not on $\mathcal{O}$ are called *exterior points* or *interior points* according as they lie on 2 or 0 tangents, by analogy with the real case. But the analogy goes no further. In the real case, every line through an interior point is a secant; this is false for finite planes. (Can you count the number of secants through a point of each type?)

STEP 2. The product of all the non-zero elements of $\mathrm{GF}(q)$ is equal to -1.

The solutions of the quadratic $x^2 = 1$ are $x = 1$ and $x = -1$; these are the only elements equal to their multiplicative inverses. So, in the product of all the non-zero elements, everything except 1 and -1 pairs off with its inverse, leaving these two elements unpaired.

For the next two steps, note that we can choose the coordinate system so that the sides of a given triangle have equations $x = 0$, $y = 0$ and $z = 0$ (and the opposite vertices are $[1,0,0]$, $[0,1,0]$, and $[0,0,1]$ respectively). We'll call this the *triangle of reference.*

STEP 3. Suppose that concurrent lines through the vertices of the triangle of reference meet the opposite sides in the points $[0,1,a]$, $[b,0,1]$, and $[1,c,0]$. Then $abc = 1$.

(The equations of the concurrent lines are $z = ay$, $x = bz$ and $y = cx$ respectively; the point of concurrency must satisfy all three equations, whence $abc = 1$.)

REMARK. This result is equivalent to the classical Theorem of Menelaus.

STEP 4. Let the vertices of the triangle of reference be chosen to be three points of $\mathcal{O}$, and let the tangents at these points have equations $z = ay$, $x = bz$ and $y = cx$ respectively. Then $abc = -1$.

Proof: There are $q-2$ further points of $\mathcal{O}$, say $p_1, \ldots, p_{q-2}$. Consider the point $[1,0,0]$. It lies on the tangent $z = ay$, meeting the opposite side in $[0,1,a]$; two secants which are sides of the triangle; and $q-2$ further secants, through $p_1, \ldots, p_{q-2}$. Let the secant through p_i meet the opposite side in $[0,1,a_i]$. Then $a\prod_{i=1}^{q-2} a_i = -1$, by Step 2. If b_i, c_i are similarly defined, we have also $b\prod_{i=1}^{q-2} b_i = c\prod_{i=1}^{q-2} c_i = -1$. Thus

$$abc \prod_{i=1}^{q-2}(a_i b_i c_i) = -1.$$

But, by Step 3, $a_i b_i c_i = 1$ for $i = 1, \ldots, q-2$; so $abc = -1$.

STEP 5. Given any three points p, q, r of $\mathcal{O}$, there is a conic $\mathcal{C}$ passing through p, q, r and having the same tangents at these points as does $\mathcal{O}$.

Proof: Choosing coordinates as in Step 4, the conic with equation

$$yz - czx + caxy = 0$$

can be checked to have the required property. (For example, $[1,0,0]$ lies on this conic; and, putting $z = ay$, we obtain $ay^2 = 0$, so $[1,0,0]$ is the unique point of the conic on this line.)

STEP 6. Now we are finished if we can show that the conic $\mathcal{C}$ of Step 5 passes through an arbitrary further point s of $\mathcal{O}$.

Let $\mathcal{C}'$ and $\mathcal{C}''$ be the conics passing through p, q, s and p, r, s respectively and having the correct tangents there. Let the conics $\mathcal{C}$, $\mathcal{C}'$ and $\mathcal{C}''$ have equations $f = 0$, $f' = 0$, $f'' = 0$ respectively. (These equations are determined up to a constant factor.) Let L_p, L_q, L_r, L_s be the tangents to $\mathcal{O}$ at p, q, r, s respectively. Since all three conics are tangent to L_p at p, we can choose the normalisation so that f, f', f'' agree identically on L_p.

Now consider the restrictions of f' and f'' to L_s. Both are quadratic functions having a double zero at s, and the values at the point $L_s \cap L_p$ coincide; so the two functions agree identically on L_s. Similarly, f and f' agree on L_q, and f and f'' agree on L_r. But then f, f' and f'' all agree at the point $L_q \cap L_r$. So the quadratic functions f' and f'' agree on L_p, L_s, and $L_q \cap L_r$, which forces them to be equal. So the three conics coincide, and our claim is proved (and with it Segre's Theorem).

9.8. Project: Proof of the Bruck–Ryser Theorem

In this section, we prove the Bruck–Ryser Theorem:

> If $n \equiv 1$ or $2 \pmod 4$ and a projective plane of order n exists, then n is a sum of two squares of integers.

The proof uses a fair amount of number theory. It also has a very *ad hoc* appearance; you may wonder how anybody ever thought of it! In fact, there are deeper and more general number-theoretic regions lying hidden here, for relating integer zeros of quadratic forms to zeros modulo primes, going by the name of *Hasse–Minkowski theory*, which have important applications in combinatorics. The argument here can be regarded as the general argument translated into a simpler form in the special case.

We need four 'facts' from number theory, Proofs and discussions of these will be given after the proof of the Bruck–Ryser Theorem.

FACT 1. The 'four-squares identity':

$$(a_1^2 + a_2^2 + a_3^2 + a_4^2)(x_1^2 + x_2^2 + x_3^2 + x_4^2) = y_1^2 + y_2^2 + y_3^2 + y_4^2,$$

where

$$y_1 = a_1 x_1 - a_2 x_2 - a_3 x_3 - a_4 x_4,$$
$$y_2 = a_1 x_2 + a_2 x_1 + a_3 x_4 - a_4 x_3,$$
$$y_3 = a_1 x_3 + a_3 x_1 + a_4 x_2 - a_2 x_4,$$
$$y_4 = a_1 x_4 + a_4 x_1 + a_2 x_3 - a_3 x_2.$$

FACT 2. If p is an odd prime, and there exist integers x_1, x_2, not both divisible by p, such that $x_1^2 + x_2^2 \equiv 0 \pmod p$, then p is the sum of two integer squares. The analogous result holds for four squares.

FACT 3. Every positive integer is the sum of four integer squares.

FACT 4. For any integer n, if the equation $x^2 + y^2 = nz^2$ has an integer solution with x, y, z not all zero, then n is the sum of two integer squares (that is, the equation has a solution with $z = 1$).

PROOF OF BRUCK–RYSER THEOREM. Suppose that there is a projective plane of order n, where $n \equiv 1$ or $2 \pmod 4$. The number of points of the plane is $N = n^2 + n + 1$; and we see that $N \equiv 3 \pmod 4$.

Let A be an *incidence matrix* of the plane, an $N \times N$ matrix with rows indexed by points and columns by lines, with (i, j) entry equal to 1 if the ith point is on the jth line, 0 otherwise. Then $AA^\top$ has (i, j) entry equal to the number of lines containing the 6th and j^{th} points, which is $n + 1$ if $i = j$, and 1 otherwise; that is,

$$AA^\top = nI + J,$$

where J is the matrix with every entry 1.

Let $x_1, \ldots, x_N$ be indeterminates, and let $x = (x_1, \ldots, x_N)$. Let $xA = z = (z_1, \ldots, z_N)$; then $z_1, \ldots, z_N$ are linear combinations of $x_1, \ldots, x_N$ with integer coefficients. We have

$$zz^\top = xAA^\top x^\top = n x x^\top + x J x^\top,$$

that is,

$$z_1^2 + \ldots + z_N^2 = n(x_1^2 + \ldots + x_N^2) + w^2,$$

where $w = x_1 + \ldots + x_N$. We take a new indeterminate x_{N+1} and add $n x_{N+1}^2$ to both sides of the above equation. Note that $N + 1$ is divisible by 4. Write $n = a_1^2 + a_2^2 + a_3^2 + a_4^2$ (by Fact 4), and use the four-squares identity (Fact 1) to write

$$n(x_{4i+1}^2 + \ldots + x_{4i+4}^2) = y_{4i+1}^2 + \ldots + y_{4i+4}^2,$$

where the y's are linear combinations of the x's. We have

$$z_1^2 + \ldots + z_N^2 + nx_{N+1}^2 = y_1^2 + \ldots + y_{N+1}^2 + w^2.$$

In the next step, we make a number of specialisations, each expressing some x_i as a rational linear combination of other x's. Note that the quadratic is positive definite, so, no matter how we do this, the resulting form will involve all the variables. To begin with, x_1 is involved in at least one y and at least one z; without loss of generality, it is involved in y_1 and z_1. If it has different coefficients in these two expressions, we impose the condition $y_1 = z_1$; otherwise, we impose $y_1 = -z_1$. In either case, we can express x_1 in terms of the other x's; and also $z_1^2 = y_1^2$, so this term can be cancelled. Now repeat this process to cancel the terms y_i^2 and z_i^2 for $i = 2, \ldots, N$, obtaining finally

$$nx_{N+1}^2 = y_{N+1}^2 + w^2,$$

where y_{N+1} and w are rational linear combinations (that is, rational multiples) of x_{N+1}. So we can choose an integer value of x_{N+1} such that y_{N+1} and w are also integers, and we have a non-zero solution of the above equation in integers. By Fact 4, n is a sum of two integer squares. The theorem is proved.

We now return to the proofs of the four 'facts'.

PROOF OF FACT 1. Straightforward calculation. But the result has a deeper significance. The *quaternions* are a number system $\mathbb{H}$ extending the complex numbers. They have the form

$$a = a_1 + a_2 \mathrm{i} + a_3 \mathrm{j} + a_4 \mathrm{k},$$

where $\mathrm{i}^2 = \mathrm{j}^2 = \mathrm{k}^2 = -1$, $\mathrm{ijk} = -1$,[3] from which it follows that $\mathrm{ij} = \mathrm{k}$, $\mathrm{ji} = -\mathrm{k}$, $\mathrm{jk} = \mathrm{i}$, $\mathrm{kj} = -\mathrm{i}$, $\mathrm{ki} = \mathrm{j}$, $\mathrm{ik} = -\mathrm{j}$. It is easily checked that

$$(a_1 + a_2 \mathrm{i} + a_3 \mathrm{j} + a_4 \mathrm{k})(x_1 + x_2 \mathrm{i} + x_3 \mathrm{j} + x_4 \mathrm{k}) = y_1 + y_2 \mathrm{i} + y_3 \mathrm{j} + y_4 \mathrm{k},$$

where $y_1, \ldots, y_4$ are as in Fact 1. There is a 'norm' defined on the quaternions by

$$\|a_1 + a_2 \mathrm{i} + a_3 \mathrm{j} + a_4 \mathrm{k}\| = a_1^2 + a_2^2 + a_3^2 + a_4^2;$$

the four-squares identity says that

$$\|a\| \cdot \|x\| = \|ax\|.$$

If we treat the complex numbers similarly, using the norm $\|a\| = |a|^2$, we obtain a 'two-squares identity'

$$(a_1^2 + a_2^2)(x_1^2 + x_2^2) = (a_1 x_1 - a_2 x_2)^2 + (a_1 x_2 + a_2 x_1)^2.$$

There is also an 'eight-squares identity', related to a number system called the *octonions* or *Cayley numbers*.

PROOF OF FACT 2. We are given that $rp = x_1^2 + x_2^2$, for some positive r; take an expression of this form in which r is as small as possible. We have to prove that $r = 1$. So suppose not. Choose u_1, u_2 such that $u_1 \equiv x_1 \pmod{r}$, $u_2 \equiv -x_2 \pmod{r}$, and $|u_i| \le r/2$ for $i = 1, 2$. Then

$$u_1^2 + u_2^2 \equiv x_1^2 + x_2^2 \equiv 0 \pmod{r},$$

say $u_1^2 + u_2^2 = rs$. Then $s < r$, because of the bounds on u_1 and u_2. We have

$$r^2 sp = (x_1^2 + x_2^2)(u_1^2 + u_2^2) = (x_1 u_1 - x_2 u_2)^2 + (x_1 u_2 + x_2 u_1)^2$$

by the two-squares identity. We have

$$x_1 u_1 - x_2 u_2 \equiv x_1^2 + x_2^2 \equiv 0 \pmod{r}$$

and

$$x_1 u_2 + x_2 u_1 \equiv x_1 x_2 - x_2 x_1 \equiv 0 \pmod{r},$$

so the equation has a factor r^2, and we obtain

$$sp = y_1^2 + y_2^2$$

for $y_1 = (x_1 u_1 - x_2 u_2)/r$, $y_2 = (x_1 u_2 + x_2 u_1)/r$. But this contradicts our choice of r, since $s < r$. The argument for four squares is very similar.

---

[3] These formulae were discovered by Hamilton, while walking by a canal in Dublin. He was so pleased with his discovery that he wrote it on a bridge he passed.

PROOF OF FACT 3. According to the four-squares identity, if two numbers are sums of four squares, then so is their product. So it will suffice to show that every prime is the sum of four squares. Clearly $2 = 1^2 + 1^2 + 0^2 + 0^2$, so we need only deal with odd primes.

We need another fact. Let p be an odd prime. A non-zero congruence class m mod p is called a *quadratic residue* (QR) if the congruence $m = x^2$ is solvable, and a *quadratic non-residue* (QNR) otherwise. Now, of the $p - 1$ congruence classes, half are QRs and half are QNRs, and the product of two QNRs is a QR. (See Exercise 12.)

Now we separate two cases.

Case 1: -1 is a QR. In other words, the congruence $x^2 + 1 \equiv 0 \pmod{p}$ has a solution. By Fact 2, p is a sum of two squares.

Case 2: -1 is a QNR. Let m be the smallest positive QNR. Then $-m$ and $m - 1$ are QRs, and so the congruences $x^2 \equiv m - 1 \pmod{p}$, $y^2 \equiv -m \pmod{p}$ are solvable. But then

$$x^2 + y^2 + 1^2 \equiv 0 \pmod{p},$$

and by Fact 2, p is a sum of four squares.

PROOF OF FACT 4. First, we argue that it suffices to prove the result for squarefree numbers n. For suppose it is true for squarefree n, and let $n = mu^2$ with m squarefree; let $x^2 + y^2 = nz^2$, where x, y, z are not all zero. Then $x^2 + y^2 = m(uz)^2$. By assumption, m is a sum of two squares, say $m = a^2 + b^2$; and then $n = (au)^2 + (bu)^2$.

So let n be squarefree, say $n = p_1 \ldots p_k$, where $p_1, \ldots, p_k$ are distinct primes; and suppose that $x^2 + y^2 = nz^2$, where x, y, z are not all zero. We may suppose that x, y, z have no common factor. Then x and y are not both divisible by p_i; for if they were, then p_i^2 divides nz^2, contradicting the facts that p_i^2 doesn't divide n and that p_i doesn't divide z. Now by Fact 2, p_i is a sum of two squares. This holds for all i. By applying the two-squares identity $k - 1$ times, we see that n is a sum of two squares, as required.

9.9. Appendix: Finite fields

This section gives an algebraic proof of the basic existence result (due to Galois) for finite fields, cited in the first section of this chapter. The details may be somewhat sketchy, but a standard algebra textbook will fill them in for you.

The proof requires a technical result, the *uniqueness of splitting field*. First, a definition. Let F be a field. We call a field containing F an *extension* of F. Let E_1, E_2 be two extensions of F. We say that E_1 and E_2 are F-*isomorphic* if there is an isomorphism from E_1 to E_2 which fixes every element of F.

Step 1. *Let F be a field, $f(x)$ an irreducible polynomial over F. Then there exists an extension E of F such that $f(x)$ has a root in E. Any two such fields which are minimal with respect to inclusion are F-isomorphic.*

An example of such a field is the quotient ring $F[x]/(f(x))$, where $F[x]$ is the polynomial ring over F and $(f(x))$ the ideal generated by $f(x)$. (Since f is irreducible, the ideal it generates is maximal, and the quotient is a field.) Now, if E_1 and E_2 are minimal extensions of F containing roots α_1 and α_2 of $f(x)$ respectively, then every element of E_i is expressible as a polynomial $g(\alpha_i)$ in α_i with coefficients in F, two polynomials representing the same element if and only if their difference is divisible by f; and the map which takes $g(\alpha_1)$ to $g(\alpha_2)$ is an F-isomorphism from E_1 to E_2.

The unique minimal extension of F containing α is denoted by $F(\alpha)$.

It follows by an easy induction that, if $f(x)$ is any polynomial over F, then there is an extension E of F such that f has all its roots in E; that is, f can be factorised into linear factors over E. (Just adjoin roots of $f(x)$ one at a time.) A minimal such extension is called a *splitting field* of $f(x)$ over F.

The *degree* of a field extension E of F is its dimension as a vector space over F (when we forget multiplication in E and remember only how to add elements of E or multiply them by elements of F).

Step 2. *Any two splitting fields of $f(x)$ over F are F-isomorphic.*

This is proved by induction on the degree of one of the splitting fields. If the degree is 1, so that $f(x)$ already splits in F, the result is clear. So suppose not. Let E_1 and E_2 be splitting fields of $f(x)$ over F. Let α_1 be a root of $f(x)$ in E_1 but not in F, and α_2 a root of *the same irreducible factor* of $f(x)$ in E_2. Then there is an F-isomorphism from $F(\alpha_1)$ to $F(\alpha_2)$ carrying α_1 to α_2, by Fact 1; so we may suppose that $\alpha_1 = \alpha_2$. Now E_1 and E_2 are splitting fields for $f(x)$ over $F(\alpha_1)$, with smaller degree than they have over F; by induction, they are $F(\alpha_1)$-isomorphic (and, *a fortiori*, F-isomorphic).

Now we turn our attention to finite fields.

Step 3. *Let F be a finite field. There exists a prime number p such that $p \cdot a = 0$ for all $a \in F$, where*

$$p \cdot a = a + a + \ldots + a \qquad p \text{ terms.}$$

The additive group of F is finite, so its elements have finite order. Suppose that the element 1 has order p; that is, $p \cdot 1 = 0$. Then p is prime; for if $p = mn$ with $m, n > 1$, then $(m \cdot 1)(n \cdot 1) = 0$, but neither $m \cdot 1$ nor $n \cdot 1$ is zero (since, by definition, p is the smallest integer k for which $k \cdot 1 = 0$). But this contradicts the fact that F has no divisors of zero.

The prime p is called the *characteristic* of F.

Step 4. *The number of elements in a finite field F is a power of the characteristic of F.*

This follows from (9.2.1), once we check that F is a vector space over $\mathbb{Z}/(p)$. (In fact, F is an extension of $\mathbb{Z}/(p)$, where $\mathbb{Z}/(p)$ consists of the elements $0, 1, \ldots, (p-1) \cdot 1$ of F.)

Step 5. *If F has q elements, then F is a splitting field of the polynomial $x^q - x$ over $\mathbb{Z}/(p)$, where p is the characteristic of F.*

For the multiplicative group of F has order $q - 1$, so all non-zero elements satisfy $x^{q-1} = 1$, whence also $x^q = x$; this polynomial is also satisfied by 0. But a polynomial of degree q cannot have more than q roots; so the elements of F are all the roots, and F is a splitting field.

Now Step 2 shows the uniqueness of the field with q elements, *if it exists*.

Step 6. *If q is a power of the prime p, then the splitting field of $x^q - x$ over $\mathbb{Z}/(p)$ has q elements.*

The derivative of the polynomial $x^q - x$ is -1 (remember that the characteristic divides q); this is coprime to $x^q - x$, so all the roots of the polynomial $x^q - x$ in its splitting field are distinct, so there are q of them. We have to show that these roots form a field. So let S be the set of roots, and $a, b \in S$; that is, $a^q = a$ and $b^q = b$. Then

$$(a + b)^q = a^q + b^q = a + b,$$
$$(ab)^q = a^q b^q = ab,$$

so $a + b, ab \in S$; similarly $1/a \in S$ if $a \neq 0$. (The first equation above is non-trivial. We have

$$(a + b)^p = \sum_{i=0}^{p} \binom{p}{i} a^{p-i} b^i = a^p + b^p,$$

since the characteristic is p and divides all the binomial coefficients $\binom{p}{i}$ for $1 \leq i \leq p - 1$. Then, by induction on k,

$$(a + b)^{p^k} = a^{p^k} + b^{p^k},$$

and the result follows since q is a power of p (Fact 4).) So S is a field of order q, completing the proof.

9.10. Exercises

1. How many additions and multiplications are needed (in the worst case) to transform an $m \times n$ matrix into reduced echelon form?

2. For fixed q, show that the probability that a random $n \times n$ matrix over $\mathrm{GF}(q)$ is non-singular tends to a limit $c(q)$ as $n \to \infty$, where $0 < c(q) < 1$.

3. Let $F_q(n)$ be the total number of subspaces of an n-dimensional vector space over $\mathrm{GF}(q)$. Prove that $F_q(0) = 1$, $F_q(1) = 2$, and

$$F_q(n+1) = 2F_q(n) + (q^n - 1)F_q(n-1)$$

for $n \geq 1$. [HINT: By (9.2.3) and (9.2.4), we have

$$\begin{bmatrix} n+1 \\ k \end{bmatrix}_q = \begin{bmatrix} n \\ k-1 \end{bmatrix}_q + \begin{bmatrix} n \\ k \end{bmatrix}_q + (q^n - 1)\begin{bmatrix} n-1 \\ k-1 \end{bmatrix}_q.$$

Now sum over k.]
     Prove that $F_q(n) \geq q^{\lfloor n^2/4 \rfloor}$.

4. Prove

$$\begin{bmatrix} n+1 \\ k \end{bmatrix}_q = \begin{bmatrix} n \\ k \end{bmatrix}_q + q^{n+1-k}\begin{bmatrix} n \\ k-1 \end{bmatrix}_q.$$

in two ways: by using (9.2.3) and (9.2.4), or by dividing the $k \times (n+1)$ matrices into two classes according to their first column.

5. Prove that the right-hand side of the q-binomial theorem (9.2.5) for $t = 1$ counts the number of $n \times n$ matrices in echelon form over $\mathrm{GF}(q)$, that is, satisfying the first two conditions in the definition of reduced echelon form. How many $n \times n$ matrices in reduced echelon form are there?

6. Prove that a set of points of a projective space is a flat if and only if it contains the line through any two of its points. [The corresponding set of vectors of the vector space is closed under scalar multiplication, since it is a union of 1-dimensional subspaces. So you must show that the set of vectors is closed under addition if and only if the set of points contains the line through any two of its points.]

7. Show that any set of $m - 2$ MOLS of order m can be enlarged to a set of $m - 1$ MOLS. [HINT: Construct the net corresponding to the given MOLS. Show that its points fall into m sets of m pairwise non-collinear points; these sets comprise the 'missing' parallel class.]
REMARK. R. H. Bruck generalised this result; he showed that any set of $m - f(m)$ MOLS of order m can be enlarged to a set of $m - 1$ MOLS, where $f(m)$ is a function of magnitude roughly $m^{1/4}$.

8. Show that there are two non-isomorphic nets of order 4 and degree 3. (The corresponding Latin squares are the multiplication tables of the two groups of order 4.) Show that one, but not the other, can be enlarged to an affine plane.

9. (a) Prove that there is a unique projective plane of order 3.
     (b) Prove that there is a unique projective plane of order 4.

10. Let $\mathcal{O}$ be an oval in a projective plane of even order q. Prove that the tangents to $\mathcal{O}$ all pass through a common point p, and that $\mathcal{O} \cup \{p\}$ is a set of $q + 2$ points which meets every line in either 0 or 2 points. (Such a set is called a *hyperoval*. Note that, if any one of its points is omitted, the resulting set is an oval.) [HINT: Let x_i be the number of points not on $\mathcal{O}$ which lie on i tangents. Show that $x_0 = 0$, and calculate $\sum(i - 1)(i - (q + 1))x_i$.]

11. Prove that, if q is a prime power, then any five points of $\mathrm{PG}(2, q)$, such that no three of them are collinear, are contained in a unique conic. Deduce that the number of conics is

$$(q^2 + 1 + 1)q^2(q - 1).$$

12. Define a geometry as follows. The points are to be all the 2-element subsets of $\{1, 2, 3, 4, 5, 6\}$; the lines are all the disjoint triples of 2-subsets. Prove that the geometry is a generalised quadrangle with $s = t = 2$, $\alpha = 1$.

13. Let p be an odd prime. Show that half the non-zero congruence classes mod p are quadratic residues and half are non-residues, and that the product of two non-residues is a residue. [HINT: Any non-zero element of $\mathbb{Z}/(p)$ has 0 or 2 square roots in $\mathbb{Z}/(p)$. Further, multiplying by a fixed non-residue is one-to-one and maps residues to non-residues.]

14. Write a quaternion formally as $a + \mathbf{x}$, where a is a real number and $\mathbf{x}$ a 3-dimensional vector (relative to the standard basis $(\mathbf{i}, \mathbf{j}, \mathbf{k})$). Show that

$$(a + \mathbf{x}) + (b + \mathbf{y}) = (a + b) + (\mathbf{x} + \mathbf{y}),$$
$$(a + \mathbf{x}) \cdot (b + \mathbf{y}) = (ab - \mathbf{x}.\mathbf{y}) + (a\mathbf{y} + b\mathbf{x} + \mathbf{x} \times \mathbf{y}),$$

where $\mathbf{x}.\mathbf{y}$ and $\mathbf{x} \times \mathbf{y}$ are the usual scalar and vector products ('dot product' and 'cross product') of vectors.

10. Ramsey's Theorem

Complete disorder is impossible

T. S. Motzkin (attr.)

TOPICS: Pigeonhole Principle; Ramsey's Theorem; estimates for Ramsey numbers; applications

TECHNIQUES: Double induction; probabilistic existence proof

ALGORITHMS:

CROSS-REFERENCES:

In 1930, F. P. Ramsey[1] proved a lemma in a paper on mathematical logic. The lemma has proved to be of greater importance than the theorem it was used to prove,[2] and has given its author's name to an area where combinatorics, logic, topology and probability interact. Roughly speaking, a theorem of Ramsey theory says that any structure of a certain type, no matter how 'disordered', contains a much more highly ordered substructure of the same type.

Several mathematicians (notably Hilbert, Schur and van der Waerden) had before 1930 proved theorems which are now regarded as part of Ramsey theory. As Kafka in Borges' essay,[3] Ramsey created his own predecessors; with the hindsight of Ramsey's Theorem, we can see that these independent results are closely connected.

10.1. The Pigeonhole Principle

The *Pigeonhole Principle* is, at first sight, not the kind of thing that you would expect to be discovered by (and named after) a mathematician. In its simplest form, it is rather obvious:

(10.1.1) Pigeonhole Principle

If $n + 1$ letters are placed in n pigeonholes, then some pigeonhole must contain more than one letter.

---

[1] Ramsey was a brilliant economist in the circle of Keynes. Though he died at the age of 29, he had already made notable contributions to this discipline. He was an atheist, but his younger brother became Archbishop of Canterbury.

[2] This theorem concerned what are now called 'indiscernible sequences'.

[3] Jorge Luis Borges, 'Kafka and his precursors', *Labyrinths* (1964)

We will see, however, that it can be quantified and generalised into some highly non-trivial mathematics. In any event, it is clear that it is a 'combinatorial' result. It bears the name of the nineteenth-century German algebraist Dirichlet. He was surely not the first person to discover it, but the first to make effective use of it, as we will soon see. (By the way, can you give a formal proof?)

Even in the basic form above, it has many applications. One of these (ordering elements in a rectangular array) is given as Exercise 1. Here is the application which Dirichlet made, and resulted in his name being attached to the principle. It concerns the existence of good rational approximations to an irrational number. The topic really belongs to Number Theory, but the argument is combinatorial.

(10.1.2) Proposition. *Let α be an irrational number. Then there are infinitely many different rational numbers p/q for which*

$$\left| \alpha - \frac{p}{q} \right| < \frac{1}{q^2}.$$

PROOF. For this proof, we let $\{x\}$ denote the *fractional part* of the real number x, that is, $\{x\} = x - \lfloor x \rfloor$.

Our strategy is to show:

> For any natural number n, there is a rational number p/q with $q \leq n$ such that $|\alpha - p/q| < 1/(nq)$.

Of course, we then have $|\alpha - p/q| < 1/(q^2)$. Moreover, since α is irrational, $\alpha \neq p/q$, and we can find n_1 with $|\alpha - p/q| > 1/n_1$. Then repeating the argument with n_1 in place of n gives another solution p_1/q_1 which is different from p/q (since $|\alpha - p_1/q_1| < 1/(n_1 q_1) \leq 1/n_1 < |\alpha - p/q|$). Continuing this process, we find infinitely many such 'good' rational approximations.

Consider the $n+1$ numbers $\{i\alpha\}$, for $i = 1, 2, \ldots, n+1$. We put these numbers into the n pigeonholes $(j/n, (j+1)/n)$, for $j = 0, \ldots, n-1$. (None of the numbers coincides with an end-point of the intervals, since α is irrational.) By the Pigeonhole Principle, some interval contains more than one of the numbers, say $\{i_1\alpha\}$ and $\{i_2\alpha\}$, which therefore differ by less than $1/n$. Putting $q = |i_1 - i_2|$, we see that there exists an integer p such that

$$|q\alpha - p| < \frac{1}{n},$$

from which the result follows on division by n. Moreover, q is the difference between two integers in the range $1, \ldots, n+1$, so $q \leq n$.

Instead of pigeonholes, we use the terminology of colouring. The Pigeonhole Principle states that, if $r+1$ objects are coloured with r different colours, then there must be two objects with the same colour. In order to move towards Ramsey's Theorem, we quantify the result further as follows.

(10.1.3) Proposition. *Suppose that $n \geq 1 + r(l-1)$. Let n objects be coloured with r different colours; then there exist l objects all with the same colour. Moreover, the inequality is best possible.*

PROOF. If the conclusion is false, then there are at most $l-1$ objects of each colour, hence at most $r(l-1)$ altogether, contrary to assumption.

When we say that the result is *best possible*, what we mean is this. If fewer than $1+r(l-1)$ objects are given, then *there is some way of colouring them* such that no l have the same colour. This too is obvious: 'fewer than $1+r(l-1)$' means 'at most $r(l-1)$', and the objects can be divided into r groups with at most $l-1$ in each group.

Still more generally, suppose that $n \geq k_1 + \ldots + k_r - r + 1$; let the points of an n-set be coloured with r colours $c_1, \ldots, c_r$. Then, for some value of i in the range $1, \ldots, r$, there exist k_i points all having colour i; and this is best possible.

10.2. Some special cases

We now consider the two-player game introduced in Chapter 1.

Mark six points on the paper, no three in line (for example, the vertices of a regular hexagon). Now the players take turns. On each player's turn, he draws a line in his colour between two of the points which haven't already been joined. (Crossings of lines other than at marked points are not significant.) The first player to create a triangle with all sides of his colour, having three of the marked points as vertices, *loses*.

The game is finite, since at most $\binom{6}{2} = 15$ edges can be drawn. If you play it with a friend, you will notice that it always ends in a win for one player; a draw is not possible. We prove that this is necessarily so.

(10.2.1) Proposition. *Suppose that the 2-element subsets of a 6-element set are coloured with two colours. Then there is a 3-element set, all of whose 2-element sets have the same colour. This is not true for fewer than six points.*

PROOF. Let us suppose that the colours are red and blue; let $1, \ldots, 6$ be the points. Consider the five 2-subsets 16, 26, 36, 46, 56. These are coloured with two colours; so there must be three of the five edges which have the same colour (by the Pigeonhole Principle with $r = 2$, $l = 3$). Let us suppose that 16, 26, and 36 are red. Now there are two possibilities: if any one of 12, 23, 31 is red (say 12), there is a red triangle (126); but if none of the three is red, then 123 is a blue triangle.

To show that six is best possible, we must colour the 2-subsets of a 5-set red and blue without creating a monocromatic (single-coloured) triangle. If the points are 1, 2, 3, 4, 5, let 12, 23, 34, 45, 51 be red and 13, 24, 35, 41, 52 blue.

Here are some more results of the same type.

(10.2.2) Proposition. *(i) If the 2-subsets of a 9-set are coloured red and blue, there is either a red 3-set or a blue 4-set.*
(ii) If the 2-subsets of a 18-set are coloured red and blue, there is a monochromatic 4-set.
(iii) If the 2-subsets of a 17-set are coloured red, blue and green, there is a monochromatic 3-set.
(iv) All the above are best possible.

PROOF. The proofs all follow the same pattern, except for one trick in the proof of (i). We prove (i) first for 10 points. Consider the nine edges joining one point x to the others. By the 'more general' form of the Pigeonhole Principle, either there are four red edges, or six blue edges. Suppose first that there are four red edges; let X be the set of their four endpoints other than x. If X contains a red edge yz, then xyz is a red triangle; else X is a blue 4-set. Now consider the other case, six blue edges; let Y be the set of their endpoints other than x. Now we use the result proved above, that Y contains a monochromatic triangle uvw. If it is red, we are done; if blue, then $xuvw$ is a blue 4-set.

Now suppose there are just nine points. The only way we can avoid the above situation is that every point x lies on exactly three red and five blue edges. But this contradicts the *Handshaking Lemma* of Chapter 2. (Could there be nine people at a convention, each of whom shakes hands exactly three times?) So the result holds for 9 points too.

(ii) Take a set of 18 points and colour the edges. Any point x lies on 17 edges; by the Pigeonhole Principle, either 9 are red or 9 are blue. Assume the former. By (i), the endpoints of these 9 edges either contain a red triangle (giving a red 4-set with x), or a blue 4-set (and we are finished).

(iii) Now take 17 points and colour the edges with three colours, red, blue and green. A point x is joined to 16 others, so by PP six of them have the same colour, say green. If the set X of endpoints of these edges contains a green edge yz, we have a green triangle xyz; otherwise all edges within X are red and blue, and there is a red or blue triangle by our earlier result.

The fact that these are best possible requires construction of colourings with 8, 17 and 16 points, not having monochromatic subsets of the specified sizes. This can be done, but I don't give details here (but see Exercise 6).

10.3. Ramsey's Theorem

The results above and their manner of proof suggest their generalisation, which is known as Ramsey's Theorem.

> **(10.3.1) Ramsey's Theorem**
>
> *Let r, k, l be given positive integers. Then there is a positive integer n with the following property. If the k-subsets of an n-set are coloured with r colours, then there is a monochromatic l-set, i.e., one all of whose k-sets have the same colour.*

More generally, let $r, k, a_1, \ldots, a_r$ be given. Then there exists n with the property that, if the k-subsets of an n-set are coloured with r colours $c_1, \ldots, c_r$, then for some i in the range $1, \ldots, r$, there is an a_i-set, all of whose subsets have colour c_i.

We denote by $R(r, k, l)$ the smallest n for which Ramsey's Theorem holds, and by $R^*(r, k; a_1, \ldots, a_r)$ the smallest n for which the 'more general' statement holds.

Clearly we have $R(r, k, l) = R^*(r, k; l, l, \ldots, l)$. To familiarise the notation, check that we proved the following results: $R(2, 2, 3) = 6$; $R^*(2, 2; 3, 4) = 9$; $R(2, 2, 4) = 18$; $R(3, 2, 3) = 17$; and

$$R^*(r, 1; a_1, \ldots, a_r) = \sum_{i=1}^{r} a_i - r + 1.$$

Moreover, there are some trivial evaluations: $R(r, k, k) = k$, $R(1, k, l) = l$. (We always assume that $k \leq l$, or that all of $a_1, \ldots, a_r$ are at least k, otherwise the assertions are trivial.)

It is also true that

$$R^*(r + 1, k; a_1, \ldots, a_r, k) = R^*(r, k; a_1, \ldots, a_r).$$

For, if there is a k-set of colour c_{r+1}, we have won; otherwise, only the first r colours occur.

The proof of Ramsey's Theorem uses induction, similar to the examples. As the arguments for (i) and (ii) suggest, we prove the 'more general' assertion. We already have the result for $k = 1$, so assume that $k > 1$. We may assume that $a_i > k$ for all i. By induction, we may assume that the numbers

$$A_i = R^*(r, k; a_1, \ldots, a_{i-1}, a_i - 1, a_{i+1}, \ldots, a_r)$$

are defined (and the statement is true for these).

Take $n = 1 + R^*(r, k - 1; A_1, \ldots, A_r)$. Let X be a set of n points, whose k-sets are coloured with r colours $c_1, \ldots, c_r$. Take a point $x \in X$, and let $Y = X \setminus \{x\}$. We define a colouring of the $(k - 1)$-subsets of Y with colours $c_1^*, \ldots, c_r^*$, by the rule that, for any $(k - 1)$-subset U, the colour of U is c_i^* if and only if the colour of $U \cup \{x\}$ is c_i. By definition of n, for some i, there is a c_i^*-monochromatic set Z_i of size A_i. By definition of A_i, the set Z_i contains either a set of size a_j with all its k-sets of colour c_j for some $j \neq i$, or a set V of size $a_i - 1$ with all its k-sets of colour c_i. In the first case, we have won. In the second case, $\{x\} \cup V$ is a set of size a_i, and all its k-sets have colour c_i — by assumption for subsets not containing x, and by the definition of the c^*-colouring and the fact that all $(k - 1)$-subsets have colour c_i^* in the case of subsets containing x.

10.4. Bounds for Ramsey numbers

It is extremely difficult to calculate exact values of Ramsey numbers. Apart from the values given in the last section, only seven values are known precisely, as of August 1995:

$R^*(2, 2; 3, l) = 14, 18, 23, 28, 36$ for $l = 5, 6, 7, 8, 9$;

$R^*(2, 2; 4, 5) = 25$; $R(2, 3, 4) = 14$.

Others, such as $R^*(2, 2; 3, 10)$, $R^*(2, 2; 4, 6)$, $R(2, 2, 5)$ and $R(4, 2, 3)$, are less than 100, but are only known to lie in fairly large ranges of integers. If you have an Internet connection and a PostScript printer, updated information is contained in a Dynamic Survey by S. Radziszowski in the Electronic Journal of Combinatorics, at http://ejc.math.gatech.edu:8080/Journal/Surveys/ds1.ps.

In the absence of exact values, we rely on inequalities, upper and lower bounds. I stress that upper bounds come from the proof of Ramsey's Theorem or a refinement of it, and lower bounds from constructions of colourings without large monochromatic sets.

The proof of Ramsey's Theorem in the last section gives us a 'recurrence inequality' for the Ramsey numbers, viz.

$$R^*(r, k; a_1, \ldots, a_r) \le 1 + R^*(r, k - 1; A_1, \ldots, A_r),$$

where

$$A_i = R^*(r, k; a_1, \ldots, a_{i-1}, a_i - 1, a_{i+1}, \ldots, a_r).$$

In general, this is a very tangled web which is difficult to disentangle into explicit bounds. We consider one case where this can be done.

(10.4.1) Proposition. *If* $a_1, a_2 \ge 2$, *then*

$$R^*(2, 2; a_1, a_2) \le \binom{a_1 + a_2 - 2}{a_1 - 1}.$$

PROOF. If $a_1 = 2$, then

$$R^*(2, 2; 2, a_2) = R^*(1, 2; a_2) = a_2 = \binom{2 + a_2 - 2}{2 - 1},$$

and the result is true; similarly, if $a_2 = 2$. So we will use induction, assuming the result is true when either a_1 or a_2 is reduced. In the notation of Ramsey's Theorem,

$$A_1 = R^*(2, 2; a_1 - 1, a_2) \le \binom{a_1 + a_2 - 3}{a_1 - 2},$$

$$A_2 = R^*(2, 2; a_1, a_2 - 1) \le \binom{a_1 + a_2 - 3}{a_1 - 1},$$

where the inequalities are the inductive hypothesis; so

$$\begin{aligned}
R^*(2, 2; a_1, a_2) &\le 1 + R^*(2, 1; A_1, A_2) \\
&= 1 + (A_1 + A_2 - 1) \\
&= A_1 + A_2 \\
&\le \binom{a_1 + a_2 - 3}{a_1 - 2} + \binom{a_1 + a_2 - 3}{a_1 - 1} \\
&= \binom{a_1 + a_2 - 2}{a_1 - 1},
\end{aligned}$$

where the second line comes from the Pigeonhole Principle (the case $k = 1$ of Ramsey's Theorem) and the last is the standard binomial coefficient identity $\binom{n}{k-1} + \binom{n}{k} = \binom{n+1}{k}$.

(10.4.2) Corollary. $R(2,2,l) \leq \binom{2l-2}{l-1}$.

PROOF. $R(2,2,l) = R^*(2,2;l,l)$ by definition.

The right-hand side here is less than $2^{2l-2} = 4^{l-1}$, since the sum of all binomial coefficients $\binom{2l-2}{i}$ is equal to 2^{2l-2}. Moreover, it is larger than $4^{l-1}/(2l-1)$, since there are $2l-1$ of these binomial coefficients, and the middle one $\binom{2l-2}{l-1}$ is the largest. So the upper bound grows exponentially with constant 4. We conclude this section by proving a lower bound for this Ramsey number, which is also exponential, but with the smaller constant $\sqrt{2}$. (The true order of magnitude is not known.) The proof uses an important combinatorial technique known as the *Probabilistic Method*.

(10.4.3) Proposition. $R(2,2,l) \geq 2^{(l-2)/2}$ for $l \geq 3$.

PROOF. Let X be a set of n points; the size of n will be specified later. We consider all possible colourings of the 2-subsets of X with two colours (red and blue, say). Since there are $\binom{n}{2} = n(n-1)/2$ pairs, there are $2^{n(n-1)/2}$ such colourings.

How many of these colourings contain a monochromatic l-subset? There are $\binom{n}{l}$ choices of an l-set L. For each choice, L is monochromatic in a proportion $2/2^{l(l-1)/2} = 2^{1-l(l-1)/2}$ of all the colourings; for, of the $2^{l(l-1)/2}$ ways in which the colours could fall on the 2-subsets of L, only two are monochromatic. So the number of colourings which contain a monochromatic l-set does not exceed a fraction $\binom{n}{l}2^{1-l(l-1)/2}$ of the total. (The number could in principle be calculated exactly, using the Principle of Inclusion and Exclusion; but this bound is good enough.)

Now suppose that $n = \lfloor 2^{(l-2)/2} \rfloor$. Then

$$\binom{n}{l}2^{1-l(l-1)/2} < n^l 2^{-l(l-2)/2}$$

$$\leq 1,$$

the first inequality holding since $\binom{n}{l} < n^l$ and $1 - l(l-1)/2 < -l(l-2)/2$, and the second by definition of n. In other words, the proportion of colourings having a monochromatic l-set is strictly less than 1. This means that there exists some colouring which has no monochromatic l-set. Hence $R(2,2,l) > n = \lfloor 2^{(l-2)/2} \rfloor$, whence $R(2,2,l) > 2^{(l-2)/2}$, as required.

The argument can be re-phrased as follows. Instead of considering the *set* of all colourings, and calculating the *proportion* that have a monochromatic n-set, we can instead speak of the *probability* that a *random* colouring has a monochromatic l-set. This probability p is bounded by the *expected number* of monochromatic l-sets in a random colouring, which is equal to $\binom{n}{l}2^{1-l(l-1)/2}$ (the number of l-sets times the probability that a given l-set is monochromatic). No mention of inclusion and exclusion is required. It is this interpretation which led to the term 'probabilistic method' for this type of argument.

In more detail:

Colour at random the set of all 2-subsets of the given n-set X, where each set has probability $1/2$ of being red and $1/2$ of being blue, with decisions about different sets independent. Now consider any l-set Y. It has $\binom{l}{2} = l(l-1)/2$ subsets of size 2. The probability that all are red is $2^{-l(l-1)/2}$, with the same probability that all are blue; so the probability that Y is monochromatic is twice this number, or $2^{1-l(l-1)/2}$.

The expected number of monochromatic l-sets is equal to this probability multiplied by the total number of l-subsets, hence $\binom{n}{l}2^{1-l(l-1)/2}$.

If n and l are such that this expected value is less than one, then it cannot occur that there is at least one monochromatic set in every colouring; hence there exists a colouring containing no monochromatic l-set.

However the argument is phrased, note that it is a non-constructive existence proof: it shows that there must be a way of doing the colouring so that no monochromatic l-sets are created, but it gives us absolutely no indication of how to find one (except, possibly, choosing the colouring at random and trusting to luck). It is generally regarded as 'better' to have an explicit construction of an object, in such a way that it is possible to verify directly that it has the required properties, than to have only an existence proof.

10.5. Applications

Here are some applications of Ramsey's Theorem. In the first case, there is a beautiful direct argument giving the exact bound.

(10.5.1) Proposition. *There is a function $f(m,n)$ with the following property:*

If $x_1, x_2, \ldots, x_N$ is any sequence of distinct real numbers with $N > f(m,n)$, then there is either a monotonic increasing sequence of length greater than m, or a monotone decreasing sequence of length greater than n.

Here is the proof using Ramsey's Theorem. We take $f(n,m) = R^*(2,2; m + 1, n+1) - 1$. Suppose that $N > f(m,n)$, and we are given a sequence of N distinct real numbers. Take $X = \{1, \ldots, N\}$, and colour the 2-subsets of X as follows: given a 2-set $\{i,j\}$, with $i < j$, colour it red if $x_i < x_j$, blue if $x_i > x_j$. Since $|X| \geq R^*(2,2; m+1, n+1)$, there is either a red $(m+1)$-set or a blue $(n+1)$-set. But a red set indexes a monotone increasing subsequence; for if $n_1 < n_2 < \ldots$ and all edges are red, then $x_{n_1} < x_{n_2} < \ldots$. Similarly a blue set indexes a decreasing subsequence.

Now here is the elegant direct proof, due to Erdős and Szekeres. We take the function $f(m,n)$ to be simply mn. So suppose that we have a sequence of $mn + 1$ distinct real numbers, and suppose that it contains no monotone increasing sequence of length $m + 1$ or greater. For $i = 1, \ldots, m$, let

$K_i = \{k :$ the longest monotone increasing sequence ending at x_k has length $i\}$.

Now we have partitioned the set $\{1, 2, \ldots, mn + 1\}$ into m subsets $K_1, \ldots, K_m$. By the Pigeonhole Principle, some one of these sets, say K_i, contains at least $n + 1$ members.

Now we claim that K_i indexes a monotone decreasing subsequence. For suppose that $k, l \in K_i$ with $k < l$ and $x_k < x_l$. Now, by definition of K_i, there is a monotone increasing sequence of length i ending at k, say $x_{j_1} < x_{j_2} < \ldots < x_k$. But then

$$x_{j_1} < x_{j_2} < \ldots < x_k < x_l$$

is a monotone increasing sequence of length $i + 1$ ending at x_l, contradicting the fact that $l \in K_i$. This claim establishes the result.

The bound $f(m,n) = mn$ is best possible. For consider the mn numbers

$$n - 1, 2n - 1, \ldots, mn - 1, n - 2, 2n - 2, \ldots, mn - 2, \ldots, 0, n, \ldots, (m-1)n.$$

It is not hard to check that the longest increasing subsequence has length m, and the longest decreasing subsequence has length n.

Another application is due to Erdős and Szekeres. A set of points in the Euclidean plane is *convex* if it contains the line segment joining any two of its points. The *convex hull* of a set S of points is the smallest convex set containing S. It can also be described as the set of linear combinations of points (x, y) in S, where the coefficients in the linear combination are restricted to being non-negative and having sum 1. A *convex polygon* is a finite set of points, none of which lies in the convex hull of the others. Another description is that each of the points lies on a line with the property that all the other points are on the same side of the line.

(10.5.2) Proposition. *There is a function f such that, given any $f(n)$ points in the plane with no three collinear, some set of n of the points form a convex polygon.*

PROOF. We need two preliminary facts:

FACT 1. Given any five points in the plane, no three collinear, some four of the points form a convex quadrilateral.[4]

This is clear if the convex hull of the points is a pentagon or quadrilateral. So suppose that it is a triangle, with vertices A, B, C, and let D and E be the remaining points. Then the line DE meets two sides of the triangle, say AB and AC; and the quadrilateral $BCDE$ (or $BCED$) is convex.

FACT 2. Given a set of n points in the plane, if every four points form a convex quadrilateral, then all n points form a convex polygon.

The proof is an exercise.

Now let $f(n) = R^*(2, 4; 5, n)$. Given $f(n)$ points in the plane, colour a 4-set red if it is a convex quadrilateral, blue otherwise. By Fact 1, there is no blue 5-set. So there is a red n-set; and, by Fact 2, it is a convex polygon with n points.

The exact value of the function $f(n)$ is unknown for n of moderate size.

---

[4] This special case of (10.5.2), due to Esther Klein, was the inspiration for the general result, which involved an independent discovery of Ramsey's Theorem by Erdős and Szekeres. See the comments by Szekeres in the introduction to the volume of selected papers by Paul Erdős, *The Art of Counting* (1973).

10.6. The infinite version

As our very last item, we mention without proof the infinite version of Ramsey's Theorem. As usual, the prototype is the Pigeonhole Principle:

> If the elements of an infinite set are coloured with finitely many colours, then there is an infinite monochromatic subset.

Ramsey's theorem generalises to colourings of the k-subsets of an infinite set with finitely many colours:

(10.6.1) Ramsey's Theorem (infinite form)

Let X be an infinite set, and k and r positive integers. Suppose that the k subsets of X are coloured with r colours. Then there is an infinite subset Y of X, all of whose k-subsets have the same colour.

We will discuss this result, and various extensions of it, in Section 19.4.

A remarkable discovery in logic is that it is possible to deduce the finite form of Ramsey's theorem from the infinite, but not *vice versa*. This fact has important consequences, notably a variant of the finite Ramsey theorem (the 'Paris–Harrington Theorem') which is true but not provable from the axioms for the natural numbers (essentially because the 'Paris–Harrington numbers' grow so fast that they are not provably computable). But we cannot follow this any further.

10.7. Exercises

1. A platoon of soldiers (all of different heights) is in rectangular formation on a parade ground. The sergeant rearranges the soldiers in each row of the rectangle in decreasing order of height. He then rearranges the soldiers in each column in decreasing order of height. Using the Pigeonhole Principle, or otherwise, prove that it is not necessary to rearrange the rows again; that is, the rows are still in decreasing order of height.

2. Show that any finite graph contains two vertices lying on the same number of edges.

3. (a) Show that, given five points in the plane with no three collinear, the number of convex quadrilaterals formed by these points is odd.
    (b) Prove Fact 2 in the proof of (10.5.2).

4. Show that, if $N > mnp$, then any sequence of N real numbers must contain either a strictly increasing subsequence with length greater than m, a strictly decreasing subsequence with length greater than n, or a constant subsequence of length greater than p. Show also that this result is best possible.

5. (a) Show that any infinite sequence of real numbers contains an infinite subsequence which is either constant or strictly monotonic.

(b) Using the Principle of the Supremum,[5] prove that every increasing sequence of real numbers which is bounded above is convergent.

(c) Hence prove the *Bolzano–Weierstrass Theorem*: Every bounded sequence of real numbers has a convergent subsequence.

6. (a) Let X be the set of residues modulo 17. Colour the 2-element subsets of X by assigning to $\{x, y\}$ the colour red if

$$x - y \equiv \pm 1, \pm 2, \pm 4 \text{ or } \pm 8 \pmod{17},$$

blue otherwise. Show that there is no monochromatic 4-set. [HINT: By symmetry, we may assume that the 4-set contains 0 and 1; this greatly reduces the number of cases to be considered!]

(b) Find a colouring of the 2-subsets of an 8-set red and blue so that there is no red 3-set and no blue 4-set.

(c) Let X consist of all subsets of $\{0, 1, 2, 3, 4\}$ of even cardinality. Colour the 2-subsets $\{x, y\}$ of X as follows:
• red if $x \triangle y = \{i, j\}$ and $i - j \equiv \pm 1 \pmod 5$;
• blue if $x \triangle y = \{i, j\}$ and $i - j \equiv \pm 2 \pmod 5$;
• green if $|x \triangle y| = 4$.
Show that there is no monochromatic 3-set.

7. (a) Prove the following theorem of Schur:

Schur's Theorem

There is a function f on the natural numbers with the property that, if the numbers $\{1, 2, \ldots, f(n)\}$ are partitioned into n classes, then there are two numbers x and y such that x, y and $x + y$ all belong to the same class.

(In other words, the numbers $\{1, 2, \ldots, f(n)\}$ cannot be partitioned into n 'sum-free sets'.)

[HINT: Colour the 2-subsets of $\{1, 2, \ldots, N + 1\}$ with n colours, according to the rule that $\{x, y\}$ has the ith colour if $|x - y|$ belongs to the ith class (where N is some suitable, sufficiently large, integer).]

(b) State and prove an infinite version of Schur's Theorem.

8. A *delta-system* is a family of sets whose pairwise intersections are all equal. (So, for example, a family of pairwise disjoint sets is a delta-system.) Prove the existence of a function f of two variables such that any family $\mathcal{F}$ of at least $f(n, k)$ sets of cardinality n contains k sets forming a delta-system.

[HINT: Construct a sequence of sets $A_1, A_2, \ldots$ in $\mathcal{F}$, and a sequence $\mathcal{F}_1, \mathcal{F}_2, \ldots$ of subfamilies, such that

---

[5] The Principle of the Supremum is the basic principle expressing the completeness of the real number system. It asserts that, if a non-empty set of real numbers has an upper bound, then it has a *supremum* or least upper bound.

- $\mathcal{F}_i \supseteq \mathcal{F}_{i+1}$ for all i;
- $A_i \cap A = A_i \cap A'$ for all $A, A' \in \mathcal{F}_i$;
- $A_j \in \mathcal{F}_i$ for all $j > i$.

Show that

- the sequence can be continued for m terms if $\mathcal{F}$ is sufficiently large (in terms of m and n);
- if the sequence continues for $(k-1)(n+1)+1$ terms, then some k of the sets A_i form a delta-system.]

State and prove an infinite version of this theorem.

Do you regard this theorem as part of 'Ramsey theory'?

9. Why are constructive existence proofs more satisfactory than non-constructive ones?

11. Graphs

Only connect!

E. M. Forster, *Howards End* (1910)

TOPICS: Graph properties related to paths and cycles, especially trees, Eulerian and Hamiltonian graphs; networks, Max-Flow Min-Cut and related theorems; [Moore graphs]

TECHNIQUES: Algorithmic proofs; approximate solutions; [Eigenvalue techniques]

ALGORITHMS: Graph algorithms; greedy algorithm; stepwise improvement; Eulerian and Hamiltonian circuits; approximate algorithms

CROSS-REFERENCES: Trees (Chapters 3, 4); Hall's Marriage Theorem (Chapter 6); [de Bruijn–Erdős theorem (Chapter 7)]

We have met graphs several times before, in various guises. Now, we return to them, and consider them more systematically. Graphs describe the connectedness of systems; typically, they model transport or communication systems, electrical networks, etc. In this chapter, we concentrate on issues related to this aspect. In Part 2, we return to graphs and look at colouring problems.

Graph theory is a cuckoo in the combinatorial nest;[1] it has grown to the status of an independent discipline, though still closely linked with other parts of combinatorics.

11.1. Definitions

In Section 2.5, we introduced a *graph* as a set V of *vertices* equipped with a set E of 2-subsets of V called *edges*. Sometimes it is necessary to broaden the definition.[2] In particular, we may want to allow *loops*, which are edges joining vertices to themselves; *multiple edges*, more than one edge between the same pair of vertices; and *directed edges*, which have an orientation so that they go *from* one vertex *to*

---

[1] This comment is not a disparagement. Graph theory has been successful because it provides mathematicians with a large supply of interesting problems, many of them related to applications.

[2] Where necessary to avoid confusion, the structure just defined is called a *simple graph*.

another.[3] The exact details of the formal mathematical machinery needed to define all these concepts is not too important; just note that directed edges are easily represented as ordered pairs rather than 2-subsets of vertices. A graph with some or all of these extended features is called a *general graph*; in particular, if it has directed edges, it is a *directed graph* or *digraph*, and if it has multiple edges, it is a *multigraph*.

Most of these concepts can be expressed in the language of relations introduced in Section 3.8. Since knowing a graph involves knowing which pairs of vertices are adjacent, we can regard a graph as a binary 'adjacency' relation on the vertex set. For a simple graph, adjacency is irreflexive and symmetric; relaxing these two conditions allows loops and directed edges respectively. However, multiple edges cannot easily be described in this language.

For the most part, we consider only undirected graphs without loops; but we sometimes need to allow multiple edges. The exception is Section 11.9; a network is most naturally based on a directed graph.

In a simple graph, we say that vertices x and y are *adjacent* if $\{x, y\}$ is an edge; they are *non-adjacent* otherwise.

We write $G = (V, E)$ for a graph G with vertex set V and edge set E.

Two simple but important kinds of graphs are *complete graphs*, in which every pair of vertices is an edge; and *null graphs*, having no edges at all. The complete and null graphs on n vertices are denoted by K_n and N_n respectively. Other important graphs will appear from time to time.

A *subgraph* of a graph $G = (V, E)$ is a graph whose vertex and edge sets are subsets of those of G. Note that, if $G' = (V', E')$ is a subgraph of G, then for every edge $e \in E'$, it must hold that both the vertices of e lie in V'.

Two kinds of subgraphs are of particular importance. An *induced subgraph* of G is a subgraph $G' = (V', E')$ whose edge set consists of *all* the edges of G which have both ends in V'. A *spanning subgraph* is one whose vertex set is the same as that of G. Thus, for example, every graph with at most n vertices is a subgraph of K_n, and every graph with exactly n vertices is a spanning subgraph; but the only induced subgraphs of K_n are complete graphs.

An induced subgraph is specified by giving its vertex set V'; we speak of the subgraph *induced on the set* V'.

Now we have to consider various kinds of routes in graphs. There are several different terms to be defined here; the differences are not very important, as you will see. My terminology is slightly different from the standard.

A *walk* in a graph is a sequence

$$(v_0, e_1, v_1, e_2, v_2, \ldots, e_n, v_n),$$

where e_i is the edge $\{v_{i-1}, v_i\}$ for $i = 1, \ldots, n$. We say that it is a walk *from v_0 to v_n*. The *length* of the walk is the number n of edges in the sequence (or one less than

---

[3] Directed edges could arise in modelling traffic flow in a town with some one-way streets, for example.

than the number of vertices). We allow $n = 0$ here. The walk is *closed* if $n > 0$ and $v_n = v_0$. Note that there are no restrictions; when walking, we may retrace our steps arbitrarily.

In a simple graph, the edges in a walk are uniquely determined by the vertices; so we often speak of the walk $(v_0, v_1, \ldots, v_n)$, defined by the condition that v_{i-1} and v_i are adjacent for $i = 1, \ldots, n$.

We define special kinds of walks: treks, trails, and paths. A *trek* is a walk in which any two consecutive edges are distinct;[4] if it is closed, we also require that the first and last edge are distinct. Thus, a trek is a bit more purposeful than a walk: we never retrace the edge we have just used. The last condition ensures that, in a closed trek, we can start at any point and describe a closed trek.

A *trail* is a walk with all its edges distinct; a *path* is a walk with all its vertices distinct (except perhaps the first and the last). The idea is that a trail might be followed by an explorer, who is not interested in revisiting an edge he has once explored; while a path proceeds efficiently from one place to another without any repetition. Further, we define a *circuit* to be a closed path.

Note that these concepts get progressively stronger; a path is a trail is a trek.[5] However, from the point of view of connections, there is no essential difference:

(11.1.1) Proposition. *(a) For any distinct vertices x, y of a graph G, the conditions that there exists a walk, trek, trail or path from x to y are all equivalent.*

(b) For any graph G, the conditions that G contains a closed trek, trail or path are all equivalent.

PROOF. Given a walk from x to y, if it is not a trek, then some two consecutive edges are repeated, so that there is a subsequence (v, e, v', e, v). Replacing this by the single vertex v gives a shorter walk. The process terminates in a trek from x to y.

Now a trek with a repeated edge must have a repeated vertex; so it suffices to show that, if there is a trek from x to y (with possibly $x = y$), then there is a path. If the vertex v is repeated (but not as the first and last vertex), there is a subsequence $(v, \ldots, v)$, which can be replaced by a single v to obtain a shorter trek. Continuing this process produces a path. Note that a closed trek cannot be reduced to the trek of length zero by this process.

Now define a relation $\equiv$ on the vertex set V by the rule: $x \equiv y$ if there is a path (or trail, or trek, or walk) from x to y. We have:

$\equiv$ *is an equivalence relation on V.*

This is straightforward: there is a walk of length 0 from x to x; reversing a walk from x to y gives a walk from y to x; and following a walk from x to y with a walk from y to z gives a walk from x to z. (Note that the proof would be untidier if we used one of the more special types of walk.)

---

[4] A trek with s edges is called an s-arc in the graph-theoretic literature; but this does not convey the sense of being intermediate in purposiveness between an walk and a trail, and also could be confused with the use of 'arc' for an edge of a directed graph.

[5] Mnemonic: a term later in the dictionary describes a wider concept.

This equivalence relation, of course, defines a partition of the vertex set of G. We define the *connected components* (or, for short, the *components*) of G to be the subgraphs induced on the equivalence classes. Note that no edge joins points in different equivalence classes; so the edge set of G is partitioned into the edge sets of its components.

A graph is *connected* if it has just one component, and *disconnected* otherwise. Note that any connected component of G is indeed a connected graph.

The *valency*, or *degree*, of a vertex x of a graph G is the number of edges containing x.[6] In a directed graph, we have to distinguish between the *out-valency* of a vertex (the number of directed edges starting at that vertex) and the *in-valency* (the number of edges ending there).

If every vertex of a graph has the same valency, the graph is called *regular*, and the common valency d is the *valency* of the graph. We call such a graph d-*valent*, and use the terms *divalent, trivalent*, etc. when $d = 2, 3$, etc.

Often we will modify a graph G by removing a vertex v and all edges containing it, or by removing an edge e, or by adding an edge e joining two vertices not previously joined. We use the shorthand notations $G - v$, $G - e$, $G + e$ for the results of these operations. (The strictly correct set-theoretic notation would be much more cumbersome, and would depend on the precise kind of graph in question.)

Sometimes our graphs will carry additional, numerical information: an edge may represent a pipeline, for example, and be labelled with its capacity, or the cost of building it. Formally, a *weight function* on a set X is a function from X to the non-negative real numbers. A *vertex-weighted*, resp. *edge-weighted*, graph is a graph with a weight function on the set of vertices, resp. edges. Edge-weighted graphs are more common, but we allow either or both types of weight function.

11.2. Trees and forests

A *tree* is a connected graph without circuits. We have met trees before, in Section 3.10 (where we proved Cayley's Theorem, that there are n^{n-2} labelled trees on n vertices) and Section 4.7 (binary trees, in connection with searching and sorting).

We might expect that a connected graph has 'many' edges, and a graph without circuits has 'few'. The next result shows that trees are extremal for both these properties. We need one piece of notation: a graph without circuits is called a *forest* — its connected components are trees!

---

[6] Both terms are commonly used. I prefer the first. The term 'degree' is over-used in mathematics, and there is no analogy between the degree of a graph and the degree of a polynomial, permutation group, etc. On the other hand, anyone who has studied chemistry will recognise the same concept. In the methane molecule CH_4, the carbon atom has valency 4 and the hydrogen atoms have valency 1. The standard representation

of the methane molecule shows the analogy clearly.

(11.2.1) Theorem. (a) *A connected graph with n vertices has at least $n-1$ edges, with equality if and only if it is a tree.*

(b) *A forest with n vertices and m connected components has $n-m$ edges. Thus, a forest has at most $n-1$ edges, with equality if and only if it is a tree.*

PROOF. We show first that a tree has $n-1$ edges. This is proved by induction; it is clear for $n=1$. The inductive step depends on the following fact:

A tree with more than one vertex has a vertex of valency 1.

Since a tree is connected, it has no isolated vertices (if $n>1$); so, arguing by contradiction, we can assume that every vertex has valency at least 2. But then there are arbitrarily long treks in the graph, since whenever we enter a vertex along one edge, we may leave along another. A trek of length greater than n must return to a vertex it has visited previously; so there is a closed trek, and hence a circuit, and we have arrived at a contradiction. So the assertion is proved.

Now let x be a vertex in the tree T which has valency 1. Let $T-v$ denote the graph obtained by removing v and the unique edge incident with it. Then $T-v$ has $n-1$ vertices, and contains no circuits. We claim that $T-v$ is connected. This holds because a path in T between two vertices $x,y \neq v$ cannot pass through v. Thus $T-v$ is a tree. By the induction hypothesis, it has $n-2$ edges; so T has $n-1$ edges.

Now (b) of the theorem follows easily. For let F be a forest with n vertices and m components $T_1,\ldots,T_m$, with $a_1,\ldots,a_m$ vertices respectively. Then $\sum_{i=1}^{m} a_i = n$. Now T_i is a tree, and so has $a_i - 1$ edges. So F has

$$\sum_{i=1}^{m}(a_i - 1) = n - m$$

edges.

To prove (a), let G be any connected graph with n vertices, and suppose that G is not a tree. Then G contains a circuit C. Let e be an edge in this circuit, and $G_1 = G - e$ the graph obtained by removing e. Then G_1 is still connected. For, if a path from x to y uses the edge e, then there is a walk from x to y not using e. (Instead of using e, we traverse the circuit the other way.) Repeating this procedure, we must reach a tree after, say, r steps. Since r edges are removed, G has $n-1+r$ edges altogether.

Let G be a graph. A *spanning forest* is a spanning subgraph of G (consisting of all the vertices and some of the edges of G) which happens to be a forest. A *spanning tree* is similarly defined.

(11.2.2) Corollary. *Any connected graph has a spanning tree.*

This follows from the argument for part (a) of the theorem above; by removing edges from G, we can obtain a spanning tree. There is another way to proceed, which will be useful later; this involves building up the spanning tree 'from below'.

(11.2.3) Spanning tree algorithm

Let $G = (V, E)$ be a connected graph.

Set $S = \emptyset$.

WHILE the graph (V, S) is not connected, let e be an edge of G joining vertices in different components, and add e to the set S.

RETURN (V, S).

To prove that this algorithm works, we have to show that the choice of e is always possible and its addition creates no circuit. Let Y be a connected component of (V, S), and $Z = V \setminus Y$; choose vertices y, z in Y, Z respectively. In G, there is a path from y to z; some edge in this path must cross from Y to Z, and this is a suitable choice for e. Now suppose that $(V, S) + e$ contains a circuit. If we start, say, in Y, and follow this circuit, at some moment we cross into Z by using the edge e; then there is no way to return to Y to complete the circuit without re-using e.

We see that there is a great deal of freedom in creating spanning trees. How many are there? Cayley's Theorem (Section 3.10) can be stated in the form:

(11.2.4) Cayley's Theorem. *The complete graph K_n has n^{n-2} spanning trees.*

For, obviously, any tree on the vertex set $\{1, \ldots, n\}$ is a spanning tree of the complete graph.

There is a general technique for counting the spanning trees in an arbitrary graph, using the adjacency matrix of the graph. This is described in the chapter on graph spectra in Beineke and Wilson, *Selected Topics in Graph Theory* (1977).

11.3. Minimal spanning trees

Suppose that n towns are to be linked by a telecommunication network. For each pair of towns, the cost of installing a cable between these two towns is known. What is the most economical way of connecting all the towns?

This is known as the *minimal connector* problem. The data can be regarded as an edge-weighted graph. (As described, the graph G in question is the complete graph; but this is not essential. We could suppose that, for various reasons, it is impossible or uneconomic to connect certain pairs of towns directly.)

The solution to the problem will be that connected spanning subgraph H of the graph G of minimal total weight (that is, the sum of the weights of the edges of H is as small as possible). Assume that all weights are positive. Then H must be a tree; for, if not, then edges could be deleted, reducing the weight, without disconnecting it. The problem is solved by a simple-minded algorithm called the *greedy algorithm*. This says: at each stage, build the cheapest link which joins two towns not already connected by a path. Formally:

(11.3.1) Greedy algorithm for minimal connector

Let $G = (V, E)$ be a connected graph, w a non-negative weight function on E.

Set $S = \emptyset$.

WHILE (V, S) is not connected, choose the edge e of minimal weight subject to joining vertices in different components.

RETURN (V, S).

This algorithm is just a specialisation of the spanning tree algorithm in the last section; so it does indeed produce a spanning tree. We have to show that this spanning tree has minimum weight.

Let $e_1, e_2, \ldots, e_{n-1}$ be the edges in S, in the order in which the Greedy Algorithm chooses them. Note that

$$w(e_1) \leq \ldots \leq w(e_{n-1}),$$

since if $w(e_j) < w(e_i)$ for $j > i$, then at the ith stage, e_j would join points in different components, and should have been chosen in preference to e_i.

Suppose, for a contradiction, that there is a spanning tree of smaller weight, with edges $f_1, \ldots, f_{n-1}$, ordered so that

$$w(f_1) \leq \ldots \leq w(f_{n-1}).$$

Thus,

$$\sum_{i=1}^{n-1} w(f_i) < \sum_{i=1}^{n-1} w(e_i).$$

Choose k as small as possible so that

$$\sum_{i=1}^{k} w(f_i) < \sum_{i=1}^{k} w(e_i).$$

Note that $k > 1$, since the greedy algorithm chooses first an edge of smallest weight. Then we have

$$\sum_{i=1}^{k-1} w(f_i) \geq \sum_{i=1}^{k-1} w(e_i);$$

hence

$$w(f_1) \leq \ldots \leq w(f_k) < w(e_k).$$

Now, at stage k, the greedy algorithm chooses e_k, and not any of the edges $f_1, \ldots, f_k$ of strictly smaller weight; so all of these edges must fail the condition that they join points in different components of (V, S), where $S = \{e_1, \ldots, e_{k-1}\}$. It follows that the connected components of (V, S'), where $S' = \{f_1, \ldots, f_k\}$, are subsets of those of (V, S); so (V, S') has at least as many components as (V, S).

But this is a contradiction, since both (V, S) and (V, S') are forests, and their numbers of components are $n - (k - 1)$ and $n - k$ respectively; it is false that $n - k \geq n - (k - 1)$.

In general, the *greedy algorithm* refers to any algorithm for constructing an object in stages, where at each stage we make the choice which locally optimises some 'objective function', subject to the condition that we move closer to our final goal. Obviously, this short-sighted local optimisation does not usually produce the best overall solution. It is quite remarkable that it does so in this case! (See Exercise 3.)

11.4. Eulerian graphs

One's first encounter with graph theory often takes the form of the familiar puzzle 'trace this figure without taking your pencil off the paper'. Euler's[7] experience was similar. He showed that it was not possible to walk round the town of Königsberg[8] crossing each of its seven bridges just once. This demonstration is commonly taken as the starting point of graph theory.[9]

In problems of this sort, we are required to traverse every edge of a graph once, but we may revisit a vertex. So the appropriate type of route is a trail (see Section 11.1). We define an *Eulerian trail* in a graph to be a trail which includes every edge. (A closed Eulerian trail is sometimes called an *Eulerian circuit*, but this conflicts with our definition of a circuit as a closed path.) Clearly an *isolated* vertex (lying on no edges) has no effect, and may be deleted. Also, it is convenient here to work in the more general class of *multigraphs*, where two vertices may be joined by more than one edge. Now Euler's result can be stated thus:

(11.4.1) Euler's Theorem. *(a) A multigraph with no isolated vertices has a closed Eulerian trail if and only if it is connected and every vertex has even valency.*

(b) A multigraph with no isolated vertices has a non-closed Eulerian trail if and only if it is connected and has exactly two vertices of odd valency.

PROOF. It's obvious that a graph with an Eulerian trail must be connected if no vertex is isolated. The other conditions are also necessary. For consider a graph with a closed Eulerian trail. As we follow the circuit, each time we reach a vertex by an edge, we must leave it by a different edge, using up two of the edges through that vertex; since every edge is used, the valency must be even. The same applies at the initial vertex of a closed Eulerian trail, since the first and last edge of the circuit play the same rôle. For a non-closed Eulerian trail, however, the valencies of the first and last vertices are odd, since the first and last edges are 'unpaired'.

REMARK. According to the Handshaking Lemma (Chapter 2), the number of vertices of odd valency in a graph is even. So, if there is a vertex of odd valency, then there are at least two.

---

[7] Euler could be claimed as the founder of combinatorics. He was not the first person to work on a combinatorial problem; but he is undoubtedly the mathematician of greatest stature who has made a serious contribution to the subject. We saw his encounter with orthogonal Latin squares in Chapter 8, and we will meet him again.

[8] Now Kaliningrad.

[9] See, for example, Biggs, Lloyd and Wilson, *Graph Theory 1736–1936* (1976).

Now we turn to the sufficiency of the conditions: we have to construct Eulerian trails in graphs satisfying them. The argument is, in some sense, algorithmic.

So let $G = (V, E)$ be a graph satisfying the condition of either (a) or (b). In case (a), let v be any vertex; in (b), let v be one of the vertices of odd valency. Now follow a trail from v, never re-using an edge, for as long as possible. Let S be the set of edges in this trail.

For any vertex x other than v (in case (a)) or the other vertex of odd valency (in case (b)), whenever the trail reaches x, there are an odd number of edges through x not yet used. This is because we reached x along an edge, and previous visits accounted for an even number of edges (except for v, where previous visits accounted for an odd number of edges). Thus, we don't get stuck at x; zero is not odd, so there is an edge by which we can leave. So the trail must end at v (in case (a)) or the other vertex of odd valency (in case (b)).

If $S = E$, we have constructed an Eulerian trail, and we are finished. So suppose not. There must be a vertex u lying on both an edge in S and an edge not in S. For otherwise, the sets X and Y of vertices lying on edges in S, not in S respectively, form a partition of V; and no edge joins vertices in different parts, contradicting connectedness.

Moreover, in the graph $(V, E \setminus S)$, every vertex has even valency. So, starting at u and using only edges of $E \setminus S$, we can find a closed trail, by the same argument as before. Now we can 'splice in' this trail to produce a longer one: start at v and follow the old trail to u; then traverse the new trail; then continue along the old trail.

After a finite number of applications of this construction, we must arrive at an Eulerian trail of the type desired.

Note that, in case (b), any Eulerian trail must start at one vertex of odd valency and finish at the other — a fact well known to anyone who has tried a 'trace without lifting the pencil' puzzle.

The map of Königsberg is easily converted into a multigraph whose edges are the bridges, as shown in Fig. 11.1. All four vertices have odd valency; so there is no Eulerian trail.

Fig. 11.1. The bridges of Königsberg

11.5. Hamiltonian graphs

There is a natural analogue for vertices of an Eulerian trail: a *Hamiltonian path*[10] is a path which passes once through each vertex (except that it may be closed, that is,

---

[10] Hamilton's claim to give his name to this concept is much weaker than Euler's claim to Eulerian trails. Hamilton demonstrated that the graph formed by the twenty vertices and thirty edges of a dodecahedron possesses a Hamiltonian circuit, and patented a puzzle based on this; but he proved

its start and finish may be the same). A closed Hamiltonian path is a *Hamiltonian circuit*. A graph possessing such a circuit is called *Hamiltonian*.

Clearly multiple edges are irrelevant here; so we may assume that our graphs are simple in this section.

For $n > 2$, there is a unique graph on n vertices which is connected and divalent (regular with valency 2). This is the so-called *n-cycle C_n*. It can be represented as the vertices and edges of an n-gon. Now a graph is Hamiltonian if and only if it has a cycle as a spanning subgraph.

Hamiltonian graphs are much harder to deal with than Eulerian ones. There is no simple necessary and sufficient condition for a graph to be Hamiltonian, and it is notoriously difficult to decide this question for a given graph of even moderate size. A lot of effort has gone into proving *sufficient* conditions. As an example, we prove one of the simplest of these conditions, *Ore's Theorem*.

(11.5.1) Ore's Theorem. *Let G be a graph with n vertices, and suppose that, for any two non-adjacent vertices x and y in G, the sum of their valencies is at least n. Then G is Hamiltonian.*

PROOF. By contrast to Euler's Theorem, this proof is non-constructive. However, it can be recast as an algorithm (see Exercise 15).

Arguing by contradiction, we suppose that G is a graph which satisfies the hypothesis of Ore's Theorem but is not Hamiltonian. We also may suppose that G is *maximal* with these properties, so that the addition of any edge to G produces a Hamiltonian graph. (We achieve this by adding new edges joining previously non-adjacent vertices as long as G remains non-Hamiltonian. Adding an edge does not decrease the valency of any vertex, and does not create any new non-adjacent pair of vertices, so the valency condition remains true. But we won't know when G is maximal unless we can test all the graphs obtained by adding an extra edge and show that they are Hamiltonian!)

Now G is certainly not complete, so it has a non-adjacent pair of vertices x and y. Since G is maximal non-Hamiltonian, the graph obtained by adding the edge $e = \{x, y\}$ is Hamiltonian; and a Hamiltonian circuit in this graph must contain e. So G itself contains a Hamiltonian path $(x = v_1, e_2, v_2, \ldots, v_n = y)$.

Now let A be the set of vertices adjacent to x; let $B = \{v_i : v_{i-1}$ is adjacent to $y\}$. (Since y is not adjacent to $v_n = y$, this set is well-defined.) By assumption, $|A| + |B| \geq n$. But the vertex $v_1 = x$ doesn't belong to either A or B; so $|A \cup B| \leq n - 1$. It follows that $|A \cap B| \geq 1$, and so there is a vertex v_i lying in both A and B.

Now we obtain a contradiction by constructing a Hamiltonian circuit in G. Starting at $x = v_1$, we follow the path $v_2, v_3, \ldots$ as far as v_{i-1}. Now v_{i-1} is adjacent to y (since $v_i \in B$), so we go to $y = v_n$ at the next step. Then we follow the path backwards through $v_{n-1}, \ldots$ as far as v_i, and then home to x (this edge exists because $v_i \in A$).

---

no general result, and there is some evidence that he got the idea from Kirkman, who made the same observation at about the same time. Also, a problem involving a Hamiltonian circuit in a different graph, the 'knight's tour' on the chessboard, had been solved earlier by (of all people) Euler.

The result is best possible in some sense. Consider the graph with $2m + 1$ vertices $x_1, \ldots, x_m, y_1, \ldots, y_{m+1}$, and having as edges all pairs $\{x_i, y_j\}$. (This is a *complete bipartite* graph.) It is not Hamiltonian; for any edge crosses between the sets $A = \{a_1, \ldots, a_m\}$ and $B = \{b_1, \ldots, b_m\}$, and so a path of odd length starting in A must finish in B and cannot be closed. But two non-adjacent vertices are both in A or both in B, and the sum of their valencies is $2m + 2 = n + 1$ or $2m = n - 1$ respectively.

Nevertheless, there are a great many results which strengthen Ore's Theorem by varying the hypotheses slightly.

11.6. Project: Gray codes

An analog-to-digital converter is a device that takes a continuously-varying real number and converts it to an integer, ideally the integer part or the nearest integer. The result is presented in the standard way, usually to base 10 (in an odometer or gas meter) or base 2 (in an electronic device connected directly to a computer).

We considered the operation of an odometer in Chapters 2 and 4. There are points in its operation where several digits must change simultaneously. Owing to mechanical limitations, the change is not quite simultaneous. Thus, a reading taken at this point may involve a considerable error. For example, in the course of changing from 36999 to 37000, the reading could be as low as 36000 or as high as 37999; and even if we assume that the digits change sequentially from the right, the low value 36000 is still a possible reading.

To eliminate this error, we need to arrange the numbers in order (different from the usual order) so that only one digit changes at each step. If this can be done, the only possible error will arise from a time delay in the mechanical operation of the device, and will be at most 1. In the case of binary representation, such a sequence is known as a *Gray code*. It has a natural graph-theoretic interpretation.

The n-cube Q_n is the graph whose vertices are all n-tuples $x_{n-1} \ldots x_0$ of zeros and ones, two vertices being adjacent if they agree in all but one position. (Note that there are 2^n vertices, which we write as the binary representations of the integers from 0 to $2^n - 1$. The n-cube consists of the vertices and edges of the familiar regular polytope of the same name in $\mathbb{R}^n$.) Now a Gray code is the same thing as a Hamiltonian path in the n-cube. For $n = 1$, the graph Q_1 is a single edge, and trivially has a Hamiltonian path. But for $n \geq 2$, we can do better:

(11.6.1) Theorem. *For $n \geq 2$, the graph Q_n is Hamiltonian.*

The proof is by induction. For $n = 2$, Q_2 is the 4-cycle C_4, and the assertion is true: we fix the Hamiltonian circuit $(00, 01, 11, 10, 00)$. Suppose that Q_n has a Hamiltonian circuit $(v_0, \ldots, v_{2^n-1})$. Then

$$(0v_0, 0v_1, \ldots, 0v_{2^n-2}, 0v_{2^n-1}, 1v_{2^n-1}, 1v_{2^n-2}, \ldots, 1v_1, 1v_0, 0v_0)$$

is a Hamiltonian circuit in Q_{n+1}.

The case $n = 3$ is shown in Fig. 11.2.

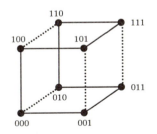

Fig. 11.2. Hamiltonian circuit in the 3-cube

For practical use, it is necessary to be able to 'encode' and 'decode' this code. That is, we have to be able to calculate the Nth term of the Gray code, and the position of (the binary representation of) N in the code. (Note that these statements make sense independent of the value of n as long as $2^n > N$; for the Gray code of length 2^{n+1} begins with the Gray code of length 2^n with its terms preceded by 0, which doesn't change the integers they represent.) The key observation is that, if 2^k is the exact power of 2 dividing N, then the digit which changes at the Nth step is the kth digit from the right. (This is easily proved by induction: in our construction, the nth digit changes only at the $(2^n)^{\text{th}}$ step.) At the same step in the odometer, the $0^{\text{th}}, \ldots, k^{\text{th}}$ digits all change. From this observation, it is not difficult to prove the following assertions; the reader is encouraged to supply proofs:

> Let $x_{n-1} \ldots x_0$ be the binary representation of N. For $i = 0, \ldots, n-1$, let $y_i = x_i + x_{i+1} \pmod 2$; that is, $y_i = 0$ if $x_i = x_{i+1}$, $y_i = 1$ otherwise. (By convention, $x_{n+1} = 0$.) Let $y_{n-1} \ldots y_0$ be the binary representation of M. Then the number in the Nth position in the Gray code is M.

> Let $y_{n-1} \ldots y_0$ be the binary representation of M. For $i = 0, \ldots, n-1$, let x_i be the number of ones in the set $\{y_i, \ldots, y_{n-1}\}$, taken mod 2, and let $x_{n-1} \ldots x_0$ be the binary representation of N. Then the number M occurs in the Nth position in the Gray code.

11.7. The Travelling Salesman

A salesman for the Acme Widget Corporation[11] has to visit all n cities in a country on business. The distance between each pair of cities is known. She wants to minimise the total distance travelled, and return to her starting point.

This is the notorious Travelling Salesman Problem (TSP). In graph-theoretic terminology, it asks for the Hamiltonian circuit of smallest weight in an edge-weighted complete graph.

In fact, there is no real loss in restricting to the complete graph. For a general edge-weighted graph, simply add new edges with ridiculously large weights, so that these cannot occur in any minimum-weight circuit.

Indeed, the Hamiltonian circuit problem for a given graph G is a special case of the Travelling Salesman problem. If G has n vertices, assign the weight 1 to an edge of the complete graph K_n if it is an edge of G, and 2 otherwise. Then the minimum weight of a Hamiltonian circuit of K_n is n if and only if G has a Hamiltonian circuit.

The existence of an algorithm to solve this problem is not in doubt.

(11.7.1) (Slow) Algorithm for Travelling Salesman
- *Generate all permutations of $\{1, \ldots, n\}$ (see Chapter 3).*
- *For each permutation π, calculate*

$$\sum_{i=1}^{n-1} w(\{i\pi, (i+1)\pi\}) + w(\{n\pi, 1\pi\}).$$

- *Return the smallest value.*

---

[11] Widgets are generic industrial products in Operational Research problems.

The disadvantage is that there are $n!$ permutations; for even moderate values of n (say, $n = 50$), this number is so large that the method cannot be contemplated.

Some small improvements can be made. For example, we can assume that the circuit starts at vertex 1, so that $1\pi = 1$; this saves a factor of n. Unfortunately, nobody knows how to do substantially better!

Because of the practical importance of the problem (not just for sales departments, but for other applications such as design of circuits), some compromises have been reached. Methods which deliver an approximate solution have been developed. Out of a huge literature, I have selected one example, chosen because it uses concepts we have already developed.

(11.7.2) Twice-round-the-Tree Algorithm
(An approximate solution to the Travelling Salesman)
- *Find a minimal connector S.*
- *In the multigraph obtained by duplicating each edge of S, find a closed Eulerian trail.*
- *Follow this trail, but whenever the next step would involve revisiting a vertex, go instead to the first unvisited vertex on the trail. When every vertex has been visited, return to the start.*

In the second step, every edge in S is duplicated, resulting in a connected graph with all valencies even; so there does indeed exist a closed Eulerian trail, and we have seen an algorithm for finding one.

It is clear that this algorithm produces a Hamiltonian circuit. How good is it?

We say that an edge-weighted complete graph satisfies the *triangle inequality* if, for any three vertices a, b, c, we have

$$w(\{a, b\}) + w(\{b, c\}) \geq w(\{a, c\}).$$

This condition certainly holds if the weights are distances between towns.[12]

(11.7.3) Theorem. *Let G be an edge-weighted complete graph, m the weight of a minimal connector, M the smallest weight of a Hamiltonian circuit. Then*
(a) $m \leq M$;
(b) if G satisfies the triangle inequality, then $M \leq 2m$.

PROOF. (a) is clear, since a Hamiltonian circuit is certainly connected. (Indeed, it remains connected when any edge is deleted, so its weight is at least the sum of m and the smallest edge weight in G. This can be further improved.)

For (b), note that the weight of the closed Eulerian trail in the second stage of the algorithm is equal to $2m$. Now, in the third stage, we take various short cuts,

---

[12] Or much more general distances. Under minimal assumptions, the shortest route from a to c cannot be longer than a route via b.

replacing a path $v_i, \ldots, v_j$ by a single edge from v_i to v_j. By the triangle inequality (and induction), the length of the edge doesn't exceed the length of the path. So we end up with a Hamiltonian circuit of weight at most $2m$, giving a constructive proof of the inequality.

Another celebrated problem bears the same relation to closed Eulerian trails as the Travelling Salesman does to Hamiltonian circuits. This is the *Chinese postman* problem: Given an edge-weighted connected graph, find the closed walk of minimum weight which uses every edge of the graph. (The postman must pass along every street delivering letters.) If the graph G is Eulerian, then a closed Eulerian trail is the solution. If not, then some edges must be traversed more than once. There is an efficient algorithm for this problem.

11.8. Digraphs

The most important variant of graphs consists of *directed graphs* or *digraphs*, where the edges are ordered pairs of vertices (rather than unordered pairs). Each edge (x, y) has an *initial vertex* x and a *terminal vertex* y. Note that (x, y) and (y, x) are different edges.

With any digraph D is associated an ordinary (undirected) graph, the *underlying graph*: it has the same vertex set as D, and its edges are those of D without the order (that is, $\{x, y\}$ for each edge (x, y) of D). The underlying graph will fail to be simple if D contains two oppositely-directed edges (such as (x, y) and (y, x)). If the underlying graph is simple, then D is called an *oriented graph*.

The definitions of the various types of route in a digraph are the same as in a graph, with the important exception that the edges must be traversed in the correct direction: so, if

$$(v_0, e_1, v_1, \ldots, e_n, v_n)$$

is a trek, trail, or path, then e_i is the edge (v_{i-1}, v_i) for $i = 1, \ldots, n$. In a digraph, we cannot immediately retrace an edge, and so every walk is a trek. (11.1.1) holds without modification for digraphs.

The situation with connected components is different, however. If, as before, we let R be the relation defined by the rule that $(x, y) \in R$ if there is a path (or trail, or trek) from x to y, then the relation R is reflexive and transitive, but not necessarily symmetric; so it is a partial preorder but not necessarily an equivalence relation. (See Sections 3.8–9 and Exercise 18 of Chapter 3 for partial preorders and their connection with equivalence relations.) Accordingly, we define two types of connectedness:

- the digraph D is *(weakly) connected* if its underlying graph is connected;
- D is *strongly connected* if, for any two vertices x, y, there is a path from x to y.

It's clear that a strongly connected digraph is weakly connected. The converse is false.[13]

The definitions of Eulerian trail and circuit and of Hamiltonian path and circuit are just what you expect. A digraph possessing a Hamiltonian circuit is obviously

---

[13] It is possible to imagine a town with one-way streets in which you can drive from x to y but not from y to x (but very impracticable!)

strongly connected; one with a Hamiltonian path is weakly (but not necessarily strongly) connected. Similar statements hold for Eulerian trails and circuits.

The analogue of Euler's theorem runs as follows:

(11.8.1) Euler's Theorem for digraphs. *A digraph with no isolated vertices has a closed Eulerian trail if and only if it is weakly connected and the in-valency and out-valency of any vertex are equal.*

You are invited to prove this, and to formulate and prove a necessary and sufficient condition for the existence of a non-closed Eulerian trail.

11.9. Networks

A *network* is an edge-weighted digraph possessing two distinguished vertices, the *source* s and the *target* t, with $s \neq t$. The weight of an edge e is referred to as its *capacity*, and denoted by $c(e)$.

A good model to keep in mind is a hydraulic network consisting of pipes and junctions. Fluid is pumped in at the source and out at the target; the capacity of an edge reflects the maximum rate of flow possible in that pipe. Of course, much wider interpretations are also possible, such as the movement of commercial product through distribution systems between factories, warehouses, etc.

In accordance with this interpretation, we define a *flow* in a network to be a function f from the edge set to the non-negative real numbers, satisfying the two properties

- $0 \leq f(e) \leq c(e)$ for all edges e;
- $\sum_{\iota(e)=v} f(e) = \sum_{\tau(e)=v} f(e)$ for all vertices $v \neq s, t$.

Here c is the capacity; s and t the source and target; and, for any edge e, $\iota(e)$ and $\tau(e)$ denote the initial and terminal vertices of e. Thus, the first condition asserts that the flow in any edge is non-negative and doesn't exceed the capacity of the edge; the second asserts that, for any vertex v other than the source and target, the flow out of v is equal to the flow into v, so there is no net accumulation at any point.

The *value* of a flow f is defined to be

$$\mathrm{val}(f) = \sum_{\iota(e)=s} f(e) - \sum_{\tau(e)=s} f(e),$$

the net flow out of the source. It turns out to be equal to the net flow into the target, as the next result shows. For any set S of vertices, we use the notation

$$S^{\rightarrow} = \{e : \iota(e) \in S, \tau(e) \notin S\},$$
$$S^{\leftarrow} = \{e : \iota(e) \notin S, \tau(e) \in S\}.$$

(11.9.1) Proposition. *Let f be a flow in a network, S a set of vertices containing the source but not the target. Then*

$$\sum_{e \in S^{\rightarrow}} f(e) - \sum_{e \in S^{\leftarrow}} f(e) = \mathrm{val}(f).$$

PROOF. To show this, we calculate

$$\sum_{v \in S} \left(\sum_{\iota(e)=v} f(e) - \sum_{\tau(e)=v} f(e) \right).$$

On one hand, this is equal to val(f), since the term of the outer sum with $v = s$ is equal to val(f), while the other terms are all zero by definition of a flow.

On the other hand, consider this as a sum over edges. Let $e = (v, w)$ be an edge. If $v \in S$, then $f(e)$ occurs in the term of the outer sum corresponding to v; if $w \in S$, then $-f(e)$ occurs in the term corresponding to w. Thus, only those edges with exactly one end in S, viz., those in $S^\rightarrow$ and $S^\leftarrow$, contribute to the sum, and their contributions are $f(e)$ and $-f(e)$ respectively.

Now take $S = V \setminus \{t\}$, where V is the vertex set; then $S^\rightarrow = \{e : \tau(e) = t\}$ and $S^\leftarrow = \{e : \iota(e) = t\}$, and so the net flow *into* t is equal to val(f).

The main question about a network is:

What is the maximum value of a flow in the network?

A *cut* in a network is a set C of edges with the property that any path from the source to the target contains an edge in C. Its *capacity* cap(C) is the sum of the capacities of its edges. Intuitively, it is clear that the capacity of a cut is an upper bound for the value of any flow. We will show this and more:

(11.9.2) Max-Flow Min-Cut Theorem
The maximum value of a flow in a network is equal to the minimum capacity of a cut.

This important theorem has a number of consequences, including Hall's Marriage Theorem and Menger's Theorem on paths in graphs. Our proof of the Max-Flow Min-Cut Theorem is in part algorithmic. More precisely, the proof is algorithmic in the case when all the capacities are integers, and we prove something more:

(11.9.3) Integrity Theorem
Suppose that the capacity of every edge in a network is an integer. Then there is a flow of maximum value, such that the flow in every edge is an integer.

The integral case is the important one; we'll see that the general case can be deduced from it by quite different (and non-constructive) methods.

Our first task is to prove:

$$\begin{pmatrix} \text{the value} \\ \text{of any flow} \end{pmatrix} \;\leq\; \begin{pmatrix} \text{the capacity} \\ \text{of any cut} \end{pmatrix}.$$

It is enough to prove this for minimal cuts (those for which, if any edge is removed, the result is not a cut). So let C be a minimal cut. Define S to be the set of vertices v for which there exists a path from s to v using no edge of C. Then $C = S^{\rightarrow}$. (If e is any edge in C then, by minimality, there is a path from s to t using the edge e and no other edge of C; so $\iota(e) \in S$ and $\tau(e) \notin S$.) Now, if f is any flow, then

$$\mathrm{val}(f) = \sum_{e \in S^{\rightarrow}} f(e) - \sum_{e \in S^{\leftarrow}} f(e)$$

$$\leq \sum_{e \in S^{\rightarrow}} c(e)$$

$$= \mathrm{cap}(C).$$

Now we treat the case where all capacities are integers. We prove the following:

If all capacities of edges in a network are integers, then there is a flow f, with integer values on all edges, and a cut C, such that

$$\mathrm{val}(f) = \mathrm{cap}(C).$$

By what we just proved, f is a maximal flow and C a minimal cut; so the Max-Flow Min-Cut Theorem (in this case) and the Integrity Theorem will be proved.

The proof involves showing the following.

Let all capacities of edges in a network be integers, and let f be an integer-valued flow. Then either
- *there is an integer-valued flow f' with $\mathrm{val}(f') = \mathrm{val}(f) + 1$; or*
- *there is a cut C with $\mathrm{cap}(C) = \mathrm{val}(f)$.*

Now we can start with any flow, and apply this result successively. As long as the first alternative holds, the value of the flow is increased. So eventually the second alternative becomes true, and we have finished. In order to prove the theorem, we can start with the zero flow (the zero function is always a flow!); but in practice it is usually possible to spot a starting flow which is close to maximal, and shorten the calculation. The proof of the assertion is algorithmic.

So let f be an integer-valued flow in a network with integer capacities. Perform the following algorithm.

> Set $S = \{s\}$.
> WHILE there is an edge $e = (v, w)$ with either
> - $f(e) < c(e)$, $v \in S$, $w \notin S$, or
> - $f(e) > 0$, $v \notin S$, $w \in S$,
>   add to S the vertex of e it doesn't already contain (w or v respectively).
> RETURN S.

Now there are two cases, according as $t \in S$ or not.

CASE 1. $t \in S$. By construction of S, it follows that there is a path from s to t in the underlying graph, say $(v_0 = s, v_1, \dots, v_d = t)$, such that, for each i, either
(a) (v_{i-1}, v_i) is an edge e of the network with $f(e) < c(e)$; or
(b) (v_i, v_{i-1}) is an edge e of the network with $f(e) > 0$.

Let A and B be the sets of edges of the digraph appearing under cases (a) and (b) respectively. Now define a new flow f' by the rule

$$f'(e) = \begin{cases} f(e) + 1 & \text{if } e \in A; \\ f(e) - 1 & \text{if } e \in B; \\ f(e) & \text{otherwise.} \end{cases}$$

We have to show that this is a flow, and that its value is one more than that of f. The first axiom for a flow, that $0 \le f'(e) \le c(e)$ for all e, holds because all capacities and flow values are integers, so (for example) if $f(e) < c(e)$, then $f(e) + 1 \le c(e)$. The second axiom requires some case checking. Let v_i be a vertex on the path (no vertex off the path is affected); suppose that $i \ne 0, d$. If (v_{i-1}, v_i) and (v_i, v_{i+1}) are both edges, then the net flow into v_i and the net flow out of v_i are both increased by 1, and the flows still balance. The other cases are similar. Also similar is the fact that the value of the flow is increased by 1.

CASE 2. $t \notin S$. Then $S^{\rightarrow}$ is a cut. Also, by definition of S, if $e \in S^{\rightarrow}$, then $f(e) = c(e)$, and if $f \in S^{\leftarrow}$, then $f(e) = 0$ (else the algorithm would enlarge S). So

$$\mathrm{val}(f) = \sum_{e \in S^{\rightarrow}} f(e) - \sum_{e \in S^{\leftarrow}} f(e)$$

$$= \sum_{e \in S^{\rightarrow}} c(e)$$

$$= \mathrm{cap}(S^{\rightarrow}),$$

as required.

This completes the proof in the integral case.

The rest of the proof of the Max-Flow Min-Cut Theorem is quite different (and of less interest). It parallels the construction of the real numbers from the integers: first we construct the rational numbers by division, and then we construct the reals by an analytic process (typically Cauchy sequences or Dedekind cuts).

So suppose first that all capacities are rational. By multiplying by the highest common factor m of the denominators of these rationals, we obtain a new network in which all capacities are integers. By the previous case, the Max-Flow Min-Cut Theorem holds for the new network; hence it also holds for the old one (on dividing the flow values by the same number m).

Finally, suppose that the capacities are real numbers. We can approximate them arbitrarily closely from below by rational numbers, and hence find flows whose values are arbitrarily close to the capacity of a minimal cut. Then the result follows by a limiting process.[14]

This concludes the proof of the Max-Flow Min-Cut Theorem.

---

[14] There is an additional subtlety here. We construct a sequence of flows whose values converge to $\mathrm{cap}(C)$, where C is a minimal cut. Now the flows can be regarded as points in a Euclidean space whose dimension is equal to the number of edges. Moreover, they lie in closed and bounded region

11.10. Menger, König and Hall

The Max-Flow Min-Cut Theorem, in combination with the Integrity Theorem, is a very powerful tool in graph theory. The key to its application is to consider an arbitrary directed graph with distinguished vertices s and t as a network in which each edge has capacity 1. Now, in an integer-valued flow in this network, the flow in any edge must be 0 or 1; so the flow 'picks out' a subset of the edges, those carrying a flow of 1. Now, if the value of the flow is m, then there are m edge-disjoint paths from s to t. (This is proved by induction on m. Starting from s and using only edges with positive flow, never re-using an edge, we eventually arrive at t, having constructed a trek from s to t. Deleting circuits between repeated vertices, we obtain a path from s to t. Now, if we reduce the flow in the edges of this path to 0, the value of the flow is decreased by 1. By induction, we can find $m - 1$ edge-disjoint paths among the remaining edges. So the claim is proved.)

This conclusion can be put in the following form, where an st-separating set of edges is a set C such that every path from s to t uses an edge of C:

(11.10.1) **Menger's Theorem.** *Let s and t be vertices of a digraph D. Then the maximum number of pairwise edge-disjoint paths from s to t is equal to the minimum number of edges in an st-separating set.*

Menger's Theorem also has a version for undirected graphs, and versions which refer to vertices instead of edges. You can read about these in Beineke and Wilson, *Selected Topics in Graph Theory*.

Further results involve more specific digraphs. A very important class of digraphs are those derived from bipartite graphs.

A graph $G = (V, E)$ is *bipartite* if there is a partition of the vertex set into two parts A and B such that every edge has one end in A and the other end in B. The partition $\{A, B\}$ is called a *bipartition* of G.

Given a bipartite graph G with bipartition $\{A, B\}$, we construct a network as follows. The vertex set is $\{s, t\} \cup A \cup B$, where the source s and target t are new vertices. The edges are

- all pairs (s, a) for $a \in A$;
- all pairs (b, t) for $b \in B$;
- all pairs (a, b), with $a \in A$, $b \in B$, for which $\{a, b\}$ is an edge of G.

I will call this network $N(G)$.

In order to interpret flows and cuts in $N(G)$, we need two definitions. In a graph G, a *matching* is a set of pairwise disjoint edges; and an *edge-cover* is a set S of vertices with the property that every edge contains a vertex of S.

Now a path from s to t in $N(G)$ has the form (s, a, b, t), where $a \in A$, $b \in B$, and $\{a, b\}$ is an edge of G. So a set of edge-disjoint paths (s, a_i, b_i, t) in $N(G)$ arises from a matching in G consisting of the edges $\{a_i, b_i\}$, $i = 1, \ldots, m$.

---

of the space. Such a region is compact; so, by the Bolzano–Weierstrass Theorem, the sequence of flows has a convergent subsequence. The limit of this subsequence is a flow whose value is equal to $\mathrm{cap}(C)$. See Chapter 10, Exercise 5, for the 1-dimensional Bolzano–Weierstrass Theorem; the general case is proved coordinatewise.

An edge-cover S in G gives rise to a cut in $N(G)$, consisting of the edges (s, a) for $a \in S \cap A$, together with the edges (b, t) for $b \in S \cap B$. (Any path from s to t must use an edge of G, and hence pass through a vertex of S, since S is an edge-cover.) Now there may be other cuts, containing some edges of the form (a, b); but *none of these can be smaller than all those of the first type*. For let S be an arbitrary cut. Replace every edge (a, b) in S ($a \in A$, $b \in B$) with the edge (s, a), deleting repetitions; the result is a cut containing edges of the form (s, a) and (b, t) only. We conclude that

> The size of the smallest cut in $N(G)$ is equal to the size of the smallest edge-cover in G.

Hence we conclude:

(11.10.2) König's Theorem. *The maximum size of a matching in the bipartite graph G is equal to the minimum size of an edge-cover in G.*

Finally, we will show that Hall's Marriage Theorem is a consequence of König's Theorem.[15]

In order to do this, we have to translate a family of sets into a bipartite graph. This is a common and important procedure.

Let $\mathcal{F} = (A_1, \ldots, A_n)$ be a family of subsets of $\{1, \ldots, m\}$. We define the *incidence graph* G of $\mathcal{F}$ as follows. The vertex set V of G is the union of two parts $A = \{1, \ldots, m\}$ and $B = \{A_1, \ldots, A_n\}$; and the vertices $i \in A$ and $A_j \in B$ are joined if and only if $i \in A_j$.

The incidence graph is clearly bipartite; the sets A and B used in its definition form a bipartition. If the dual rôle played by the vertices (which are also sets or elements of sets) is confusing, you may take A to be a set in one-to-one correspondence with $\{1, \ldots, m\}$, and B a set in one-to-one correspondence with $\mathcal{F}$.

Now a matching in the incidence graph G is a set of disjoint edges $\{i, A_j\}$; thus, each point i lies in its corresponding set A_j, and the points are all distinct, as are the sets. This is just a system of distinct representatives for a subfamily of $\mathcal{F}$. So $\mathcal{F}$ has an SDR if and only if there is a matching whose edges contain all vertices of B.

Recall that, for $J \subseteq \{1, \ldots, n\}$, we set $A(J) = \bigcup_{j \in J} A_j$. We say that $\mathcal{F}$ satisfies *Hall's Condition* if $|A(J)| \geq |J|$ for all $J \subseteq \{1, \ldots, n\}$.

(11.10.3) Hall's Marriage Theorem. *The family $(A_1, \ldots, A_n)$ possesses a SDR if and only if it satisfies Hall's Condition.*

PROOF. As in Chapter 8, the necessity of the condition is clear: if a SDR exists, then $A(J)$ must contain representatives of all the sets A_j for $j \in J$, and so must have size at least as great as J. So suppose that Hall's Condition is satisfied. Let G be the incidence graph of the family. We have to show that there is a matching of size n in G. By König's Theorem, we must show that any edge-cover in G has size at least n.

---

[15] In fact, König's Theorem was proved before Hall's, but this implication was not noticed until afterwards. (Hall was a group theorist, König a graph theorist.)

The set of all vertices in B is an edge-cover of G of size n. Let S be any edge-cover, and let $J = B \setminus S$. Each vertex in $A(J)$ is joined to a vertex of J by an edge; so the edge-cover S must contain $A(J)$. Thus

$$|S| \geq |B| - |J| + |A(J)| \geq n,$$

by Hall's Condition.

REMARK. We have, in some sense, given a constructive proof of Hall's Theorem. Given a family $\mathcal{F}$ of sets satisfying Hall's Condition, construct its incidence graph G, and the network $N(G)$. Use the algorithm of the last section to find a maximum flow in $N(G)$. Then the edges from A from B carrying non-zero flow define the required SDR.

The network algorithm can be translated into more graph-theoretic language for this purpose. A formulation of the algorithm for König's Theorem is given in Exercise 10.

11.11. Diameter and girth

We know what it means for a graph to be connected. How do we decide in practice? Here is an algorithm which computes the connected component of a graph G containing a vertex x (and more besides, as we will see).

(11.11.1) Algorithm: Component containing x

Mark x with the integer 0. Set $d = 0$.

WHILE *any vertex was marked at the preceding stage,*
- *look at all vertices marked d; mark all unmarked neighbours of such vertices with $d + 1$;*
- *replace d by $d + 1$.*

At the termination of this algorithm, the marked vertices comprise the connected component containing x, and the mark of each vertex is the length of the shortest path from x to that vertex.

In a connected graph G, the *distance* $d(x, y)$ from x to y is the length of the shortest path from x to y. (Sometimes, distance is defined in a general graph, so that the distance between two vertices in different components is ∞; we ignore this complication.) The algorithm above gives a method for computing the distance between two vertices of a graph.

The distance function satisfies
- $d(x, y) \geq 0$, and $d(x, y) = 0$ if and only if $x = y$;
- $d(x, y) = d(y, x)$;
- $d(x, y) + d(y, z) \geq d(x, z)$.

The first two properties are clear. For the third, note that there is a walk of length $d(x, y) + d(y, z)$ from x to z via y; this can be converted into a path by removing repetitions in the usual way, so the length of the shortest path cannot be greater than this.

The third condition is the *triangle inequality*, which we met already in Section 11.7. If you have studied introductory topology, you will recognise the three properties as the axioms for a *metric*. So, in this language, a connected graph, equipped with its distance function, is a metric space.

The *diameter* of a connected graph G is the maximum value attained by the distance function.

The number of vertices of a graph is bounded in terms of the diameter and the maximum valency of a vertex:

(11.11.2) Theorem. *In a connected graph with diameter d and maximum valency k, the number of vertices is at most*

$$1 + k + k(k-1) + k(k-1)^2 + \ldots + k(k-1)^{d-1} = 1 + k\frac{(k-1)^d - 1}{k-2}.$$

PROOF. We show by induction that there are at most $k(k-1)^{i-1}$ vertices at distance i from a given vertex x, for $i \geq 1$. This is clear for $i = 1$. For the inductive step, we double-count pairs (y, z), where y and z are adjacent and lie at distances i and $i+1$ from x respectively. There are at most $k(k-1)^{i-1}$ choices for y; each is joined to one point at distance $i-1$ from x (lying on a shortest path from x to y), and so for given y there are at most $k-1$ choices for z. On the other hand, for each z, there is at least one y (again on a shortest path to x); so there are at most $k(k-1)^i$ such z.

Now the result is obtained by summation.

In the next section, we examine graphs meeting this bound. First, however, we prove a 'dual' result.

The *girth* of a graph G is the length of the shortest closed path in G. Thus, forests don't have a girth (or we could say the girth of a forest is infinite). Alternatively, the girth is the smallest $n \geq 3$ for which the graph contains the n-cycle C_n as an induced subgraph. (A closed path of length n is a subgraph isomorphic to C_n; if it is not an induced subgraph, then there must be an edge of G joining two non-consecutive vertices, in which case the circuit is cut into two shorter circuits.)

(11.11.3) Theorem. *Let G be a graph of girth g, and let $e = \lfloor (g-1)/2 \rfloor$. Suppose that the minimum valency of G is k. Then G has at least*

$$1 + k + k(k-1) + \ldots + k(k-1)^{e-1} = 1 + k\frac{(k-1)^e - 1}{k-2}$$

vertices.

PROOF. The argument is similar to the previous theorem: we show that, for $1 \leq i \leq e$, the number of vertices at distance i from x is at least $k(k-1)^{i-1}$. Again, the induction begins trivially. Now consider the double count.

For each y with $d(x, y) = i < e$, there is one neighbour of y at distance $i-1$ from x, and none at distance i from x. (Otherwise, we could start from x, trek to y, and return a different way, to create a closed trek of length $2i$ or $2i + 1$; so there would be a closed path of length at most $2i + 1$. Since $2i + 1 < g$, this is impossible.) Thus, at least $k - 1$ neighbours of y lie at distance $i + 1$ from y.

In the same way, given z with $d(x, z) = i + 1$, there can be only one neighbour y of z at distance i from x (since $2(i + 1) \leq 2e < g$ by assumption). So the induction goes through.

Close inspection of the argument shows the following:

Theorem. *Of the following conditions on a graph G, any two imply the third:*
- *G is connected with maximum valency k and diameter d;*
- *G has minimum valency k and girth $2d + 1$;*
- *G has $1 + k((k - 1)^d - 1)/(k - 2)$ vertices.*

A graph satisfying these three conditions is called a *Moore graph* of diameter d and valency k. (The first two conditions show that a Moore graph is regular.) In the next section, we examine Moore graphs of diameter 2.

It turns out that Moore graphs are very rare. So the next question is: how close to these bounds can we get (for general values of k and d, or asymptotically)? A lot of work has been done on this question, but the results will not be described here.

11.12. Project: Moore graphs

In this section, we decide (almost completely) for which values of k there exists a Moore graph of diameter 2 and valency k.

Let G be a graph with vertex set $\{1, 2, \ldots, n\}$. The *adjacency matrix* $A(G)$ of G is the $n \times n$ matrix whose (i, j) entry is equal to 1 if $\{i, j\}$ is an edge of G, 0 otherwise. It is a real symmetric matrix, and thus it can be diagonalised. The argument involves calculating the eigenvalues of $A(G)$ and their multiplicities.

Let G be a Moore graph of valency k and diameter 2. From the argument in the last section, we see that G has

$$n = 1 + k + k(k - 1) = k^2 + 1$$

vertices, and that G has girth 5. This means that, if x and y are adjacent, then no vertex is adjacent to both; and, if x and y are non-adjacent, they have exactly one common neighbour.

Let A be the adjacency matrix of G, and J the $n \times n$ matrix with every entry 1. If I is the $n \times n$ identity matrix, we claim that

$$A^2 = kI + (J - I - A).$$

To see this, we prove:

> For any graph G, the (i, j) entry of $A(G)^2$ is equal to the number of common neighbours of i and j.

For the (i, j) entry is

$$\sum_{h=1}^{n} (A)_{ih}(A)_{hj};$$

and every entry of A is zero or 1, so the sum counts the number of vertices h for which $(A)_{ih} = (A)_{hj} = 1$, that is, the number of vertices h joined to both i and j. This proves the assertion.

Now, in the case of our Moore graph G, the number of common neighbours of i and j is k if $i = j$ (since the valency is k), and is 0 if $\{i, j\}$ is an edge and 1 otherwise. This means that A^2 has diagonal entries k, and off diagonal entries 0 or 1 according as A has entries 1 or 0 (in other words, off the diagonal, it coincides with $J - I - A$). this proves the claim.

Now we examine the spectrum of A. Let $\mathbf{j}$ be the vector with all its entries 1. Then the ith entry of $A\mathbf{j}$ is just the row sum of the ith row of A, which is equal to k since G is regular with valency k. Thus, $A\mathbf{j} = k\mathbf{j}$, and $\mathbf{j}$ is an eigenvector of A with eigenvalue k.

Since A is symmetric, the subspace W of $\mathbb{R}^n$ consisting of vectors perpendicular to $\mathbf{j}$ is preserved by A. Also, for any $\mathbf{w} \in W$, the sum of the entries of $\mathbf{w}$ is 0, and so $J\mathbf{w} = 0$. Thus, for $\mathbf{w} \in W$, we have

$$A^2\mathbf{w} = k\mathbf{w} + (-I - A)\mathbf{w},$$

whence

$$(A^2 + A - (k-1)I)\mathbf{w} = 0.$$

Let α be any eigenvalue of A (acting on W). If $\mathbf{w}$ is the corresponding eigenvector, then the above equation shows that

$$\alpha^2 + \alpha - (k-1) = 0.$$

So α is a root of this quadratic equation, whence

$$\alpha = \frac{1}{2}\left(-1 \pm \sqrt{4k-3}\right).$$

Now we distinguish two cases.

CASE 1. $4k - 3$ is not a perfect square. Then the eigenvalues are irrational. So the multiplicity of the two roots of the quadratic, as eigenvalues of A, are equal, and so each is $(n-1)/2 = k^2/2$. Now we use the fact that the sum of the eigenvalues of a matrix is equal to its *trace* (the sum of the diagonal elements). A has the eigenvalue k with multiplicity 1, and $(-1 \pm \sqrt{4k-3})/2$ each with multiplicity $(n-1)/2$; and its trace is zero, since all its diagonal elements are zero. Thus, we have

$$k + \left(\frac{k^2}{2}\right) \cdot \left(\frac{-1 + \sqrt{4k-3}}{2}\right) + \left(\frac{k^2}{2}\right) \cdot \left(\frac{-1 - \sqrt{4k-3}}{2}\right) = 0,$$

from which we find that $k = k^2/2$, or $k = 2$.

Now there is a unique graph of valency 2, diameter 2, and girth 5: the 5-cycle or *pentagon*.

CASE 2. $4k - 3$ is a square. Since it is odd, so is its square root; say $4k - 3 = (2s+1)^2$ for some integer s, from which we find that $k = s^2 + s + 1$. The eigenvalues of A are k (with multiplicity 1), s, and $-s - 1$. The multiplicities of the last two eigenvalues are, say, f and g; we know that $f + g = n - 1 = k^2$. Since the trace of A is equal to 0, we also find that

$$k + fs + g(-s-1) = 0.$$

From these two linear equations, it is possible to solve for f and g. We find that

$$f = \frac{s(s^2 + s + 1)(s^2 + 2s + 2)}{2s + 1}.$$

Now the multiplicity of an eigenvalue of a matrix must be an integer; so we conclude that

$$2s + 1 \text{ divides } s(s^2 + s + 1)(s^2 + 2s + 2).$$

Multiplying this expression by 32 and doing some manipulation, we find that $2s + 1$ divides

$$((2s+1) - 1)((2s+1)^2 + 3)((2s+1)(2s+3) + 5).$$

From this, it is clear that $2s + 1$ divides 15, so that $2s + 1 = 1, 3, 5$ or 15. This gives the possible values

$s = 0, 1, 3$ or 7;

$k = 1, 3, 7$ or 57;

$n = 2, 10, 50$ or 3250.

The case $n = 2$ is spurious, since G would have a single edge and would not have diameter 2. So we conclude:

(11.12.1) Theorem. *If there is a Moore graph of diameter 2 and valency k, then $k = 2, 3, 7$ or 57, and the number of vertices is $5, 10, 50$ or 3250.*

For $k = 2$, we saw that the pentagon is the only graph. In a moment, we will construct the unique Moore graph of valency 3. There is also a unique Moore graph of valency 7, though this is harder to construct. Nobody knows whether one of valency 57 exists or not!

THE PETERSEN GRAPH.

Let G be a Moore graph of valency 3 and diameter 2, with 10 vertices. Let $\{a, b\}$ be an edge of G. Then each of a and b has two further neighbours, with no vertex joined to both. Let b, c, d be the neighbours of a, and a, e, f the neighbours of b. There are no edges within the set $\{c, d, e, f\}$, for any such edge would create a circuit of length 3 or 4.

Now c and e have a unique common neighbour, since they are not adjacent; let g be this neighbour. Similarly, let h be the common neighbour of c and f; i that of d and e; and j that of d and f. These vertices are all distinct and are joined to none of $a, \ldots, f$ except where specified. Now we have all vertices. The first six have three neighbours each, and the last four have two each (so far); so we need two more edges to complete the graph, with each of g, h, i, j on one edge. But g is not joined to h or i; so we have edges $\{g, j\}$ and $\{h, i\}$.

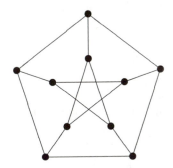

Fig. 11.3. Uniqueness of a Moore graph

This completes the unique Moore graph of diameter 2 and valency 3 (see Fig. 11.3). It can be drawn in other, more symmetrical ways (as in Fig. 11.4, for example).

Fig. 11.4. The Petersen graph

This graph is the notorious *Petersen graph*. Its fame stems from the fact that it is a counterexample to a large number of conjectures in graph theory. If you discover an assertion you believe to be true of all graphs, test it first on the Petersen graph! It is now the star of a book in its own right: Holton and Sheehan, *The Petersen Graph* (1993).

To complete the story of Moore graphs, here are the facts. As noted above, there is a unique Moore graph of diameter 2 and valency 7, the *Hoffman–Singleton graph*; the remaining case for diameter 2 is unknown. For larger diameter, Damerell, and Bannai and Ito, independently showed the following result.

(11.12.2) Theorem. *For $d \geq 3$, the only Moore graph of diameter d is the $(2d+1)$-cycle C_{2d+1} (with valency 2).*

11.13. Exercises

1. There are 34 non-isomorphic graphs on 5 vertices (compare Exercise 6 of Chapter 2). How many of these are (a) connected, (b) forests, (c) trees, (d) Eulerian, (e) Hamiltonian, (f) bipartite?

2. Show that the Petersen graph (Section 11.12) is not Hamiltonian, but does have a Hamiltonian path.

3. Show that the greedy algorithm does not always succeed in finding the path of least weight between two given vertices in a connected edge-weighted graph.

4. Consider the modification of the greedy algorithm for minimal connector. Choose the edge e for which $w(e)$ is minimal subject to the conditions that $S + e$ contains no cycle *and e shares a vertex with some previously chosen edge (unless $S = \emptyset$)*. Prove that the modified algorithm still correctly finds a minimal connector.

5. Let $G = (V, E)$ be a multigraph in which every vertex has even valency. Show that it is possible to direct the edges of G (that is, replace each unordered pair $\{x, y\}$ by the ordered pair (x, y) or (y, x)) so that the in-valency of any vertex is equal to its out-valency.

6. Let G be a graph on n vertices. Suppose that, for all non-adjacent pairs x, y of vertices, the sum of the valencies of x and y is at least $n - 1$. Prove that G is connected.

7. (a) Prove that a connected bipartite graph has a unique bipartition.
   (b) Prove that a graph G is bipartite if and only if every circuit in G has even length.

8. Choose ten towns in your country. Find from an atlas (or estimate) the distances between all pairs of towns. Then
   (a) find a minimal connector;
   (b) use the 'twice-round-the-tree' algorithm to find a route for the Travelling Salesman.
How does your route in (b) compare with the shortest possible route?

9. Consider the result of Chapter 6, Exercise 7, viz.

> Let $\mathcal{F} = (A_1, \ldots, A_n)$ be a family of sets having the property that $|A(J)| \geq |J| - d$ for all $J \subseteq \{1, \ldots, n\}$, where d is a fixed positive integer. Then there is a subfamily containing all but d of the sets of $\mathcal{F}$, which possesses a SDR.

Prove this by modifying the proof of Hall's Theorem from König's given in the text.

REMARK. This extension of Hall's Theorem is in fact 'equivalent' to König's theorem. Can you deduce König's Theorem from it?

10. König's Theorem is often stated as follows:

> *The minimum number of lines (rows or columns) which contain all the non-zero entries of a matrix A is equal to the maximum number of independent non-zero entries,*

where a set of matrix entries is *independent* if no two are in the same row or column. Show the equivalence of this form with the one given in the text. [HINT: if A is $m \times n$, let G be the bipartite graph with vertices $a_1, \ldots, a_m, b_1, \ldots, b_n$, in which $\{a_i, b_j\}$ is an edge whenever $(A)_{ij} \neq 0$. Show that sets of independent non-zero entries correspond to matchings in G, and sets of lines containing all non-zero entries correspond to edge-covers of G.]

11. In this exercise, we translate the 'stepwise improvement' algorithm in the proof of the Max-Flow Min-Cut Theorem into an algorithm for König's Theorem.

Let G be a bipartite graph with bipartition $\{A, B\}$. We observed in the text that an integer-valued flow f in $N(G)$ corresponds to a matching M in G, consisting of those edges $\{a, b\}$ for which the flow in (a, b) is equal to 1. Now consider the algorithm in the proof of the Max-Flow Min-Cut Theorem, which either increases the value of the flow by 1, or finds a cut. Suppose that we are in the first case, where there is a path

$$(s, a_1, b_1, a_2, b_2, \ldots, a_r, b_r, t)$$

in the underlying graph of $N(G)$ along which the flow can be increased. Then

> $(a_1, b_1, \ldots, a_r, b_r)$ *is a path in G, such that all the edges $\{b_i, a_{i+1}\}$ but none of the edges $\{a_i, b_i\}$ belong to M; moreover, no edge containing a_1 or b_r is in M.*

Such a path in G is called an *alternating path* with respect to M. (An alternating path starts and ends with an edge not in M, and edges not in and in M alternate. Moreover, since no edge of M contains a_1 or b_r, it cannot be extended to a longer such path.)

Show that, if we delete the edges $\{b_i, a_{i+1}\}$ from M $(i = 1, \ldots, r-1)$, and include the edges $\{a_i, b_i\}$ $(i = 1, \ldots, r)$, then a new matching M' with $|M'| = |M| + 1$ is obtained.

So the algorithm is:

> WHILE there is an alternating path, apply the above replacement to find a larger matching.
> When no alternating path exists, the matching is maximal.

12. Let G be a graph with adjacency matrix A. Prove that the (i, j) entry of A^d is equal to the number of walks of length d from i to j.

13. This exercise proves the 'friendship theorem': *in a finite society in which any two members have a unique common friend, there is somebody who is everyone else's friend.* In graph-theoretic terms, a graph on n vertices in which any two vertices have exactly one common neighbour, possesses a vertex of valency $n - 1$, and is a 'windmill' (Fig. 11.5).

Fig. 11.5. A windmill

STEP 1. Let the vertices be $1, \ldots, n$, and let A_i be the set of neighbours of i. Using the de Bruijn–Erdős Theorem (Chapter 7), or directly, show that *either* there is a vertex of valency $n - 1$, *or* all sets A_i have the same size (and the graph is regular). In the latter case, the sets A_i are the lines of a projective plane (Chapter 9).

STEP 2. Suppose that G is regular, with valency k. Use the eigenvalue technique of Section 11.11 to prove that $k = 2$.

14. The 'Trackwords' puzzle in the *Radio Times* consists of nine letters arranged in a 3×3 array. It is possible to form an English word from all nine letters, where consecutive letters are adjacent horizontally, vertically or diagonally. Consider the problem of setting the puzzle; more specifically, of deciding in how many ways a given word (with all its letters distinct) can be written into the array.

(a) Formulate the problem in graph-theoretic terminology.
(b) (COMPUTER PROJECT.) In how many ways can it be done?

15. PROJECT. The following algorithm, due to Peter Johnson, makes the proof of Ore's theorem (11.5.1) constructive. Verify the details.

Let G satisfy the hypotheses of Ore's theorem. Choose any circuit C on the vertex set of G. If $E(C) \subseteq E(G)$, we are done. Otherwise, let $d = |E(C) \setminus E(G)|$; we find a new circuit C' with $|E(C') \setminus E(G)| < d$. A finite number of such steps finds the required Hamiltonian circuit.

Take any edge $e \in E(C) \setminus E(G)$. The graph G' with edge set $E(G) \cup E(C) \setminus \{e\}$ satisfies Ore's condition and has a Hamiltonian path $E(C) \setminus \{e\}$. Now proceed as in the proof of (11.5.1) to find a Hamiltonian circuit C' in G'.

12. Posets, lattices and matroids

... good order and military discipline

Army regulations

TOPICS: Posets, lattices; distributive lattices; (propositional logic); chains and antichains; product and dimension; Möbius inversion; matroids; (Arrow's Theorem)

TECHNIQUES: Möbius inversion

ALGORITHMS: Calculating the Möbius function; minimum-weight basis

CROSS-REFERENCES: PIE (Chapter 5); Hall's Theorem (Chapter 6); q-binomial theorem (Chapter 9)

Order is fundamental to the process of measurement: representing objects by numbers presupposes that we can arrange them in order. Often, however, we have only enough information to decide the order of some pairs of elements; in this case, partial order may be a more relevant concept. In this chapter, we introduce some of the many themes of the theory of order.

12.1. Posets and lattices

First, we recall the definitions, from Chapter 3. A *partial order* on X is a relation R on X which is

- *reflexive*: $(x, x) \in R$ for all $x \in X$;
- *antisymmetric*: $(x, y), (y, x) \in R$ imply $x = y$; and
- *transitive*: $(x, y), (y, z) \in R$ imply $(x, z) \in R$.

(Thus, order models the relation 'less than or equal'. For the connection with 'less than', see Exercise 17 of Chapter 3.) As usual, we write $x \leq y$ for $(x, y) \in R$. The pair (X, R) is called a *partially ordered set*, or *poset* for short.

Here are some examples of posets. In each case, the point set is $\{1, \ldots, n\}$, for some n; we list some elements of R, and the rest follow by reflexivity and transitivity.

Two comparable points: $n = 2$, $1 < 2$ (so $R = \{(1, 1), (1, 2), (2, 2)\}$).

Two incomparable points: $n = 2$, $R = \{(1, 1), (2, 2)\}$.

The poset N: $n = 4$, $1 < 3$, $2 < 3$, $2 < 4$.

The pentagon: $n = 5$, $1 < 2 < 5$, $1 < 3 < 4 < 5$.

The three-point line: $n = 5$, $1 < 2 < 5$, $1 < 3 < 5$, $1 < 4 < 5$.

A convenient way of representing a poset is by its *Hasse diagram*. We say that y *covers* x if $x < y$ but no element z satisfies $x < z < y$. (In the list above, we gave all pairs $x < y$ for which y covers x.) Now the Hasse diagram of a poset P is a graph drawn in the Euclidean plane, such that each vertex corresponds to a point of the poset, and for each covering pair $x < y$, the points representing x and y are joined by an edge and the point representing x is 'below' the point representing y (in the sense that it has smaller Y-coordinate).

The figure below gives the Hasse diagrams of the five posets described above. Note that the Hasse diagram determines the entire poset: $u < v$ if and only if there is a path from u to v, every edge of which goes 'upward'.

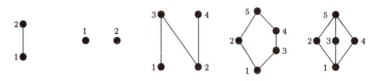

Fig. 12.1. Some Hasse diagrams

Two specialisations of posets are important. A *total order* is a partial order satisfying

• *trichotomy*: for any $x, y \in X$, $(x, y) \in R$ or $x = y$ or $(y, x) \in R$.

(With the definition here, the middle alternative $x = y$ is actually covered by the other two; but this would not have been so if we had used the 'strict' definition of partial order.) In any poset, we say that elements x and y are *comparable* if either $(x, y) \in R$ or $(y, x) \in R$. Thus, a total order is a partial order in which any two elements are comparable. A total order is sometimes called a *linear order*,[1] and a totally ordered set is called a *chain*.

A *maximal element* of a poset $(X, \leq)$ is an element x such that, if $x \leq y$, then $x = y$. (We do not require that $y \leq x$ for all x, so there may be more than one maximal element.) *Minimal elements* are defined dually.

(12.1.1) Lemma. *Any (non-empty) finite poset contains a maximal element.*

PROOF. Choose any $x_1 \in X$. If x_1 is not maximal, there exists $x_2 \in X$ with $x_1 < x_2$ (which means, of course, that $x_1 \leq x_2$ and $x_1 \neq x_2$). Continue this process, either until a maximal element is found, or we reach an element previously encountered. In fact, the second alternative cannot occur; for, if $i < j$, then

$$x_i < x_{i+1} < \ldots < x_{j-1} < x_j,$$

so $x_i = x_j$ is impossible. So eventually a maximal element will be found.

This argument obviously fails in infinite posets: there is no maximal integer, for example.

---

[1] This usage comes from geometry, where the points on a line in Euclidean space are linearly ordered, as opposed to the points of a line in projective space, which are circularly ordered.

In a poset, we say that z is a *lower bound* of x and y if $z \le x$ and $z \le y$. A *greatest lower bound* (g.l.b.) of x and y is a maximal element in the set of lower bounds. By (12.1.1), if two elements of a finite poset have a lower bound, they have a greatest lower bound; but it may not be unique. *Upper bounds* and *least upper bounds* (l.u.b.s) are defined similarly.

A *lattice* is a poset in which each pair of elements has a unique greatest lower bound and a unique least upper bound. (We are considering only finite lattices here.) A lattice has a unique minimal element 0, which satisfies $0 \le x$ for any element x. (For let 0 be any minimal element, and x any element. If z is the g.l.b. of 0 and x, then $z \le 0$, so $z = 0$ by minimality, whence $0 \le x$. If x happened to be a minimal element also, then $x \le 0$, whence $x = 0$ by antisymmetry.) Dually, a lattice has a unique maximal element 1, satisfying $x < 1$ for all x.[2]

We use the notation $x \wedge y$ and $x \vee y$ for the g.l.b. and l.u.b. of x and y in a lattice. These are also called the *meet* and *join* of x and y.

Any totally ordered set is a lattice: if $x \le y$, then $x \wedge y = x$ and $x \vee y = y$. Other examples of lattices include:

- The *power-set lattice* $\mathcal{P}(X)$, whose elements are the subsets of a set X, ordered by inclusion. It has $x \wedge y = x \cap y$ and $x \vee y = x \cup y$.
- The lattice $D(n)$ of (positive) divisors of the positive integer n, ordered by divisibility: $x \le y$ if x divides y. The g.l.b. and l.u.b. of x and y are their greatest common divisor (x, y) and least common multiple $xy/(x, y)$ respectively.
- The lattice of subspaces of a finite vector space $V = V(n, q)$, ordered by inclusion: this is the *projective geometry* $\mathrm{PG}(n, q)$, looked at in a different way. We have $x \wedge y = x \cap y$ and $x \vee y = \langle x, y \rangle = x + y$ (sum of subspaces!) respectively.

Following the nineteenth-century tendency towards abstraction and axiomatisation in mathematics, a lattice can be regarded as a set on which are defined two binary operations $\wedge$ and $\vee$ and two elements 0 and 1. The next result gives the axiomatisation of lattices from this point of view.

(12.1.2) Proposition. *Let X be a set, $\wedge$ and $\vee$ two binary operations defined on X, and 0 and 1 two elements of X. Then $(X, \wedge, \vee, 0, 1)$ is a lattice if and only if the following axioms are satisfied:*

- *Associative laws: $x \wedge (y \wedge z) = (x \wedge y) \wedge z$ and $x \vee (y \vee z) = (x \vee y) \vee z$;*
- *Commutative laws: $x \wedge y = y \wedge x$ and $x \vee y = y \vee x$;*
- *Idempotent laws: $x \wedge x = x \vee x = x$;*
- *$x \wedge (x \vee y) = x = x \vee (x \wedge y)$;*
- *$x \wedge 0 = 0$, $x \vee 1 = 1$.*

PROOF. Verifying that the axioms hold in a lattice is not difficult — try it yourself. The converse is a little harder. We have to recover the partial order from the lattice operations. If $x \le y$, then the g.l.b. of x and y is obviously x; we reverse this and define the relation $\le$ by the rule that $x \le y$ if $x \wedge y = x$. We have to show that this really is a partial order, and that $x \wedge y$ and $x \vee y$ are the g.l.b. and l.u.b., and 0 and 1 the least and greatest elements, in this order.

---

[2] In an infinite lattice, the existence of 0 and 1 cannot be deduced, and must be postulated.

First, note that $x \wedge y = x$ implies $x \vee y = y \vee (y \wedge x) = y$, using the commutative laws; and conversely. So the 'dual' definition of the order is equivalent to the one we used.

Now we show that $\leq$ is a partial order. The idempotent laws show that it is reflexive. Suppose that $x \leq y$ and $y \leq x$. Then

$$x = x \wedge y = y \wedge x = y,$$

so the relation is antisymmetric. Finally, suppose that $x \leq y$ and $y \leq z$. Then $x \wedge y = x$ and $y \wedge z = y$. So

$$x \wedge z = (x \wedge y) \wedge z = x \wedge (y \wedge z) = x \wedge y = x,$$

so $x \leq z$.

Now, for any x and y,

$$(x \wedge y) \wedge y = x \wedge (y \wedge y) = x \wedge y,$$

so $(x \wedge y) \leq y$. By commutativity, also $(x \wedge y) \leq x$. Thus, $x \wedge y$ is a lower bound for x and y. If z is any lower bound, then

$$z \wedge (x \wedge y) = (z \wedge x) \wedge y = z \wedge y = z,$$

so $z \leq (x \wedge y)$. It follows that $x \wedge y$ is the unique greatest lower bound. The proof that $x \vee y$ is the unique least upper bound is dual.

Finally, the last axiom shows that 0 is the unique minimal element and 1 the unique maximal element.

12.2. Linear extensions of a poset

As in the introduction to this chapter, we can regard a partial order as expressing our partial knowledge of some underlying total order. This suggests that every partial order is a subset of a total order. This is indeed true:

(12.2.1) Theorem. Let R be a partial order on X. Then there is a total order R^* on X such that $R \subseteq R^*$.

A total order containing the partial order R is called a *linear extension* of R (the word 'linear' coming from the alternative term 'linear order' for a total order). If X is finite, this result can be expressed in the form:

> Let $(X, \leq)$ be a poset. Then we can label the elements of X as $x_1, \ldots, x_n$ such that, if $x_i \leq x_j$, then $i \leq j$.

Our proof will, as usual, assume that X is finite. The idea of the proof is that, if R is not itself a total order, then some pair of elements is incomparable; intuitively, we don't yet know the order of these elements. We enlarge R by specifying the order of the two elements, and adding various consequential information. The resulting relation R' is still a partial order. After a finite number of steps, there are no more incomparable elements, and we have a total order.

So let a, b be incomparable. If we specify $a < b$, then everything below a must become less than everything above b. So we put

$$R' = R \cup (\downarrow a \times \uparrow b),$$

where $\downarrow a = \{x : (x, a) \in R\}$ and $\uparrow b = \{y : (b, y) \in R\}$. We claim that R' is a partial order. It is clearly reflexive, since R is. Also, note that $\downarrow a \cap \uparrow b = \emptyset$; for, if x lies in this intersection, then $(b, x), (x, a) \in R$, so $(b, a) \in R$ by transitivity, contradicting the incomparability of a and b.

Suppose that $(x, y), (y, x) \in R'$. If both pairs lie in R, then $x = y$ by antisymmetry of R. The remark in the last paragraph shows that we cannot have $(x, y), (y, x) \in \downarrow a \times \uparrow b$. The remaining case is that (without loss of generality) $(x, y) \in R$, $(y, x) \in \downarrow a \times \uparrow b$. Then $(b, x), (x, y), (y, a) \in R$, again contradicting the choice of a and b.

The proof of transitivity is very similar. If $(x, y), (y, z) \in R$, then $(x, z) \in R$; we cannot have $(x, y), (y, z) \in \downarrow a \times \uparrow b$; and, if $(x, y) \in R$, $(y, z) \in \downarrow a \times \uparrow b$, then $x \in \downarrow a$, so $(x, z) \in R'$.

The proof is complete.

12.3. Distributive lattices

A lattice L is *distributive* if it satisfies the two distributive laws

$$x \vee (y \wedge z) = (x \vee y) \wedge (x \vee z),$$
$$x \wedge (y \vee z) = (x \wedge y) \vee (x \wedge z).$$

Two of our examples of lattices are distributive: the lattice $\mathcal{P}(X)$ of subsets of a set X, and the lattice of divisors of a positive integer n. (In the first case, the distributive laws are familiar equations connecting unions and intersections of sets, easily checked with a Venn diagram. The second is a little harder to see; try it for yourself.)

In view of the first example, any sublattice of the lattice $\mathcal{P}(X)$ of subsets of X (that is, any family of subsets of X which is closed under union and intersection) is a distributive lattice. We could ask whether, conversely, any distributive lattice can be represented in this way. This is indeed true, and we prove a stronger version.

Let $P = (X, \leq)$ be a poset. A subset Y of X is called a *down-set* if $y \in Y$, $z \leq y$ imply $z \in Y$; that is, anything lying below an element of Y is in Y.[3] There are two trivial down-sets in any poset: the empty set, and the whole of X.

(12.3.1) Proposition. *The union or intersection of two down-sets is a down-set.*

PROOF. Let Y_1 and Y_2 be down-sets. Suppose that $y \in Y_1 \cup Y_2$ and $z \leq y$. Then $y \in Y_1$ or $y \in Y_2$; so $z \in Y_1$ or $z \in Y_2$, whence $z \in Y_1 \cup Y_2$. The argument for intersections is similar.

---

[3] The term 'ideal' is often used. But it has another, conflicting, meaning.

Thus, the set of all down-sets of P, with the operations of union and intersection, is a distributive lattice (whose 0 and 1 are the trivial down-sets). We denote this lattice by $L(P)$. For example, if P is the poset N of Fig. 12.1, then $L(P)$ is shown in Fig. 12.2. Now every finite distributive lattice has a canonical representation of this form.

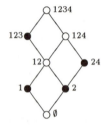

Fig. 12.2. The lattice $L(N)$

(12.3.2) Theorem. *Let L be a finite distributive lattice. Then there is a finite poset P (uniquely determined by L) such that L is isomorphic to $L(P)$.*

PROOF. How can we recover the elements of P from L? For any point x of P, the set $\downarrow x = \{y : y \leq x\}$ is a down-set, the *principal down-set* determined by x. We have to recognise elements of L corresponding to principal down-sets.

An element $a \neq 0$ of a lattice L is called *join-indecomposable*, or JI for short, if $a = b \vee c$ implies $a = b$ or $a = c$. Now, in $L(P)$, any principal down-set is JI. For, if $\downarrow x = b \vee c$, then $x \in b$ or $x \in c$, whence $\downarrow x = b$ or $\downarrow x = c$ (if b and c are down-sets). Conversely, any JI in $L(P)$ is a principal down-set. (In Fig. 12.2, the JI elements of $L(N)$ are represesnted by solid circles. Note that they form a sub-poset isomorphic to N.)

So, in any finite distributive lattice L, we let $P(L)$ be the set of JI elements, with order inherited from L. Then $P(L)$ is the only possible candidate for a poset P such that $L(P) = L$; we show that, indeed, $L(P(L)) = L$. The proof is in a number of steps.

STEP 1. Every non-zero element of L is a join of JI elements.

PROOF. If $a \in L$ is JI, we are done. Otherwise, a is a join of two elements strictly below it in the lattice. By induction (for example, on the number of elements below a), these two elements are joins of JI elements; so the same is true of a.

STEP 2. Every non-zero element $a \in L$ is the join of all the JI elements below it.

PROOF. We know that a is the join of some of these elements. The join of all of them is no smaller, but is still no larger than a.

These two steps apply also to 0, if we interpret the join of the empty set as 0.

Now let X be the set of all JI elements (the elements of the poset $P(L)$); for any $a \in L$, let $s(a) = \{x \in X : x \leq a\}$. We show that s is an isomorphism from L to $L(P(L))$.

STEP 3. $s(a)$ is a down-set.

PROOF. Clear from the definition.

STEP 4. s is a bijection.

PROOF. That s is one-to-one follows from the fact that a is the join of the elements in $s(a)$. Now let Y be any down-set in $P(L)$, and let a be the join of the elements in Y. Then each $y \in Y$ satisfies $y \leq a$. Suppose that $x \notin Y$ and $x \leq a$. If $Y = \{y_1, \ldots, y_n\}$, then we have $x \leq y_1 \vee \ldots \vee y_n$, so $x \wedge (y_1 \vee \ldots \vee y_n) = x$. By the distributive law, $(x \wedge y_1) \vee \ldots \vee (x \wedge y_n) = x$. But x is JI; so, for some i, we have $x \wedge y_i = x$, whence $x \leq y_i$. But this contradicts the facts that $x \notin Y$, $y_i \in Y$, and Y is a down-set. We conclude that $Y = s(a)$. So s is onto.

STEP 5. s is an isomorphism, i.e.
(a) $s(a \wedge b) = s(a) \cap s(b)$,
(b) $s(a \vee b) = s(a) \cup s(b)$.

PROOF. (a) For $x \in X$, we have $x \leq a \wedge b$ if and only if $x \leq a$ and $x \leq b$.
   (b) Take $x \in s(a) \cup s(b)$. Then either $x \in s(a)$ or $x \in s(b)$; so $x \leq a$ or $x \leq b$, whence $x \leq a \vee b$. Conversely, suppose that $x \in s(a \vee b)$, so $x \leq a \vee b$. Then

$$x = x \wedge (a \vee b) = (x \wedge a) \vee (x \wedge b).$$

Since x is JI, $x = x \wedge a$ or $x = x \wedge b$, whence $x \in s(a)$ or $x \in s(b)$.
   This completes the proof.

   Among distributive lattices, a special class are the *Boolean lattices*. These are the distributive lattices L possessing a unary operation $x \mapsto x'$ called *complementation*, satisfying
• $(x \vee y)' = x' \wedge y'$, $(x \wedge y)' = x' \vee y'$;
• $x \vee x' = 1$, $x \wedge x' = 0$.

(12.3.3) **Theorem.** *A finite Boolean lattice is isomorphic to the lattice of all subsets of a finite set X, with x' interpreted as $X \setminus x$.*

PROOF. Let L be a finite Boolean lattice. We have an embedding of L into $P(X)$, where X is the set of JI elements of L. To show that $L = P(X)$, we show that any two JI elements are incomparable — then any set of JI elements is a down-set.
   So suppose that a and b are distinct JI elements with $a \leq b$. Then

$$a \vee (b \wedge a') = (a \vee b) \wedge (a \vee a') = b \wedge 1 = b.$$

Since b is JI and $a \neq b$, we must have $b = b \wedge a' \leq a'$. Then

$$a = a \wedge b = a \wedge (b \wedge a') = b \wedge (a \wedge a') = b \wedge 0 = 0,$$

a contradiction.
   Now, if s is the lattice-isomorphism from L to $P(X)$ as in Theorem (12.3.2), we have $s(a) \cap s(a') = \emptyset$, $s(a) \cup s(a') = X$; so $s(a') = X \setminus s(a)$, as claimed.

Another interesting class consists of the *free distributive lattices*. These are generated (in the algebraic sense) by a set $X = \{x_1, \ldots, x_n\}$, and have the property that two expressions in the generators are unequal unless the definition of a distributive lattice forces them to be equal. I will identify the free distributive lattice as $L(P)$, as in (12.3.2), but with a bit of hand-waving; a rigorous proof has to use properly the formal algebraic definition of freeness.

Using the distributive laws, any element other than 0 and 1 can be written as a join of terms, each of which is a meet of some elements of X. So the only possible join-indecomposables apart from 1 are the meets of the non-empty subsets of X. The JI element 1 corresponds to the empty set. The order in the lattice of these meets is the reverse of the inclusion order of the subsets.

Moreover, a down-set in the poset of meets of subsets of X corresponds to an up-set in $\mathcal{P}(X)$. Since $\mathcal{P}(X)$ is 'self-dual', we have:

(12.3.4) Proposition. *The free distributive lattice generated by an n-set X is isomorphic to $L(\mathcal{P}(X))$, in other words, to $L(L(A))$, where A is an antichain with n elements.*

However, nobody knows a formula for the number of elements in this lattice for arbitrary n. This is a famous unsolved problem. The answer is known only for very small values of n.

12.4. Aside on propositional logic

The name of Boole is familiar to every computer scientist today, as a result of his project to turn set theory and logic into algebra. We now sketch the details.

Expressions in Boole's system are built from variables, just as polynomials are; but a Boolean variable can take only the two values TRUE and FALSE. (Think of these variables as elementary statements or propositions out of which more complicated expressions can be built.)

We start with a set P of propositional or Boolean variables. A *formula* is an expression involving variables, parentheses, and the *connectives* $\vee$ (disjunction, 'or'), $\wedge$ (conjunction, 'and'), and $\neg$ (negation, 'not'), defined by the rules
- any propositional variable is a formula;
- if ϕ and ψ are formulae, so are $(\phi \vee \psi)$, $(\phi \wedge \psi)$, and $(\neg\phi)$;
- any formula is obtained by these two rules.

In other words, the set of formulae is the smallest set of strings of variables, parentheses and connectives which contains the variables and is closed under the three constructions specified in the second rule.

A *valuation* is a function v from the set of variables to the set $\{\text{TRUE}, \text{FALSE}\}$. By induction, v defines a function from the set of formulae to the set $\{\text{TRUE}, \text{FALSE}\}$, which is also called a valuation and denoted by v, such that the usual 'truth table rules' for the connectives apply:
- if $v(\phi) = \text{TRUE}$ then $v((\neg\phi)) = \text{FALSE}$, and *vice versa*;
- $v((\phi \vee \psi)) = \text{TRUE}$ unless $v(\phi) = v(\psi) = \text{FALSE}$, in which case $v((\phi \vee \psi)) = \text{FALSE}$;
- $v((\phi \wedge \psi)) = \text{FALSE}$ unless $v(\phi) = v(\psi) = \text{TRUE}$, in which case $v((\phi \wedge \psi)) = \text{TRUE}$.

Further connectives can be defined in terms of the ones already given. For example, $(\phi \rightarrow \psi)$ is shorthand for $((\neg\phi) \vee \psi)$, and $(\phi \leftrightarrow \psi)$ for $((\phi \rightarrow \psi) \wedge (\psi \rightarrow \phi))$. Truth tables for these can be calculated. For example, $v((\phi \leftrightarrow \psi)) = \text{TRUE}$ if and only if $v(\phi) = v(\psi)$.

A formula ϕ is called a *tautology* if $v(\phi) = \text{TRUE}$ for all valuations v, a *contradiction* if $v(\phi) = \text{FALSE}$ for all v (that is, if $(\neg\phi)$ is a tautology). Two formulae ϕ, ψ are *equivalent* if $v(\phi) = v(\psi)$ for all valuations v; that is, if $(\phi \leftrightarrow \psi)$ is a tautology.

Now it can be checked that the 'equivalence' just defined is an equivalence relation, and that the connectives induce operations on the set of equivalence classes: if $[\phi]$ denotes the equivalence class of ϕ, then we can set

$$[\phi] \vee [\psi] = [(\phi \vee \psi)],$$
$$[\phi] \wedge [\psi] = [(\phi \wedge \psi)],$$
$$[\phi]' = [(\neg\phi)],$$

and the objects defined don't depend on the choice of representatives of the equivalence classes. Now Boole's observation can be summarised as follows:

(12.4.1) Proposition. *The set of equivalence classes of propositional formulae, with the above operations, is a Boolean lattice.*

Suppose that there are n propositional variables. The number of valuations is 2^n. Any formula ϕ defines a function $v \mapsto v(\phi)$ from valuations to $\{\text{TRUE}, \text{FALSE}\}$, and two formulae are equivalent if and only if they define the same function. Any function is represented by some formula, so the number of equivalence classes is 2^{2^n}. So the Boolean lattice has 2^{2^n} elements.

By (12.3.3), any Boolean lattice is isomorphic to $\mathcal{P}(X)$ for some set X. Can we identify such an X here? It must have cardinality 2^n. An answer is given by the *disjunctive normal form*:

(12.4.2) Disjunctive normal form. *Any formula in the variables $p_1, \dots, p_n$ which is not a contradiction is equivalent to a unique disjunction of terms $(q_1 \wedge \dots \wedge q_n)$, where each q_i is either p_i or $(\neg p_i)$.*

There are 2^n 'terms' of the form described in the proposition, and each equivalence class of formulae corresponds to a subset of the set of terms. (The equivalence class of contradictions corresponds to the empty set of terms.) Moreover, the operations $\vee, \wedge, '$ on equivalence classes correspond to union, intersection, and complementation on sets of terms. So the set of terms is the required X.

Another approach to the question gives an even more obvious answer: take X to be the set of valuations, and identify an equivalence class with the subset consisting of valuations which give the formulae in that class the value TRUE. To see the correspondence between the two approaches, note that there is a unique valuation which gives the term $q_1 \wedge \dots \wedge q_n$ the value TRUE, namely the one defined by

$$v(p_i) = \begin{cases} \text{TRUE} & \text{if } q_i = p_i, \\ \text{FALSE} & \text{if } q_i = (\neg p_i). \end{cases}$$

The disjunctive normal form theorem can be used to show that the lattice of equivalence classes of propositional formulae in n variables is the free Boolean lattice on n generators (compare the remarks at the end of the last section on free distributive lattices).

12.5. Chains and antichains

A *chain* C in a poset P is a subset of P such that any two of its points are comparable. In other words, it is a sub-poset which is a total order. An *antichain* A is a subset such that any two of its points are incomparable.

We have met these concepts before. Sperner's Theorem (7.2.1) describes the largest antichains in the lattice $\mathcal{P}(X)$ of subsets of X. Our proof of this by the LYM technique involved covering the poset by chains. A crucial point in the argument was:

If C is a chain and A an antichain in a poset, then $|C \cap A| \leq 1$.

For two points in this intersection would be both comparable and incomparable!

From this, we immediately see:

(12.5.1) Proposition. *(a) If a poset P has a chain of size r, then it cannot be partitioned into fewer than r antichains.*

(b) If a poset P has an antichain of size r, then it cannot be partitioned into fewer than r chains.

The proof is trivial, since two points in the same chain must lie in different members of a partition into antichains, and 'dually'. The main goal of this section is to prove a pair of results in the reverse direction. The first is straightforward:

(12.5.2) Theorem. *Suppose that the largest chain in the poset P has size r. Then P can be partitioned into r antichains.*

PROOF. We define the *height* of an element x of P to be one less than the greatest number of elements in a chain whose greatest member is x. (The 'one less' is conventional: the height of x is the greatest number of 'steps' up from the bottom of the poset to x.) Let A_i be the set of elements of height i. Then, by hypothesis, $A_i = \emptyset$ for $i \geq r$, so $P = A_0 \cup \ldots \cup A_{r-1}$; and each A_i is an antichain, since if $x \in A_i$ and $x < y$, then there is a chain $x_0 < \ldots < x_i = x < y$, so y has height greater than i.

The 'dual' result looks similar, but the proof is much more involved.[4]

(12.5.3) Dilworth's Theorem. *Suppose that the largest antichain in the poset P has size r. Then P can be partitioned into r chains.*

PROOF. The proof is by induction on the number of points of P. Clearly the result holds for one-element posets. So suppose that it is true for all posets with fewer points than P. Let x be a minimal element of P.

CASE 1. x is incomparable with everything else in P. Then the largest antichain in $P \setminus \{x\}$ has size $r - 1$, since adjoining x gives a larger antichain. By induction, $P \setminus \{x\}$ can be partitioned into $r - 1$ chains; we add the singleton chain $\{x\}$ to produce the required partition.

CASE 2. Some other points are comparable with x. By induction, we can partition $P \setminus \{x\}$ into r chains $C_1, \ldots, C_r$. For each i, let T_i be the set of elements of C_i which are comparable with x, and $B_i = C_i \setminus T_i$; let $B = B_1 \cup \ldots \cup B_r$. Then every element of T_i is greater than x, since x is minimal; T_i is above B_i for each i, and B is the set of all elements incomparable with x. We colour the points of B with r colours $c_1, \ldots, c_r$, by the rule that y has colour c_i if $y \in C_i$.

By the argument of Case 1, B can be written as the union of $r - 1$ chains $C'_1, \ldots, C'_{r-1}$. Each of these chains can be partitioned into 'runs' of elements of the same colour. We are about to do some rearranging of these chains, which may have to be repeated an unspecified number of times. But each rearrangement

---

[4] The result is uniformly known as Dilworth's Theorem. It was published by Dilworth in 1950. It had been found a few years earlier by Gallai and Milgram, but publication was delayed because Gallai wanted the paper translated into English, and Milgram, a topologist, did not fully appreciate its importance.

strictly decreases the total number of colour runs which occur; so we know that the rearrangement process will terminate after a finite number of moves.

A move is as follows. Suppose that the greatest elements of two or more of the chains $C'_1, \ldots, C'_{r-1}$ have the same colour c_i. Take the union of all the runs of colour c_i which lie at the top of their chains. This union U is itself a chain, since it is a subset of C_i. If y is the smallest element of U, and $y \in C'_j$, then we move all the elements of U to C'_j, where they sit at the top, forming a single run. So the number of runs has decreased, as claimed; and the new C'_i are still chains.

At some stage, it is no longer possible to apply a move of this type. This must be because the greatest elements of the chains all have different colours. Renumbering if necessary, we may assume that the greatest element of C'_i has colour c_i for $i = 1, \ldots, r-1$. Now $C'_i = T_i \cup C'_i$ is a chain for $i = 1, \ldots, r-1$, since the greatest element of C'_i lies below T_i. Finally, $C'_r = T_r \cup \{x\}$ is a chain, since x lies below all T_i. So we have the required partition into chains $C'_1, \ldots, C'_r$.

Perhaps the relative difficulty of this theorem is more understandable when you realise that it contains Hall's Marriage Theorem (6.2.1) as a special case!

Suppose that $A_1, \ldots, A_n$ are subsets of X satisfying Hall's Condition (HC):

$$|A(J)| \geq |J| \quad \text{for} \quad J \subseteq \{1, \ldots, n\},$$

where $A(J) = \bigcup_{j \in J} A_j$. We construct a poset P as follows. The elements of P are the points of X and symbols $y_1, \ldots, y_n$, with $x < y_i$ if $x \in A_i$, and no other comparabilities. We set $Y = \{y_1, \ldots, y_n\}$. Now X is an antichain in this poset. We claim that there is no larger antichain. For let S be an antichain, and set $J = \{j : y_j \in S\}$. Then S contains no element of $A(J)$; so

$$|S| \leq |J| + |X| - |A(J)| \leq |X|,$$

by (HC).

Now Dilworth's Theorem implies that P can be partitioned into $|X|$ chains. Each of these chains must contain a point of X. Let the chain through y_i be $\{x_i, y_i\}$. Then $(x_1, \ldots, x_n)$ is a system of distinct representatives for $(A_1, \ldots, A_n)$: for $x_i \in A_i$ (since $x_i < y_i$), and $x_i \neq x_j$ for $i \neq j$ (since the chains are disjoint).

12.6. Products and dimension

Suppose that a number of objects are being compared on several different numeric attributes. If x is better than y on all these attributes, we are justified in saying that x beats y. But if x is better on some attributes and y on others, then, depending how the attributes are scaled or weighted, we might come to different conclusions about their ordering, and it seems safest to say that x and y are incomparable in this case.

Accordingly, let $(X_1, \leq_1), \ldots, (X_n, \leq_n)$ be posets. The *direct product* of these posets is the poset $(X, \leq)$, where

$$X = X_1 \times \ldots \times X_n = \{(x_1, \ldots, x_n) : x_1 \in X_1, \ldots, x_n \in X_n\},$$

and

$$(x_1, \ldots, x_n) \leq (y_1, \ldots, y_n) \quad \text{if and only if} \quad x_i \leq_i y_i \text{ for } i = 1, \ldots, n.$$

It is a simple matter to show that $(X, \leq)$ is indeed a poset. Moreover, a direct product of lattices is a lattice, with meet and join defined by

$$(x_1, \ldots, x_n) \wedge (y_1, \ldots, y_n) = (x_1 \wedge_1 y_1, \ldots, x_n \wedge_n y_n),$$
$$(x_1, \ldots, x_n) \vee (y_1, \ldots, y_n) = (x_1 \vee_1 y_1, \ldots, x_n \vee_n y_n),$$

and $0 = (0_1, \ldots, 0_n)$, $1 = (1_1, \ldots, 1_n)$.

Some familiar posets are direct products. Notably:

(12.6.1) Proposition. *(a) If $|X| = n$, then the power-set lattice $\mathcal{P}(X)$ is the direct product of n copies of the two-element lattice $\{0, 1\}$.*

(b) If $n = p_1^{a_1} \ldots p_r^{a_r}$, where $p_1, \ldots, p_r$ are distinct primes, then the lattice $D(n)$ of divisors of n is isomorphic to the direct product of the lattices $D(p_1^{a_1}), \ldots, D(p_n^{a_n})$.

PROOF. (a) Let $X = \{x_1, \ldots, x_n\}$. We identify any subset Y of X with its *characteristic function* $(e_1, \ldots, e_n)$, where $e_i = 1$ if $x_i \in Y$, $e_i = 0$ otherwise. This is a bijection between $\mathcal{P}(X)$ and $\{0, 1\}^n$, Moreover, if Y and Z have characteristic functions $(e_1, \ldots, e_n)$ and $(f_1, \ldots, f_n)$ respectively, then

$$Y \subseteq Z \Leftrightarrow (\forall i)\, (x_i \in Y \Rightarrow x_i \in Z)$$
$$\Leftrightarrow (\forall i)\, (e_i = 1 \Rightarrow f_i = 1)$$
$$\Leftrightarrow (\forall i)\, (e_i \leq f_i),$$

so the map is an isomorphism.

(b) is an exercise.

The concept of direct product gives us a measure of how far a poset is from being totally ordered. Essentially, this is the smallest number of different numerical attributes required to produce the partial order by the recipe at the start of this section. Formally, we define the *dimension* of a poset P to be the smallest integer d such that P can be embedded as a sub-poset of the direct product of d totally ordered sets.

(12.6.2) Proposition. *The poset $\mathcal{P}(X)$ has dimension $|X|$. The dimension of the poset $D(n)$ is equal to the number of distinct prime divisors of n.*

PROOF. We found isomorphisms from these posets to products of the stated number of totally ordered sets. It is necessary to show that they cannot be embedded in products of fewer total orders. More generally, we claim that the product of n total orders, each with more than one point, has dimension n. The result is clear if $n \leq 2$, so we may suppose that $n \geq 3$.

We consider a special two-level poset, the *standard poset*, with $2n$ vertices

$$a_1, \ldots, a_n, b_1, \ldots, b_n;$$

the comparabilities are $a_i < b_j$ if (and only if) $i \neq j$.

STEP 1. If P is a sub-poset of Q, then $\dim(P) \leq \dim(Q)$.

STEP 2. If P is the direct product of n total orders, each with at least two points, then P contains the $2n$-point standard poset. For suppose that u_i, v_i are elements of the ith factor, with $u_i < v_i$, for $i = 0, \ldots, n-1$. Now let a_i be the n-tuple with i^{th} entry v_i and j^{th} entry u_j for $j \neq i$; and let b_i be the n-tuple with i^{th} entry u_i and j^{th} entry v_j for $j \neq i$. It is readily checked that these elements form a standard poset.

STEP 3. The dimension of a $2n$-point standard poset is n. Clearly it is not greater than n. Suppose that the standard poset is embedded in the product of m total orders. For each i, there exists a j such that the jth coordinate of b_i is strictly smaller than that of any other point b_k, since otherwise a_i (whose coordinates are all smaller than the corresponding coordinates of b_k for $k \neq i$) would lie below b_i. Clearly this requires at least n coordinates.

EXAMPLE. The poset N has dimension 2: it can be represented by the four points $(2,0)$, $(0,1)$, $(3,2)$, $(1,3)$.

It's not obvious that a finite poset has finite dimension; but this is indeed true.

(12.6.3) Theorem. *The dimension of a finite poset P is finite, and is not greater than the number of linear extensions of P.*

PROOF. Let $P = (X, R)$, and let $(X, R_1), \ldots, (X, R_k)$ be the linear extensions of P. We map X to the direct product of these total orders by the *diagonal embedding*: $x \mapsto (x, x, \ldots, x)$. Now, if $(x, y) \in R$, then $(x, y) \in R_i$ for $i = 1, \ldots, k$; so $(x, \ldots, x) \leq (y, \ldots, y)$ in the direct product. Suppose that x and y are incomparable. The proof of Theorem 12.2.1 shows that there is a linear extension R_i of R with $(x, y) \in R_i$, and another linear extension R_j with $(y, x) \in R_j$; thus, $(x, \ldots, x)$ and $(y, \ldots, y)$ are incomparable. So the diagonal embedding is an isomorphism.

12.7. The Möbius function of a poset

An $n \times n$ real matrix $A = (a_{ij})$ can be regarded as a function a from $N \times N$ to $\mathbb{R}$, where $N = \{1, 2, \ldots, n\}$, whose values are given by $a(i, j) = a_{ij}$. From this point of view, the fact that N is an ordered set leads us to consider the matrices or functions 'supported' by the order, that is, functions which satisfy $a_{ij} = 0$ unless $i \leq j$: these are precisely the upper triangular matrices. They form an algebra: that is, they are closed under matrix multiplication as well as addition and scalar multiplication. In particular, an upper triangular matrix is invertible if and only if its diagonal entries are all non-zero. We will extend this point of view to an arbitrary finite poset.

Let $P = (X, \leq)$ be a finite poset. The *incidence algebra* $I(P)$ of P is the set of functions $f : X \times X \to \mathbb{R}$ which satisfy $f(x, y) = 0$ unless $x \leq y$. Addition and scalar multiplication are defined pointwise, and two functions are multiplied by the rule

$$f \cdot g(x, y) = \sum_{x \leq z \leq y} f(x, z) g(z, y).$$

(12.7.1) Proposition. *If* $|X| = n$, *the incidence algebra* $I(P)$ *is isomorphic to a subalgebra of the algebra of upper triangular matrices. A function* f *is invertible if and only if* $f(x,x) \neq 0$ *for all* $x \in X$.

PROOF. We take a linear extension of P (Theorem 12.2.1); that is, we number the elements of X as $x_1, \ldots, x_n$ so that, if $x_i \leq x_j$, then $i \leq j$. Now we map $f \in I(P)$ to the matrix $A = (a_{ij})$ where $a_{ij} = f(x_i, x_j)$. Clearly, A is upper triangular. Also, the map is an isomorphism, since if matrices A and B correspond to f and g, then the matrix corresponding to fg has (i,j) entry

$$\sum_{x_i \leq x_k \leq x_j} f(x_i, x_k) g(x_k, x_j) = \sum_{i \leq k \leq j} a_{ik} b_{kj}$$

$$= \sum_{1 \leq i \leq n} a_{ik} b_{kj}$$

the last inequality holding because, unless $i \leq k \leq j$, either a_{ik} or b_{kj} is zero. In particular, $fg(x,y) = 0$ unless $x \leq y$ (since there are no terms in the sum); so $fg \in I(P)$.

Finally, note that the values $f(x,x)$ are the diagonal elements of the matrix corresponding to f. So a function satisfying the condition $f(x,x) \neq 0$ corresponds to an invertible matrix. We need to know that the inverse function does lie in $I(P)$. For this purpose, we give an algorithm to compute an inverse function; the fact that the inverse is unique then implies the result.

For $x \leq y$, we define the *interval* $[x,y]$ to be the set $\{z : x \leq z \leq y\}$, or the poset induced on this set. Now suppose that $f(x,x) \neq 0$ for all $x \in X$. We calculate the values $g(x,y)$ of a function $g \in I(P)$ by induction on the cardinality of $[x,y]$, as follows:

If $||[x,y]|| = 0$, then $x \nleq y$, and we set $g(x,y) = 0$.
If $||[x,y]|| = 1$, then $x = y$, and we set $g(x,x) = f(x,x)^{-1}$.
If $||[x,y]|| > 1$, we set

$$g(x,y) = -f(x,x)^{-1} \left(\sum_{x < z \leq y} f(x,z) g(z,y) \right).$$

The function g is well-defined, because the values of g on the right-hand side of the last equation have the form $g(z,y)$, where $x < z \leq y$; so the interval $[z,y]$ is properly contained in $[x,y]$, and the values are defined by induction. Clearly $g \in I(P)$. A short calculation shows that, indeed, $fg(x,x) = 1$ and $fg(x,y) = 0$ if $x \neq y$; so g is the inverse of f.

Three particular elements of $I(P)$ are specially important. The first is the function e, the characteristic function of equality:

$$e(x,y) = \begin{cases} 1 & \text{if } x = y, \\ 0 & \text{otherwise;} \end{cases}$$

this is the identity element of $I(P)$, corresponding to the identity matrix. Next is the function i, the characteristic function of the partial order:

$$i(x, y) = \begin{cases} 1 & \text{if } x \le y, \\ 0 & \text{otherwise.} \end{cases}$$

Finally, the *Möbius function* μ of the poset is the inverse of the function i. That is, it is characterised by the equation

$$\sum_{x \le z \le y} \mu(x, z) = \begin{cases} 1 & \text{if } x = y, \\ 0 & \text{otherwise.} \end{cases}$$

(12.7.2) Proposition. *The Möbius function is integer-valued.*

PROOF. Examine the proof of (12.7.1), which gives a method for calculating the inverse of a function: take $f = i$ there. Since $i(x, x) = 1$ for all x, the factor $i(x, x)^{-1}$ is equal to 1. Now $\mu(x, y)$ is a linear combination of values of $\mu(z, y)$ with integer coefficients (in fact, all equal to -1), where $x < z \le y$; by induction, $\mu(x, y)$ is an integer. (The induction starts with $\mu(x, x) = 1$.)

Note that the value of the Möbius function at (x, y) depends only on the poset $[x, y]$; points outside this interval don't affect the value. For the record, we translate the defining property of the Möbius function as follows. This result is referred to as *Möbius inversion* in the poset P.

(12.7.3) Proposition. *Let f, g be elements of $I(P)$. Then the following are equivalent:*
(a) $f(x, y) = \displaystyle\sum_{x \le z \le y} g(x, z)$;

(b) $g(x, y) = \displaystyle\sum_{x \le z \le y} f(x, z)\mu(z, y)$.

For a simple but important example, we have

(12.7.4) Proposition. *Let P be a totally ordered set. Then the Möbius function of P is*

$$\mu(x, y) = \begin{cases} 1 & \text{if } x = y, \\ -1 & \text{if } y \text{ covers } x, \\ 0 & \text{otherwise.} \end{cases}$$

PROOF. Indeed, in any poset, if y covers x, then $\mu(x, y) = -1$, since only the term $z = y$ occurs in the sum in (12.7.1). Now, if $x \le y$ and y is not the unique element z which covers x, a simple induction shows that $\mu(x, y) = 0$. (This induction begins with the case where y covers z; then $\mu(x, y) = -(\mu(z, y) + \mu(y, y)) = 0$.)

Conveniently, the Möbius function of a direct product of posets is equal to the product of the Möbius functions of the factors:

(12.7.5) Proposition. *Let $P_1, \ldots, P_k$ be posets, and let $P = P_1 \times \ldots \times P_k$. Then the Möbius function of P is defined by*

$$\mu((x_1, \ldots, x_k), (y_1, \ldots, y_k)) = \prod_{i=1}^{k} \mu(x_i, y_i).$$

PROOF. Since the Möbius function is unique, it suffices to prove that the right-hand side does have the property that

$$\sum_{x \le z \le y} \mu(z, y) = \begin{cases} 1 & \text{if } x = y, \\ 0 & \text{otherwise.} \end{cases}$$

If $z = (z_1, \ldots, z_k)$, then $x \le z \le y$ if and only if $x_i \le z_i \le y_i$ for $i = 1, \ldots, k$; so the sum on the left is over the product of the intervals $[x_i, y_i]$. Then this sum factorises as shown in the proposition.

From this, we can calculate the Möbius functions of two important posets.

(12.7.6) Theorem. *(a) The Möbius function of the Boolean lattice $\mathcal{P}(X)$ is given by*

$$\mu(Y, Z) = \begin{cases} (-1)^{|Z|-|Y|} & \text{if } Y \subseteq Z, \\ 0 & \text{otherwise.} \end{cases}$$

(b) The Möbius function of the lattice $D(n)$ of divisors of n is given by

$$\mu(y, z) = \begin{cases} (-1)^d & \text{if } z/y \text{ is the product of } d \text{ distinct primes,} \\ 0 & \text{otherwise.} \end{cases}$$

This is immediate from (12.7.4), (12.7.5) and (12.6.1).

REMARK 1. Both $\mathcal{P}(X)$ and $D(n)$ have the property that any interval is isomorphic to a lattice of the same form: $[Y, Z] \cong \mathcal{P}(Z \setminus Y)$ in case (a), and $[y, z] \cong D(z/y)$ in (b). Thus, in these cases, we can regard the Möbius function as having a single argument, setting $\mu(Y) = \mu(\emptyset, Y)$ in $\mathcal{P}(X)$, and $\mu(y) = \mu(1, y)$ in $D(n)$. The values of these functions are then given by

$$\mu(Y) = (-1)^{|Y|} \quad \text{in } \mathcal{P}(X)$$

$$\mu(y) = \begin{cases} (-1)^d & \text{if } y = p_1 \cdots p_d \\ 0 & \text{otherwise} \end{cases} \quad \text{in } D(n)$$

where $p_1, \ldots, p_d$ denote distinct primes. The latter function is the classical Möbius function met with in number theory.

REMARK 2. Using the form of the Möbius function for $\mathcal{P}(X)$, the statement of (12.6.3) translates precisely into (5.2.2), an equivalent form of the Principle of Inclusion and Exclusion. Thus Möbius inversion is a generalisation of PIE.

REMARK 3. The 'classical' form of Möbius inversion reads as follows.

> Let f, g be functions on the positive integers. Then the following are equivalent:

(a) $f(n) = \sum_{d|n} g(d)$;

(b) $g(n) = \sum_{d|n} f(d)\mu(n/d)$.

Here is an application. In Section 4.7, we found that the number a_n of monic irreducible polynomials of degree n over a field with q elements satisfies the recurrence relation

$$\sum_{d|n} da_d = q^n.$$

By Möbius inversion (applied with $f(n) = q^n$, $g(n) = na_n$), we find a formula for a_n:

$$a_n = \frac{1}{n} \sum_{d|n} q^d \mu(n/d),$$

where μ is here the classical Möbius function.

12.8. Matroids

The notion of independence shows up in many different places in mathematics: some familiar, some less so. We'll see that it obeys the same laws in these different guises. Among them are:

- *Linear independence* in vector spaces.
- A closely related notion occurs in projective or affine spaces. Any set of k points in such a space lies in a flat of dimension at most $k - 1$; it is called *independent* if it lies in no flat of dimension smaller than $k - 1$.
- In a graph (V, E), a set E' of edges is *acyclic* if (V, E') is a forest.[5]
- Let $\mathcal{F}$ be a family of subsets of X. A set $\{x_1, \ldots, x_k\}$ of points of X is a *partial transversal* for $\mathcal{F}$ if there are distinct sets $A_1, \ldots, A_k \in \mathcal{F}$ such that $x_i \in A_i$ for $i = 1, \ldots, k$; in other words, $(x_1, \ldots, x_k)$ is a SDR for a subfamily of $\mathcal{F}$.

The common concept here is that of a *matroid*.[6] A matroid is a pair $(X, \mathcal{I})$, where $\mathcal{I}$ is a non-empty family of subsets of X having the properties:

- *Hereditary property*: if $Y \in \mathcal{I}$ and $Z \subseteq Y$, then $Z \in \mathcal{I}$;
- *Exchange axiom*: if $Y, Z \in \mathcal{I}$ and $|Z| > |Y|$, then there exists $z \in Z$ such that $Y \cup \{z\} \in \mathcal{I}$.

The members of $\mathcal{I}$ are called *independent sets*. In fact, there are many other ways to define a matroid, and the beginner is often bemused by the many axiom systems. As a compromise, I will describe some other structures which are equivalent to the notion of a matroid, but without giving all the axiomatisations.

It follows immediately from the second matroid axiom that any two maximal independent sets have the same cardinality. This number is called the *rank* of the matroid, and a maximal independent set is called a *basis*. Dually, a minimal dependent set is called a *cycle*. If Y is any subset of the point set X of a matroid, then the members of $\mathcal{I}$ contained in Y clearly satisfy the matroid axioms, so define a matroid on Y. Let $\rho(Y)$ denote its rank, so that ρ is a function from $\mathcal{P}(X)$ to the non-negative integers. A set Y is called *closed* if $\rho(Y \cup \{x\}) > \rho(Y)$ for all $x \notin Y$. The *closure* $\sigma(Y)$ of an arbitrary subset Y is the smallest closed set containing it.

(12.8.1) Proposition. *A matroid on X is determined by any of the following: the bases; the rank function; the cycles; the closed sets; the closure operator on $\mathcal{P}(X)$.*

---

[5] The graph may contain loops or multiple edges. By convention, a forest has no loops, and contains at most one edge joining any pair of vertices.

[6] An alternative term is 'combinatorial pregeometry'. To the surprise of its proponents, but perhaps of nobody else, this term has not become standard.

PROOF. As we have explained, each of these structures is determined by the independent sets. We must show the converse.

The axioms imply that a set is independent if and only if it is contained in a basis. Obviously, a set is independent if and only if it contains no cycle. Also, a set is independent if and only if its rank is equal to its cardinality.

For the last two, we first observe that a set is closed if and only if it is equal to its closure, so the closed sets and the closure operator carry the same information. Moreover, the rank of a set is equal to the rank of its closure, so it is enough to determine the rank of the closed sets. Now the rank of a closed set Y is the length of any maximal chain of closed sets with greatest element Y.

(12.8.2) Proposition. *Each of the following examples defines a matroid:*
- *X is a subset of a vector space, $\mathcal{I}$ the set of linearly independent subsets of X;*
- *X is a subset of a projective or affine space, $\mathcal{I}$ is the set of independent subsets of X;*
- *X is the edge set of a graph, $\mathcal{I}$ the set of acyclic subsets of X;*
- *$\mathcal{I}$ is the set of partial transversals of a family of subsets of X.*

PROOF. The proofs show various similarities and differences, so I will sketch the first, third and fourth. (The second is almost the same as the first.)

1. Let X be a set of vectors. Clearly, any subset of a linearly independent subset is linearly independent. Suppose that Y and Z are linearly independent, with $|Z| > |Y|$. Then $\dim\langle Z\rangle > \dim\langle Y\rangle$, so $Z \not\subseteq \langle Y\rangle$. Thus, there is a vector $z \in Z$ not contained in $\langle Y\rangle$, and $Y \cup \{z\}$ is linearly independent.

3. Let X be the edge set of a graph on the vertex set V. Clearly a subset of an acyclic subset is acyclic. If Y is acyclic, then the number of connected components of (V, Y) is $|V| - |Y| + 1$, by (11.2.1). Thus, if $|Z| > |Y|$, then (V, Z) has fewer components than (V, Y), and so some edge $z \in Z$ is not contained within a component of (V, Y); thus $Y \cup \{z\}$ is acyclic.[7]

4. Any subset of a partial transversal is clearly a partial transversal. Suppose that Y and Z are partial transversals, with $|Y| < |Z|$. Let A_y be the set represented by $y \in Y$, and B_z the set represented by $z \in Z$. We consider, for each $z \in Z$, the set $X' = Y \cup \{z\}$, and the subsets $A'_y = A_y \cap X'$ and $B'_z = B_z \cap X'$. If this family of sets has a SDR, its elements must be all the points of X', which is thus a partial transversal, and we are done. So we can suppose that this fails for all z. But this means that some $n + 1$ of these sets contain only n elements of X'. These $n + 1$ sets must include B'_z, since any subfamily of the A'_y has a SDR. This means that $z \in Y$ for all $z \in Z$, a contradiction, since $|Z| > |Y|$.

As usual with abstract concepts, the point of this result is that a single argument suffices to prove a theorem applicable in several different fields. We should look to these fields for results which can be formulated in terms of independent sets. One such is the greedy algorithm for the minimal connector (Section 11.3), which

---

[7] A cycle in this matroid is the edge set of a circuit in the graph (possibly a loop or two parallel edges) — hence the name.

extends to the minimum-weight basis in a matroid whose elements have weights (see Exercise 13).

The closed sets of a matroid form a lattice, where meet is intersection, and the join of two sets is the closure of their union. Boolean lattices and finite projective and affine geometries form special cases of these so-called *geometric lattices*, which have been axiomatised and studied in their own right. We make just one observation.

A matroid is called *geometric* if the empty set and all singletons are closed. Now it is possible to pass from any matroid to a geometric matroid in a canonical way, which parallels exactly the procedure for passing from a vector space to the corresponding projective space (Chapter 9).

STEP 1. By removing all points in the closure of the empty set, we produce a matroid in which the empty set is closed.

STEP 2. Now write $x \sim y$ if $x = y$ or $\{x, y\}$ is dependent (in other words, if $\{x, y\}$ has rank 1). It follows from the exchange axiom that this is an equivalence relation. There is a matroid induced in a natural way on the set of equivalence classes. (Any closed set is a union of equivalence classes.)

In the case of a vector space V, Step 1 removes the zero vector, and Step 2 calls two vectors equivalent if one is a scalar multiple of the other; so the equivalence classes are the 1-dimensional subspaces, that is, the points of the projective geometry. In the case of a graph, Step 1 removes loops and Step 2 removes multiple edges, leaving a simple graph.

Now, in general, geometric matroids and geometric lattices are equivalent concepts: the points of the geometric matroid are the elements of the lattice which cover 0; an arbitrary element of the lattice can be identified with the set of points lying below it; and, as explained earlier, we can recover the rank function, and hence the independent sets, from the closed sets.

We conclude this section with a generalisation which pulls itself up by the bootstraps. Our third example of a matroid arose from the partial transversals of a family $\{A_1, \ldots, A_n\}$ of subsets of X, that is, the sets of points supporting SDRs of subsets of the family. Now we suppose that there is already a matroid $(X, \mathcal{I})$ defined on the point set. We ask: Is there an *independent* transversal? The answer is formally similar to Hall's Theorem (of which it is a generalisation).

(12.8.4) Theorem. *Let $A_1, \ldots, A_n$ be subsets of X, and let $(X, \mathcal{I})$ be a matroid. Then there is an independent transversal to the family if and only if, for every $J \subseteq \{1, \ldots, n\}$,*

$$\rho(A(J)) \geq |J|.$$

REMARK. Hall's Theorem corresponds to the case where the matroid is trivial (every set independent), so that $\rho(Y) = |Y|$ for any subset Y of X.

PROOF. If there is an independent transversal, then for any $J \subset \{1, \ldots, n\}$, $A(J)$ contains an independent set of size $|J|$, so its rank is at least this large. The converse is an exercise, which can be solved by re-writing (with care) the proof of Hall's Theorem given in Section 6.2 (or, indeed, almost any other of the standard textbook proofs).

12.9. Project: Arrow's Theorem

One of the problems of politics involves 'averaging out' individual preferences to reach decisions acceptable to society as a whole. In this section, we prove Arrow's Theorem, which shows that this is indeed a difficult task!

We suppose that I is a society consisting of a set of n individuals. These individuals are to be offered a choice among a set X of options, for example, by a referendum. We assume that each member i of the society has made up her/his mind about the relative worth of the options. We can describe this by a total order $\leq_i$ on X, for each $i \in I$. A *social choice function* is a rule which, given the 'individual preferences' $\leq_i$ for each $i \in I$, comes up with a 'social preference' $\leq$ on X, subject to four conditions listed (and justified) below. In other words, it is a function from the set of all n-tuples of X to the set of total orders, satisfying Axioms (A1)–(A4) below. Arrow's Theorem asserts that, if there are at least three options, then no social choice function is possible.

> *(A1) If $x \leq y$ (in the social preference), then the same remains true if the individual preferences are changed in y's favour.*

(This means that, if $\leq_i'$ $(i \in I)$ are another system of individual preferences satisfying
$u \leq_i v \Leftrightarrow u \leq_i' v$ for all $u, v \neq y$, and
$u \leq_i y \Rightarrow u \leq_i' y$ for all u,
and $\leq'$ is the corresponding social preference, then $x \leq' y$ holds.

> *(A2) If $Y \subseteq X$ and two sets $\{\leq_i\}, \{\leq_i'\}$ of individual preferences on X have the property that $\leq_i$ and $\leq_i'$ induce the same ordering on Y for each $i \in I$, then the corresponding social preferences $\leq$ and $\leq'$ induce the same ordering on Y.*

(This is the *principle of irrelevant options*, and asserts that the working of social choice should not be affected if some of the options are struck out.)

> *(A3) For any distinct $x, y \in X$, there is some system of individual preferences for which the corresponding social preference has $x \leq y$.*

(In other words, it should be possible for society to prefer y to x if enough individuals do so. In fact, it follows from (A1)–(A3) that, if $x \leq_i y$ for all $i \in I$, then $x \leq y$: that is, if everybody prefers y to x, then society does too.)

> *(A4) There is no individual i such that $\leq_i$ coincides with $\leq$ for all systems of individual preferences.*

This axiom requires that there should not be a dictator whose opinions prevail against all opposition!

(12.9.1) Arrow's Theorem. *If $|X| \geq 3$, then no social choice function exists.*

PROOF. Suppose that we have a social choice function. If (x, y) is an ordered pair of distinct options, we say that a set J of individuals is (x, y)-*decisive* if, whenever all members of J prefer y to x, then so does the social order; formally, if $x <_i y$ for all $i \in J$, then $x < y$. Further, we say that J is *decisive* if it is (x, y)-decisive for some distinct x, y. We claimed after the statement of (A3) that the whole society I is (x, y)-decisive for all x, y; let us first prove this. By (A2), we can suppose that x and y are the only options. Now by (A3), there is some system of individual preferences which causes $x < y$ to hold; and by (A1), this remains true if we alter them so that all individuals prefer y to x.

Let J be a minimal decisive set. Then $J \neq \emptyset$, by (A3). Suppose that J is (x, y)-decisive, and let i be a member of J.

CLAIM. $J = \{i\}$. For let $J' = J \setminus \{i\}$ and $K = I \setminus J$. Let z be a member of X different from x and y (remember that $|X| \geq 3$). Consider the individual preferences for which
$$x <_i y <_i z;$$

$z <_j x <_j y$ for all $j \in J'$;

$y <_k z <_k x$ for all $k \in K$.

Then

$x < y$, since all members of the (x, y)-determining set J think so;

$y < z$, since if $z < y$ then J' is (z, y)-decisive, contradicting the minimality of J.

Hence $x < z$. But then $\{i\}$ is (x, z)-decisive, since nobody else agrees with this order. By minimality of J, we have $J = \{i\}$.

The proof shows, in fact, that $\{i\}$ is (x, z)-decisive for all $z \neq x$.

CLAIM. i is a dictator.

Choose $w \neq x$, and $z \neq w, x$. Consider the individual preferences in which

$w <_i x <_i z$,

$z <_k w <_k x$ for all $k \neq i$.

Then $w < x$ (because everybody thinks so) and $x < z$ (because i thinks so); so $w < z$, and $\{i\}$ is (w, z)-decisive. Finally, a similar argument (left to the reader) shows that $\{i\}$ is (w, x)-decisive for any $w \neq x$. The claim is proved; and so Axiom (A4) is violated, proving the Theorem.

12.10. Exercises

1. Describe the lattice $L(P)$ for each of the posets P of Fig. 12.1 (other than N, see Fig. 12.2).

2. Show that the pentagon and the three-point line are lattices, but are not distributive.

REMARK. It can be shown that a lattice is distributive if and only if it contains neither the pentagon nor the three-point line as a sublattice.

3. A poset P is a *two-level poset* if it is the union of two antichains U and L with no element of L greater than any element of U (so that the only comparabilities which occur are of the form $l < u$ for $l \in L$, $u \in U$). In the deduction of Hall's Theorem from Dilworth's, we used a two-level poset. Show, conversely, that the truth of Dilworth's theorem for two-level posets can be deduced from Hall's Theorem. [HINT: you may find the form of Hall's Theorem given in Exercise 7 of Chapter 6 useful.]

4. Prove Proposition 12.5.1(b).

5. (a) Find the dimension of the pentagon and the three-point line.
   (b) Find all linear extensions of N, the pentagon, and the three-point line.

6. (a) Show that any antichain (containing more than one point) has dimension 2.
   (b) The *incidence poset* of a graph Γ consists of the vertices and edges of Γ ordered by inclusion, where an edge is regarded as a set of two vertices. Calculate the dimensions of the incidence posets of some small graphs. Show that the only connected graphs whose incidence posets have dimension 2 are the paths.

7. Prove Theorem 12.8.4.

8. Calculate the Möbius functions of the posets whose Hasse diagrams appear in Fig. 12.1.

9. Prove that the Möbius function of the lattice of subspaces of a vector space over $GF(q)$ is given by

$$\mu(Y, Z) = \begin{cases} (-1)^k q^{k(k-1)/2} & \text{if } Y \subseteq Z, \\ 0 & \text{otherwise}, \end{cases}$$

where $k = \dim(Z) - \dim(Y)$. [HINT: It suffices to consider the case when $Y = \{0\}$. Now put $t = -1$ in the q-binomial Theorem (9.2.5).]

10. Let a, b be elements of a poset P. Prove that $\mu(a, b) = \sum_{i \geq 0}(-1)^i c_i$, where c_i is the number of chains

$$a = x_0 < \ldots < x_i = b.$$

[HINT: Calling the right-hand side $\rho(a, b)$, it suffices to show that $\sum_{a \leq x \leq b} \rho(a, x) = 0$ for $a < b$. Now the displayed chain contributes $(-1)^i$ to $\rho(a, b)$, and also $(-1)^{i-1}$ to $\rho(a, x_{i-1})$.]

11. Let $(X, \mathcal{I})$ be a matroid.
    (a) Let $Y \subseteq X$. Prove that any basis for Y can be 'extended' to a basis for X.
    (b) Let $Y \subseteq X$ and let C be a cycle in Y. Prove that, for any $x \in C$, we have $\rho(Y \setminus \{x\}) = \rho(Y)$.
    (c) Show that the rank function satisfies

$$\rho(Y \cup Z) + \rho(Y \cap Z) \leq \rho(Y) + \rho(Z).$$

[HINT: Recall from linear algebra the argument which proves this (with equality) for subspaces of a vector space.]
    (d) Give an example where strict inequality holds in (c).

12. Let $(X, \mathcal{I})$ be a matroid, and $I \in \mathcal{I}$. Show that $(X \setminus I, \{J : J \cup I \in \mathcal{I}\})$ is a matroid. Prove that its rank function ρ' is given by $\rho'(Y) = \rho(Y \cup I) - \rho(I)$. Hence show that any interval in a geometric lattice is a geometric lattice.

13. Prove that the greedy algorithm succeeds in finding a basis of minimum weight in a weighted matroid.

14. Show that Arrow's Theorem is false if there are just two options and at least three individuals in the society. [HINT: try democracy!]
    How is this result related to the contents of Section 7.1?

15. Exploit the connection between terms in the disjunctive normal form and valuations to prove the disjunctive normal form theorem (12.4.2).

16. (a) Show that the free distributive lattice with 3 generators has cardinality 20.
    (b) COMPUTING PROJECT. Calculate the cardinality of the free distributive lattice for larger numbers of generators.

13. More on partitions and permutations

More and more I'm aware that the permutations are not unlimited.

Russell Hoban, *Turtle Diary* (1975)

TOPICS: Partition numbers; conjugacy classes of permutations; diagrams and tableaux; symmetric polynomials

TECHNIQUES: Generating functions; proof of identities by counting

ALGORITHMS: Robinson–Schensted–Knuth correspondence

CROSS-REFERENCES: Permutations and partitions (Chapter 3); partial order (Chapter 12); [Catalan numbers, involutions (Chapter 4), Gaussian coefficients (Chapter 9); cycle index (Chapter 15)]

In Chapter 3, we considered partitions and permutations of a finite set. Here, we look at the 'unlabelled' versions. These are partitions of an integer n, and conjugacy classes of permutations in the symmetric group S_n. It turns out that there are equal numbers of these objects, and a rich interplay between them. The story also involves symmetric functions and the character theory of S_n.

13.1. Partitions, diagrams, and conjugacy classes

Let n be a positive integer. A *partition* of n is an expression for n as a sum of positive integers, where the order of the summands is unimportant.[1] We can arrange the parts in order, with the largest first. Thus, there are five partitions of 4:

$$4 = 3 + 1 = 2 + 2 = 2 + 1 + 1 = 1 + 1 + 1 + 1.$$

As well as this obvious notation, a partition of n is sometimes written in the form $1^{a_1} 2^{a_2} \ldots n^{a_n}$. where a_i is the number of parts equal to i, that is, the number of occurrences of i as a term in the sum. The 'factor' i^{a_i} is not an exponential; the integer i is merely a placeholder for the term a_i. If $a_i = 0$, the 'factor' can be omitted. In this notation, the five partitions of 4 are

$$4^1, \ 3^1 1^1, \ 2^2, \ 2^1 1^2, \ 1^4.$$

---

[1] If the order of the summands is significant, then the number of partitions of n is 2^{n-1} for $n \geq 1$. See Exercise 9(b) of Chapter 4.

We also use the notation $\lambda \vdash n$ to mean 'λ is a partition of n'.

In addition, we use a pictorial representation of partitions by means of *diagrams*[2] $D(\lambda)$, defined as follows. Let λ be the partition $n = n_1 + \ldots + n_k$, with $n_1 \geq \ldots \geq n_k$. The diagram of λ has k rows; the ith row (numbering from the top) contains n_i cells, aligned at the left.[3] Cells may be represented either by dots or by empty squares, whichever is convenient; I will make use of both in appropriate places. Thus, the diagram of the partition $7 = 3 + 2 + 2$ or $3^1 2^2$ is shown in Fig. 13.1.

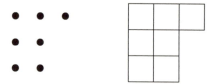

Fig. 13.1. The diagram of a partition

Let $\lambda \vdash n$. The *conjugate* or *dual partition* λ^* of λ is the partition of n whose diagram is the transpose (in the sense of matrices, that is, interchanging rows and columns) of that of λ. For example, if $\lambda = 3^1 2^2$, as above, then $\lambda^* = 3^2 1^1$. In general, if $\lambda = 1^{a_1} 2^{a_2} \ldots n^{a_n}$, then $\lambda^* = 1^{b_1} 2^{b_2} \ldots n^{b_n}$, where b_i is the number of indices j for which $a_j \geq i$. Obviously, $(\lambda^*)^* = \lambda$.

Let $p(n)$ be the number of partitions of n, the nth *partition number*. (Check that, for $n = 1, 2, 3, 4, 5$, we have $p(n) = 1, 2, 3, 5, 7$ respectively.) The function p is sometimes called the *partition function*. We prove first an expression for its generating function. By convention, $p(0) = 1$; the unique partition of 0 has no parts.

(13.1.1) Theorem. $\displaystyle\sum_{n \geq 0} p(n)t^n = \prod_{i \geq 1}(1 - t^i)^{-1}$.

PROOF. The right-hand side is

$$\prod_{i \geq 1}(1 + t^i + t^{2i} + \ldots) = (1 + t + t^2 + \ldots)(1 + t^2 + t^4 + \ldots)\ldots .$$

A term in t^n in this product is obtained by selecting, say, t^{a_1} from the first factor, t^{2a_2} from the second, and so on, with $a_1 + 2a_2 + \ldots = n$ (so that $1^{a_1} 2^{a_2} \ldots \vdash n$). Each partition of n gives a contribution of 1 to the coefficient of t^n, so this coefficient is equal to $p(n)$.

This expression for $\Pi(t) = \sum p(n)t^n$ is not much use as it stands. But in the next section, we'll see that it gives a recurrence relation for the partition numbers.[4]

---

[2] These are also called *Ferrers diagrams* or *Young diagrams*.

[3] This convention corresponds to the indexing of matrices, where rows are numbered down the page and columns from left to right. An alternative convention is based on Cartesian coordinates, where the independent variable increases from left to right, and the dependent variable from bottom to top. According to Ian Macdonald, *Symmetric Functions and Hall Polynomials*, p. 2, 'Readers who prefer this convention should read this book upside down in a mirror'. Computer users will recognise the difference between text and graphics output.

[4] For analysts, we note that $\Pi(t)$ is an analytic function of the complex variable t for $|t| < 1$, but has a singularity at every root of unity, so it cannot be analytically continued outside the unit disc. (The unit circle is a *natural boundary*.)

There are two convenient orderings defined on the set of all partitions of n. Let $\lambda, \mu \vdash n$; say, $\lambda : n = n_1 + \ldots + n_k$ and $\mu : n = m_1 + \ldots + m_l$, with the convention that undefined parts are zero.

(a) We say that λ precedes μ in the *reverse lexicographic order* (r.l.o.) if, for some i, we have $n_j = m_j$ for $j < i$ and $n_i > m_i$. (If we regard a partition as a 'word', whose 'letters' are positive integers, this is the dictionary order of words with the convention that large integers precede small ones in the 'alphabet'.) This is a total order.

(b) We say that λ precedes μ in the *natural partial order* (n.p.o.) (written $\lambda < \mu$) if $\lambda \neq \mu$ and

$$n_1 + \ldots + n_i \geq m_1 + \ldots + m_i$$

for all $i \geq 1$.

For $n \leq 5$, these two orders coincide. They differ first for $n = 6$, where $3^1 1^3$ and 2^3 are incomparable in the n.p.o. (though the first precedes the second in the r.l.o.). However, it is always true that r.l.o. is a linear extension (see Section 12.2) of n.p.o.:

(13.1.2) Proposition. *If $\lambda < \mu$, then λ precedes μ in the reverse lexicographic order.*

PROOF. With the notation as before, choose i such that $n_j = m_j$ for $j < i$ but $n_i \neq m_i$. Since $n_1 + \ldots + n_i \geq m_1 + \ldots + m_i$, we must have $n_i > m_i$.

Conjugation reverses the n.p.o.:

(13.1.3) Proposition. *If $\lambda \leq \mu$ then $\mu^* \leq \lambda^*$.*

PROOF. Suppose that $\mu^* \not\leq \lambda^*$, where μ^* is the partition $n = n_1^* + n_2^* + \ldots$, etc. (so that n_i^* is the number of j such that $n_j \geq i$, by definition of conjugation). Then, for some i, we have

$$m_1^* + \ldots + m_j^* \geq n_1^* + \ldots + n_j^* \quad \text{for } j < i$$
$$\text{and} \quad m_1^* + \ldots + m_i^* < n_1^* + \ldots + n_i^*,$$

so $t = m_i^* < n_i^* = s$.

Now $n_{i+1}^* + n_{i+2}^* + \ldots$ is the number of cells in the diagram of λ which lie to the right of the ith column; so

$$n_{i+1}^* + n_{i+2}^* + \ldots = \sum_{j=1}^{s}(n_j - i).$$

Similarly,

$$m_{i+1}^* + m_{i+2}^* + \ldots = \sum_{j=1}^{t}(m_j - i).$$

So

$$\sum_{j=1}^{t}(m_j - i) > \sum_{j=1}^{s}(n_j - i) \geq \sum_{j=1}^{t}(n_j - i),$$

the right-hand inequality holding because $s > t$ and $n_j \geq i$ for $j \leq s$. Hence

$$m_1 + \ldots + m_t > n_1 + \ldots + n_t,$$

and so $\lambda \not\leq \mu$.

Now we turn to permutations. We saw in Section 3.5 that any permutation of $\{1, \ldots, n\}$ can be expressed as the product of disjoint cycles, uniquely up to the order of the factors and the starting point of each cycle. If the cycle lengths are $n_1, \ldots, n_k$, then $n = n_1 + \ldots + n_k$, and so we have a partition of n, which is called the *cycle structure* of the permutation. Thus, cycle structure defines a map from the symmetric group S_n to the set of partitions of n.

Two permutations $g_1, g_2 \in S_n$ are said to be *conjugate* if $g_2 = h^{-1} g_1 h$ for some $h \in S_n$. Conjugacy is an equivalence relation on S_n, whose equivalence classes are called *conjugacy classes*.[5]

(13.1.4) Proposition. *Two permutations have the same cycle structure if and only if they are conjugate.*

PROOF. Suppose that $g_2 = h^{-1} g_1 h$. Let $(x_1\ x_2\ \ldots\ x_k)$ be a cycle of g_1, so that $x_i g_1 = x_{i+1}$ for $i = 1, \ldots, k-1$, and $x_k g_1 = x_1$. Let $y_i = x_i h$ for $i = 1, \ldots, k$. Then, for $i = 1, \ldots, k-1$, we have

$$y_i g_2 = y_i h^{-1} g_1 h = x_i g_1 h = x_{i+1} h = y_{i+1},$$

and similarly $y_k g_2 = y_1$. Thus, $(y_1\ y_2\ \ldots\ y_k)$ is a cycle of g_2. Thus, we obtain the cycle decompositon of g_2 from that of g_1 by replacing each point by its image under h. So the cycle structures are equal.

Conversely, let g_1 and g_2 have the same cycle structure. Calculate the cycle decomposition of each, and write that of g_2 under that of g_1 so that cycles of the same length correspond vertically. Now let h be the permutation obtained by mapping each point in the decomposition of g_1 to the point vertically below it. (So, if we forget all the brackets, what is written down is the two-line form of h.) Then $h^{-1} g_1 h = g_2$, by the same calculation as before.

For example, if $g_1 = (1\ 2\ 3)(4\ 5)(6)$ and $g_2 = (2\ 5\ 3)(4\ 6)(1)$, then $g_2 = h^{-1} g_1 h$, where $h = \begin{pmatrix} 1\ 2\ 3\ 4\ 5\ 6 \\ 2\ 5\ 3\ 4\ 6\ 1 \end{pmatrix} = (1\ 2\ 5\ 6)(3)(4)$ (in cycle notation).

It is clear that every partition of n is realised as the cycle structure of some permutation; so

the number of conjugacy classes in S_n is $p(n)$.

But we can do better, and calculate the conjugacy class sizes:

(13.1.5) Proposition. *Let* $\lambda = 1^{a_1} 2^{a_2} \ldots n^{a_n}$ *be a partition of* n. *Then the number of permutations with cycle structure* λ *is*[6]

$$\frac{n!}{\prod_{i=1}^{n} i^{a_i} a_i!}.$$

PROOF. If we write out the brackets for the cycle decomposition of such a permutation, there are $n!$ ways of entering the numbers $1, \ldots, n$ into the spaces. But we can start each of the a_i cycles of length i in any position in the cycle, in i^{a_i} ways, and permute these cycles arbitrarily, in $a_i!$ ways, for each i; so we have to divide $n!$ by the product of all these numbers.

---

[5] Conjugacy is an equivalence relation in any group. (Prove this.)

[6] In this expression, i^{a_i} has its usual mathematical meaning.

13.2. Euler's Pentagonal Numbers Theorem

A *pentagonal number* is a number of the form $k(3k-1)/2$ or $k(3k+1)/2$ for some non-negative number k. Alternatively, it is a number of the form $k(3k-1)/2$ for some (positive, negative, or zero) integer k. The second description is preferable, since it generates zero once only, whereas the first produces zero twice. The reason for the name is shown by the pictures of pentagonal numbers for small positive k.

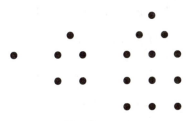

Fig. 13.2. Small pentagonal numbers

The next theorem, due to Euler, is quite unexpected, as is its application: it will enable us to derive an efficient recurrence relation for the partition numbers.

(13.2.1) Euler's Pentagonal Numbers Theorem
(a) *If n is not a pentagonal number, then the numbers of partitions of n into an even and an odd number of distinct parts are equal.*
(b) *If $n = k(3k-1)/2$ for some $k \in \mathbb{Z}$, then the number of partitions of n into an even number of distinct parts exceeds the number of partitions into an odd number of distinct parts by one if k is even, and vice versa if k is odd.*

For example, if there are four partitions of $n = 6$ into distinct parts, viz. $6 = 5+1 = 4+2 = 3+2+1$, two of each parity; while if $n = 7$, there are five such partitions, viz. $7 = 6+1 = 5+2 = 4+3 = 4+2+1$, three with an even and two with an odd number of parts.

PROOF. To demonstrate Euler's Theorem, we try to produce a bijection between partitions with an even and an odd number of distinct parts; we succeed unless n is a pentagonal number, in which case a unique partition is left out.

Let λ be any partition of n into distinct parts. We define two subsets of the diagram $D(\lambda)$ as follows:
- The *base* is the bottom row of the diagram (the smallest part).
- The *slope* is the set of cells starting at the east end of the top row and proceeding in a south-westerly direction for as long as possible.

Note that any cell in the slope is the last in its row, since the row lengths are all distinct. See Fig. 13.3.

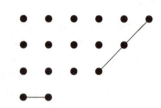

Fig. 13.3. Base and slope

Now we divide the set of partitions of n with distinct parts into three classes, as follows:

- *Class 1* consists of the partitions for which *either* the base is longer than the slope and they don't intersect, *or* the base exceeds the slope by at least 2;
- *Class 2* consists of the partitions for which *either* the slope is at least as long as the base and they don't intersect, *or* the slope is strictly longer than the base;
- *Class 3* consists of all other partitions with distinct parts.

Given a partition λ in Class 1, we create a new partition λ' by removing the slope of λ' and installing it as a new base, to the south of the existing diagram. In other words, if the slope of λ contains k cells, we remove one from each of the largest k parts, and add a new (smallest) part of size k. This is a legal partition with all parts distinct. Moreover, the base of λ' is the slope of λ, while the slope of λ' is at least as large as the slope of λ, and strictly larger if it meets the base. So λ' is in Class 2.

In the other direction, let λ' be in Class 2. We define λ by removing the base of λ' and installing it as a new slope. Again, we have a partition with all parts distinct, and it lies in Class 1. (If the base and slope of λ meet, the base is one greater than the second-last row of λ', which is itself greater than the base of λ', which has become the slope of λ. If they don't meet, the argument is similar.)

The partition shown in Fig. 13.3 is in Class 2; the corresponding Class 1 partition is shown in Fig. 13.4.

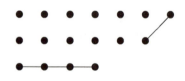

Fig. 13.4. A Class 1 partition

These bijections are mutually inverse. Thus, the numbers of Class 1 and Class 2 partitions are equal. Moreover, these bijections change the number of parts by 1, and hence change its parity. So, in the union of Classes 1 and 2, the numbers of partitions with even and odd numbers of parts are equal.

Now we turn to Class 3. A partition in this class has the property that its base and slope intersect, and either their lengths are equal, or the base exceeds the slope by 1. So, if there are k parts, then $n = k^2 + k(k-1)/2 = k(3k-1)/2$ or

$n = k(k + 1) + k(k - 1)/2 = k(3k + 1)/2$. Fig. 13.5 shows the two possibilities.

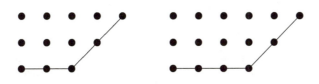

Fig. 13.5. Two Class 3 partitions

So, if n is not pentagonal, then Class 3 is empty; and, if $n = k(3k - 1)/2$, for some $k \in \mathbb{Z}$, then it contains a single partition with $|k|$ parts. Euler's Theorem follows.

(13.2.2) Corollary. $\prod_{n \geq 1}(1 - t^n) = \sum_{k=-\infty}^{\infty}(-1)^k t^{k(3k-1)/2}$.

PROOF. By Euler's Pentagonal Numbers Theorem, the right-hand side is the generating function for $\text{even}(n) - \text{odd}(n)$, where $\text{even}(n)$ and $\text{odd}(n)$ are the numbers of partitions having all parts distinct and having an even or odd number of parts respectively. We must show that the same is true for the left-hand side.

The coefficient of t^n is made up of contributions from factors $(1 - t^{n_1}), \ldots, (1 - t^{n_k})$, where $n_1 + \ldots + n_k = n$ and $n_1, \ldots, n_k$ are distinct; the contribution from this choice of factors is $(-1)^k$. So each term counted by $\text{even}(n)$ contributes 1, and each term counted by $\text{odd}(n)$ contributes -1. So the theorem is proved.

The right-hand side can be written as

$$1 + \sum_{k>0}(-1)^k \left(t^{k(3k-1)/2} + t^{k(3k+1)/2}\right),$$

using the first 'definition' of the pentagonal numbers. From this, we deduce the promised recurrence for the partition numbers. This illustrates the general principle that finding a linear recurrence relation for a sequence is equivalent to finding the inverse of its generating function (see Chapter 4, Exercise 12).

(13.2.3) Corollary. *For* $n > 0$,

$$p(n) = \sum_{k>0}(-1)^{k-1}(p(n - \tfrac{1}{2}k(3k - 1)) + p(n - \tfrac{1}{2}k(3k + 1)))$$
$$= p(n - 1) + p(n - 2) - p(n - 5) - p(n - 7) + p(n - 12) + \ldots,$$

with the convention that $p(n) = 0$ *for* $n < 0$.

PROOF. Since

$$\sum_{n \geq 0} p(n)t^n = \prod_{n>0}(1 - t^n)^{-1},$$

we have

$$\left(\sum_{n \geq 0} p(n)t^n\right) \cdot \left(1 + \sum_{k>0}(-1)^k(t^{k(3k-1)/2} + t^{k(3k+1)/2})\right) = 1.$$

For $n > 0$, the coefficient of t^n in the product is zero. Thus,

$$0 = p(n) + \sum_{k>0}(-1)^k(p(n - \tfrac{1}{2}k(3k-1)) + p(n - \tfrac{1}{2}k(3k+1))),$$

from which the result follows.

This is a linear recurrence relation in which the number of terms grows with n, but relatively slowly: there are about $\sqrt{8n/3}$ pentagonal numbers below n. Thus, it permits efficient calculation: $p(n)$ can be evaluated with $O(n^{3/2})$ additions or subtractions.

13.3. Project: Jacobi's Identity

In this section, I give a delightful proof, due to Richard Borcherds, of an identity of Jacobi.[6] The proof has the appearance of physics, although it is pure combinatorics; it involves double-counting states of Dirac electrons!

Jacobi's Identity asserts:

(13.3.1) Jacobi's Triple Product Identity

$$\prod_{n>0}(1 + q^{2n-1}z)(1 + q^{2n-1}z^{-1})(1 - q^{2n}) = \sum_{l\geq 0} q^{l^2} z^l.$$

It is an identity between formal power series in the indeterminates q and z. By replacing q by $q^{1/2}$ and moving the third term in the product to the right-hand side, the identity takes the form

$$\prod_{n>0}(1 + q^{n-1/2}z)(1 + q^{n-1/2}z^{-1}) = \left(\sum_{l\geq 0} q^{l^2/2}z^l\right)\left(\prod_{n>0}(1 - q^n)^{-1}\right) \qquad (*),$$

in which form we will prove it.

A *level* is a number of the form $n + \tfrac{1}{2}$, where n is an integer. A *state* is a set of levels which contains all but finitely many negative levels and only finitely many positive levels. The state consisting of all the negative levels and no positive ones is called the *vacuum*. Given a state S, we define the *energy* of S to be

$$\sum\{l : l > 0, l \in S\} - \sum\{l : l < 0, l \notin S\},$$

while the *particle number* of S is

$$|\{l : l > 0, l \in S\}| - |\{l : l < 0, l \notin S\}|.$$

Although it is not necessary for the proof, a word about the background is in order!

Dirac showed that relativistic electrons could have negative as well as positive energy. Since they jump to a level of lower energy if possible, Dirac hypothesised that, in a vacuum, all the negative energy levels are occupied. Since electrons obey the exclusion principle, this prevents further electrons from occupying these states. Electrons in negative levels are not detectable. If an electron gains enough energy to jump to a positive level, then it becomes 'visible'; and the 'hole' it leaves behind behaves like a particle with the same mass but opposite charge to an electron. (A few years later, positrons were discovered filling these specifications.) If the vacuum has no net particles and zero energy, then the energy and particle number of any state should be relative to the vacuum, giving rise to the definitions given.

---

[6] Jacobi's Identity implies Euler's Pentagonal Numbers Theorem: see Exercise 10.

We show that the coefficient of $q^m z^l$ on either side of $(*)$ is equal to the number of states with energy m and particle number l. This will prove the identity.

For the left-hand side this is straightforward. A term in the expansion of the product is obtained by selecting $q^{n-\frac{1}{2}}z$ or $q^{n-\frac{1}{2}}z^{-1}$ from finitely many factors. These correspond to the presence of an electron in positive level $n - \frac{1}{2}$ (contributing $n - \frac{1}{2}$ to the energy and 1 to the particle number), or a hole in negative level $-(n - \frac{1}{2})$ (contributing $n - \frac{1}{2}$ to the energy and -1 to the particle number). So the coefficient of $q^m z^l$ is as claimed.

The right-hand side is a little harder. Consider first the states with particle number 0. Any such state can be obtained in a unique way from the vacuum by moving the electrons in the top k negative levels up by $n_1, n_2, \ldots, n_k$, say, where $n_1 \geq n_2 \geq \ldots \geq n_k$. (The monotonicity is equivalent to the requirement that no electron jumps over another.) The energy of the state is thus $m = n_1 + \ldots + n_k$. Thus, the number of states with energy m and particle number 0 is equal to the number $p(m)$ of partitions of m, which is the coefficient of q^m in $\Pi(q) = \prod_{n>0}(1 - q^n)^{-1}$, by (13.1.1).

Now consider states with positive particle number l. There is a unique *ground state*, in which all negative levels and the first l positive levels are filled; its energy is $\frac{1}{2} + \frac{3}{2} + \ldots + \frac{2l-1}{2} = \frac{1}{2}l^2$, and its particle number is l. Any other state with particle number l is obtained from this one by 'jumping' electrons up as before; so the number of such states with energy m is $p(m - \frac{1}{2}l^2)$, which is the coefficient of $q^m z^l$ in $q^{l^2/2}z^l\Pi(q)$, as required.

The argument for negative particle number is similar.

13.4. Tableaux

Our definition of a tableau is not the most general one possible; what is defined here is usually called a *standard tableau*, but I will not talk about any non-standard tableaux![7]

Let λ be a partition of n, with diagram $D(\lambda)$. A *tableau*, or *Young tableau*, with shape λ, is an assignment of the numbers $1, 2, \ldots, n$ to the cells of $D(\lambda)$, in such a way that the numbers in any row or column are strictly increasing. For example, the three tableaux with shape $3^1 1^1$ are shown in Fig. 13.6.

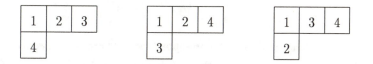

Fig. 13.6. Tableaux

The number of tableaux with shape λ is denoted by f_λ. Clearly, we have $f_\lambda = f_{\lambda^*}$, the corresponding tableaux being related by transposition.

There is a somewhat unexpected formula for f_λ. Given a cell (i, j) of the diagram $D(\lambda)$, the *hook* $H(i, j)$ associated with it is the set consisting of this cell and all those cells to the south or east of it; that is, all cells (i, j') in the diagram with $j' \geq j$, and all cells (i', j) with $i' \geq i$. The *hook length* $h(i, j)$ is the number of cells in the hook $H(i, j)$.

(13.4.1) Theorem. $f_\lambda = \dfrac{n!}{\prod_{(i,j)\in D(\lambda)} h(i, j)}.$

_____

[7] Plural of *tableau.*

In his book *Symmetric Functions and Hall Polynomials*, Ian Macdonald refers on p. 53 to 'The fact that the number of standard tableaux of shape λ is equal to $n!/h(\lambda)$', and says 'No direct combinatorial proof seems to be known.' The note refers to a proof of this *hook length formula* at the end of a series of exercises, quoting earlier results on symmetric functions. I do not plan to trace through the argument here!

The numbers f_λ have another combinatorial interpretation. Let λ be the partition $n = n_1 + \ldots + n_k$, where (as usual) $n_1 \geq \ldots \geq n_k$. Suppose that, in an election, n voters cast their votes for k candidates, with the ith candidate receiving n_i votes for $i = 1, \ldots, k$. Then the number of ways in which the votes can be counted, so that at no stage in the count is the jth candidate ahead of the ith, for any $j > i$, is f_λ. To see this, record the count by writing the numbers $1, \ldots, n$ in the cells of $D(\lambda)$, where m is put in the ith row (immediately to the right of the entries already there) if the mth vote goes to the ith candidate. By assumption, we have a tableau with shape λ; and every tableau corresponds to a possible count.

In particular, if λ is the partition $2n = n + n$, then f_λ is the Catalan number C_{n+1} — this interpretation of f_λ is in exact agreement with that for the Catalan number given in Exercise 15(b) of Chapter 4. So the numbers f_λ generalise the Catalan numbers. We can check the hook length formula (13.4.1) in this case. The hook lengths for this partition λ are $n+1, n, \ldots, 2$ in the first row, and $n, n-1, \ldots, 1$ in the second; so

$$C_{n+1} = \frac{(2n)!}{(n+1)!n!} = \frac{1}{n+1}\binom{2n}{n},$$

in agreement with (4.5.2).

Another important property of tableaux is the *Robinson–Schensted–Knuth correspondence*:

(13.4.2) Robinson–Schensted–Knuth correspondence. *There is a bijection between the set of permutations of $\{1, \ldots, n\}$, and the set of ordered pairs of tableaux of the same shape. Under this bijection, if g corresponds to the pair (S, T) of tableaux, then g^{-1} corresponds to (T, S). In particular, the two tableaux corresponding to a permutation $g \in S_n$ are equal if and only if $g^2 = 1$.*

PROOF. We give a constructive proof, of course! We build a pair (S, T) of tableaux from a permutation g, which we take in passive form $(a_1, \ldots, a_n)$. The construction proceeds step by step. Before the first step, S and T are empty. At the start of the ith step, S and T are 'partial tableaux' with i cells, having the same shape. (This means that their entries are distinct but not necessarily the first i natural numbers, and the rows and columns are strictly increasing. In fact, T is a genuine standard tableau, but S is not in general.) In step i, we add a new cell to the shape, and add entries a_i to S and i to T, in a manner to be described. The procedure is recursive; we define a 'subroutine' called INSERT, which puts an integer a in the jth row of a partial tableau T.

> **Subroutine: INSERT** a into the jth row
>
> *If a is greater than the last element of the jth row, then append it to this row. (If the jth row is empty, put a in the first position.) Otherwise, let x be the smallest element of the jth row for which $a \not> x$. 'Bump' x out of the jth row, replacing it with a; then INSERT x into the $(j+1)^{\text{st}}$ row.*

Now we can give a complete specification of the RSK algorithm:

> **RSK algorithm**
>
> *Start with S and T empty.*
> *For $i = 1, \ldots, n$, do the following:*
> - *INSERT a_i into the first row of S. This causes a cascade of 'bumps', ending with a new cell being created and a number (not less than a_i) written into it.*
> - *Now create a new cell in the same position in T and write i into it.*

We have to check that, after the ith stage, S and T are partial tableaux. The fact that rows and columns are increasing is, for S, a consequence of the way INSERT works; for T, it is because i is greater than any element previously in the tableau. The point of substance is that the newly created cell doesn't violate the condition that the row *lengths* are non-increasing; that is, there should be a cell immediately above it. This is because the element 'bumped' is smaller than the element to the right of the position it is 'bumped' out of, and so it comes to rest to the left of this position.

At the end of the algorithm, we have two tableaux of the same shape.

We illustrate the algorithm with the permutation $(2, 3, 1)$.

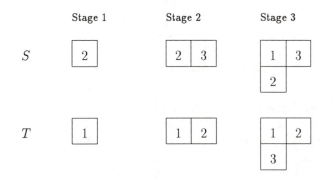

Fig. 13.7. The RSK algorithm

At stage 3, 2 is 'bumped' by 1 into the second row.

The procedure can be reversed, to construct a permutation from a pair of standard tableaux of the same shape. To see this, note that we can locate n in the tableau T, and then reconstruct the cascade of 'bumps' required to move the corresponding element of S to that position; the insertion triggering this cascade is a_n. Working back in the same way, we recover the entire permutation. (A few worked examples make this clearer than pages of explanation!)

Now we come to the final claim that, if (S, T) corresponds to g, then (T, S) corresponds to g^{-1}. My argument here will be somewhat 'hand-waving'. Let g and g^{-1} have passive forms $(a_1, \ldots, a_n)$ and $(b_1, \ldots, b_n)$ respectively. Thus, $a_i = j$ if and only if $b_j = i$. For the permutation g, stage i in the construction inserts a_i into S and i into T; a_i goes into the first row, and i into a position determined by a cascade of 'bumps' in S. Subsequently, i keeps its place in T, but a_i may be 'bumped' down by subsequent insertions corresponding to values of s with $s > i$ but $a_s < a_i$. Each 'bump' moves it down one row.

Now, corresponding to g^{-1}, at stage j, we insert $b_j = i$ into the first row of S, and j into T, in a position determined by a sequence of bumps in S. One can check that these are the same bumps that moved a_i before, but all in a single cascade rather than one at a time. Dually, the bumps which subsequently move b_j down are those which determine the position of i in the previous case. So the resulting tableaux S and T are precisely the T and S corresponding to g, and the claim is proved.

(13.4.3) Corollary. (a) $\sum_{\lambda \vdash n} (f_\lambda)^2 = n!$.

(b) $\sum_{\lambda \vdash n} f_\lambda = s(n)$, where $s(n)$ is the number of solutions of $g^2 = 1$ in S_n.

The function $s(n)$ was considered in Section 4.4, where we proved a lower bound for it. We can now re-do this and give an upper bound too.

(13.4.4) Corollary. $\sqrt{n!} \leq s(n) \leq \sqrt{p(n)n!}$.

PROOF. (a) $(\sum f_\lambda)^2 \geq \sum f_\lambda^2$, since the right-hand side omits all 'product' terms $2f_\lambda f_\mu$.

(b) The vectors $(1, 1, \ldots, 1)$ and $(f_{\lambda_1}, f_{\lambda_2}, \ldots, f_{\lambda_{p(n)}})$ in the Euclidean space of dimension $p(n)$ have lengths $\sqrt{p(n)}$ and $\sqrt{n!}$, and inner product $s(n)$.

13.5. Symmetric polynomials

Let $x_1, \ldots, x_N$ be indeterminates. A polynomial $f(x_1, \ldots, x_N)$ is called *symmetric* if it is left unchanged by any permutation of its arguments: $f(x_{1g}, \ldots, x_{Ng}) = f(x_1, \ldots, x_N)$ for all $g \in S_N$. (The older term 'symmetric functions' is often used; I will avoid this since it has at least two more general meanings.)

Any symmetric polynomial can be written uniquely as a sum of parts which are *homogeneous* (that is, every term has the same total degree). These homogeneous parts are themselves symmetric. So we may restrict our attention to homogeneous symmetric polynomials, of degree n, say.

We now define some special classes of symmetric polynomials. Let λ be the partition $n = n_1 + \ldots + n_k$.

(a) The *basic polynomial* m_λ is the sum of the term $x_1^{n_1} \ldots x_k^{n_k}$ and all the other terms which can be obtained from this one by permuting the indeterminates. (If some of the parts n_i are equal, the same term will come up more than once; but each term is only included once.)

(b) The *elementary symmetric polynomial* e_n is the sum of all products of n distinct indeterminates; the *complete symmetric polynomial* h_n is the sum of all products of n indeterminates (repetitions allowed); and the *power sum polynomial* p_n is $x_1^n + \ldots + x_N^n$.

(c) If z is one of the symbols e, h or p, then we define $z_\lambda = z_{n_1} \ldots z_{n_k}$.

For example, if there are three indeterminates, and λ is the partition $3 = 2 + 1$, then

$$m_\lambda = x_1^2 x_2 + x_2^2 x_1 + x_1^2 x_3 + x_3^2 x_1 + x_2^2 x_3 + x_3^2 x_2,$$
$$e_\lambda = (x_1 x_2 + x_1 x_3 + x_2 x_3)(x_1 + x_2 + x_3),$$
$$p_\lambda = (x_1^2 + x_2^2 + x_3^2)(x_1 + x_2 + x_3),$$
$$h_\lambda = e_\lambda + p_\lambda.$$

(13.5.1) Theorem. *For $N \geq n$, if z is one of the symbols m, e, h or p, then any homogeneous symmetric polynomial f of degree n in $x_1, \ldots, x_N$ can be written uniquely as a linear combination $\sum_{\lambda \vdash n} c_\lambda z_\lambda$. Moreover, in all cases except $z = p$, if f has integer coefficients, then the numbers c_λ are integers.*

PROOF. For $z = m$, this is clear: if one term of m_λ occurs in f, then all the other terms appear with the same coefficient.

To show the rest of the theorem, we have to demonstrate that the m_λ can be expressed as linear combinations of the z_μ (with integer coefficients if $z \neq p$). I will consider $z = e$ now; the others will emerge naturally later. The key fact is:

Suppose that $e_\lambda = \sum_{\mu \vdash n} a_{\lambda\mu} m_\mu$. Then $a_{\lambda\lambda^*} = 1$, and $a_{\lambda\mu} = 0$ unless $\mu \geq \lambda^*$ in the natural partial order.

For, if λ is the partition $n = n_1 + \ldots + n_k$, then e_λ contains the term

$$(x_1 x_2 \ldots x_{n_1})(x_1 \ldots x_{n_2}) \ldots (x_1 \ldots x_{n_k}),$$

which occurs in m_{λ^*}; so $a_{\lambda\lambda^*} = 1$. Any other monomial in e_λ corresponds to a partition greater than this one.

Thus, if the e_λ are ordered according to the reverse lexicographic order, and the m_λ according to the r.l.o. of their duals, then the matrix expressing the es in terms of the ms is upper triangular, with diagonal entries 1 and all entries integers. (Recall that the r.l.o. is a linear extension of the n.p.o.) So it is invertible, and its inverse has the same form. (Compare the Möbius inversion algorithm in Section 12.7.)

(13.5.2) Corollary. *Any symmetric polynomial $f(x_1, \ldots, x_N)$ can be written as a polynomial $g(z_1, \ldots, z_N)$, where z is one of the symbols e, h, p. In the first two cases, if f has integer coefficients, then so does g.*

This holds because the z_λ are all possible monomials of degree n which can be formed from $z_1, \ldots, z_N$.

In the case $z = e$, this is a version of *Newton's Theorem* on symmetric functions. The particular significance of this case is that, if $\alpha_1, \ldots, \alpha_N$ are the roots of the polynomial $\phi(t) = t^N + a_1 t^{N-1} + \ldots + a_N = 0$, then

$$a_i = (-1)^i e_i(\alpha_1, \ldots, \alpha_N),$$

so any symmetric polynomial in the roots of ϕ can be written as a polynomial in its coefficients. (Newton's Theorem extends to larger classes of functions, such as rational functions.)

Further results about symmetric polynomials can be expressed conveniently in terms of their generating functions. Define

$$E(t) = \sum_{n \geq 0} e_n t^n,$$

$$H(t) = \sum_{n \geq 0} h_n t^n,$$

$$P(t) = \sum_{n \geq 1} p_n t^{n-1}.$$

(These series of course also involve the indeterminates $x_1, \ldots, x_N$.) Now we have

$$E(t) = \prod_{r=1}^{N} (1 + x_r t),$$

$$H(t) = \prod_{r=1}^{N} (1 - x_r t)^{-1},$$

as is shown by expanding the products on the right in the usual way. In particular:

(13.5.3) Proposition. *(a)* $H(t) = E(-t)^{-1}$.

(b) $\sum_{r=0}^{n} (-1)^r e_r h_{n-r} = 0$ *for* $n \geq 1$.

Here (b) comes from expanding $E(-t)H(t) = 1$. It is a recursive relation expressing e_n in terms of $e_0, \ldots, e_{n-1}$ and $h_0, \ldots, h_n$. By induction, e_n can be expressed as a polynomial in $h_0, \ldots, h_n$ with integer coefficients. This is equivalent to the assertion that the polynomials e_λ are linear combinations of the h_μ with integer coefficients. This proves the case $z = h$ of Theorem 13.5.1.

The situation for $P(t)$ is a little less obvious:

(13.5.4) Proposition. *(a)* $\dfrac{d}{dt} H(t) = P(t)H(t)$ *and* $\dfrac{d}{dt} E(t) = P(-t)E(t)$.

(b) $nh_n = \sum_{r=1}^{n} p_r h_{n-r}$ *and* $ne_n = \sum_{r=1}^{n} (-1)^{r-1} p_r e_{n-r}$.

PROOF. (a)

$$P(t) = \sum_{r \geq 1} p_r t^{r-1}$$

$$= \sum_{r \geq 1} \sum_{i=1}^{N} x_i^r t^{r-1}$$

$$= \sum_{i=1}^{N} \frac{x_i}{1 - x_i t}$$

$$= \frac{d}{dt} \sum_{i=1}^{N} \log(1 - x_i t)^{-1}$$

$$= \frac{d}{dt} \log H(t);$$

the argument for the other part is similar.

(b) comes from (a) by expanding and equating coefficients.

The result of (b) allows us to express e_n or h_n as polynomials in $p_1, \ldots, p_n$, and hence e_λ or h_λ as linear combinations of the p_μ. But this time the coefficients are rational numbers, not integers, because of the terms ne_n, nh_n in (b). For example,

$$e_2 = \tfrac{1}{2}(p_1^2 - p_2), \qquad h_2 = \tfrac{1}{2}(p_1^2 + p_2).$$

There are several further reasons for combinatorialists to be interested in symmetric polynomials. One is the fact that we have the indeterminates $x_1, \ldots, x_N$ at our disposal; substitutions of particular values lead to interesting specialisations. For example (taking $n = N$):

(a) putting $x_1 = \ldots = x_n = 1$, we have $E(t) = (1 + t)^n$ and $e_r = \binom{n}{r}$ giving the Binomial Theorem (3.3.1). Similarly, $H(t) = (1 - t)^{-n}$ and $h_r = \binom{n+r-1}{r}$.

(b) Putting $x_i = q^{i-1}$ for $i = 1, \ldots, n$, we find that $E(t) = \prod_{i=1}^{n}(1 + q^{i-1}t)$ is the left-hand side of the q-binomial Theorem (9.2.4); so

$$e_r(1, q, \ldots, q^{n-1}) = q^{r(r-1)/2} \begin{bmatrix} n \\ r \end{bmatrix}_q,$$

the Gaussian coefficient.

Secondly, we have now four bases for the space of symmetric polynomials of degree n, namely (m_λ), (e_λ), (h_λ) and (p_λ). A further important basis consists of the *Schur functions* s_λ. The transition matrices between these bases define interesting arrays of numbers indexed by pairs of partitions. In many cases, these have combinatorial significance, or specialise to more familiar numbers, including the numbers f_λ of standard tableaux (Section 13.4), Stirling and Bell numbers (Sections 4.5, 5.3), and cycle indices of symmetric and alternating groups (see Section 15.3). For algebraists, I mention the fact that the transition matrix from (p_λ) to (s_λ) is the character table of the symmetric group S_n. See Macdonald, *Symmetric Functions and Hall Polynomials*, for an overview of this material. Reading it, one can appreciate the view held by some people, that if it isn't related to symmetric polynomials, then it isn't combinatorics!

13.6. Exercises

1. In the spirit of Section 3.12, devise an algorithm for generating the partitions of n, one at a time, in reverse lexicographic order.

2. Use the recurrence relation (13.2.3) to calculate $p(n)$ for $n \leq 20$.

3. Prove that $p(n) < F_n$ for $n \geq 5$, where F_n is the nth Fibonacci number.

4. Show that conjugation of partitions does not reverse the r.l.o.

5. Define two operations $\circ$, $\bullet$ on partitions as follows. Let $\lambda : n = n_1 + \ldots + n_k$, $\mu : m = m_1 + \ldots + m_l$ be partitions of n and m respectively; undefined parts are zero. Then $\lambda \circ \mu$ and $\lambda \bullet \mu$ are the partitions of $m + n$ defined thus: for $\lambda \circ \mu$ we add the parts of λ and μ, viz.

$$\lambda \circ \mu : (n + m) = (n_1 + m_1) + (n_2 + m_2) + \ldots,$$

while the parts of $\lambda \bullet \mu$ are the parts of λ and μ together (arranged in non-increasing order). Prove that

$$(\lambda \circ \mu)^* = \lambda^* \bullet \mu^*.$$

6. (a) Prove that, if $k > n/2$, the number of permutations in S_n having a cycle of length k is $n!/k$.

(b) If $t(n)$ is the *proportion* of permutations in S_n which have a cycle of length greater than $n/2$, show that

$$\lim_{n \to \infty} t(n) = \log 2.$$

7. Let $\Pi(t) = \sum_{n \geq 0} p(n)t^n$ be the generating function for the partition numbers. Let $\sigma(n)$ be the sum of the divisors of n, and $\Sigma(t) = \sum_{n > 0} \sigma(n)t^{n-1}$ its generating function. Prove that

$$\frac{d}{dt}\Pi(t) = \Sigma(t)\Pi(t),$$

and deduce that

$$np(n) = \sum_{k=1}^{n} \sigma(k)p(n - k).$$

8. Prove that $h_r(1, q, \ldots, q^{n-1}) = \begin{bmatrix} n+r-1 \\ r \end{bmatrix}_q$.

9. Let $x_i = 1/N$ for $1 \leq i \leq N$, and let $N \to \infty$. Show that the limiting values of $E(t)$ and $H(t)$ are both equal to e^t.

10. Deduce Euler's Pentagonal Numbers Theorem from Jacobi's Triple Product Identity. [HINT: put $q = t^{3/2}$, $z = -t^{-1/2}$.]

11. Let A be a matrix of zeros and ones, with row sums $n_1 \geq \ldots \geq n_k > 0$ and column sums $m_1 \geq \ldots \geq m_l > 0$; let λ and μ be the partitions $n = n_1 + \ldots + n_k$ and $n = m_1 + \ldots + m_l$. Show that the polynomial e_λ contains a term $x_1^{m_1} \ldots x_l^{m_l}$. Show further that, if

$$e_\lambda = \sum_{\mu \vdash n} a_{\lambda\mu} m_\mu,$$

then $a_{\lambda\mu}$ is equal to the number of matrices A which satisfy the above conditions.

14. Automorphism groups and permutation groups

> There is transitive motion and there is intransitive motion: the motion of a galloping horse is transitive, it passes through our field of vision and continues on to wherever it is going; the motion in a tile pattern is intransitive, it moves but it stays in our field of vision.
>
> Russell Hoban, *Pilgermann* (1983)

TOPICS: Permutation groups, automorphism groups; orbits, transitivity, primitivity, generation

TECHNIQUES: Group theory

ALGORITHMS: Schreier–Sims algorithm

CROSS-REFERENCES: Labelled and unlabelled structures (Chapter 2), permutations (Chapter 3), STS(7), [STS(9)] (Chapter 8), Petersen graph (Chapter 11), [Möbius function (Chapter 12)], cycle structure (Chapters 3, 13, 15)

Groups perform two main functions in combinatorics, paradoxically opposed. On the one hand, they measure order. Any combinatorial object has an automorphism group; the larger the group, the more symmetrical the object. On the other, they measure disorder. The most familiar example of this is Rubik's cube, whose possible configurations (more than 10^{19}) are the elements of a group, only the identity of which corresponds to the completely ordered state. We'll see in this chapter that the same basic principles underlie the study of groups in both these rôles.

14.1. Three definitions of a group

In this section, we'll re-write history a bit, tracing in idealised form the path from the definition of a group as 'all symmetries of an object' to the modern axiomatic definition. The point of this journey is to see how the various concepts are related.

By an *object* I will mean a pair $(X, \mathcal{S})$, where X is a set, and $\mathcal{S}$ any structure on X, whose exact nature needn't be specified: it may be a set of unordered or ordered pairs (i.e., a graph or digraph), a set of subsets or partitions of X, or something more recondite (such as a set of paths of length 3 using vertices of X, or a set of weight functions on the edges of the complete graph on X). The point is that, given any permutation g on X, there should be a natural way of applying g to $\mathcal{S}$. For example, if $(X, \mathcal{S})$ is a graph, we apply g to each edge in $\mathcal{S}$ to obtain the edge set $\mathcal{S}g$. If $\mathcal{S}$ is a set of sets of ..., we apply this construction recursively.

The permutation g of X is an *automorphism* of (X, S) if $Sg = S$. The *automorphism group* of (X, S) is the set $\text{Aut}(X, S)$ of all automorphisms of (X, S). A subset G of $\text{Sym}(X)$ is an *automorphism group* if $G = \text{Aut}(X, S)$ for some structure S on X. This is our first 'definition' of a group.

An automorphism group G has the following properties:
(P1) it contains the identity permutation;
(P2) it contains the inverse of each of its elements;
(P3) it contains the composition of each pair of its elements.
(The first condition is clear. For the second, if $S = Sg$, we can apply g^{-1} to both sides, yielding $Sg^{-1} = S$. For the third, if $Sg = Sh = S$, then $S(gh) = (Sg)h = S$.)

These facts form the basis of our second definition. A set G of permutations of X is a *permutation group* on X if it satisfies (P1), (P2) and (P3). We observed that every automorphism group is a permutation group; is the converse true, or have we strictly enlarged the domain of groups?

It turns out that, indeed, every permutation group is the automorphism group of some object. (See Exercise 1 for a proof.)

However, this is not the end of the story. Not every permutation group is the automorphism group of a graph, for example. (There are just two different graphs on the vertex set $\{1, 2\}$, and both have two automorphisms. So the permutation group on this set which contains only the identity permutation is not the automorphism group of any graph. Note that the construction of Exercise 1 shows that it is the automorphism group of the digraph with edge $(1, 2)$.) The problem of deciding which permutation groups are automorphism groups of graphs is unsolved.

The next step is in the spirit of nineteenth-century axiomatic mathematics. It was decided that the important thing about a group is the operation of composition. In terms of this, for example, we can characterise the identity permutation e by the fact that $eg = ge = g$ for all permutations g, and the inverse g^{-1} of a permutation g by $gg^{-1} = g^{-1}g = e$. Let us temporarily write the composition of g and h as $g \circ h$. Now a permutation group G satisfies the following conditions:
(A1) *Associativity:* $g \circ (h \circ k) = (g \circ h) \circ k$ for all $g, h, k \in G$;
(A2) *Identity:* there exists $e \in G$ with $e \circ g = g \circ e$ for all $g \in G$;
(A3) *Inverses:* for any $g \in G$, there exists $g^{-1} \in G$ with $g \circ g^{-1} = g^{-1} \circ g = e$.
Associativity is a general property of composition of functions:

$$x(g \circ (h \circ k)) = (xg)(h \circ k) = ((xg)h)k = (x(g \circ h))k = x((g \circ h) \circ k).$$

We observed that the identity and inverse permutations have the required properties, and they are contained in G by (P2) and (P3).

Dyck defined an *abstract group* to be a set G with a binary operation $\circ$ defined on it satisfying (A1), (A2) and (A3). Thus, every permutation group is an abstract group. Again, we must ask whether the converse is true. The fact that it is, is the content (and the *raison d'être*) of Cayley's Theorem:

(14.1.1) Cayley's Theorem. *Every abstract group is isomorphic to a permutation group.*

PROOF. We are given an abstract group G, with operation $\circ$, and are required to find a permutation group G' on a set X, whose elements are in one-to-one correspondence with those of G, such that the element of G' corresponding to $g \circ h$ is the composition of the elements corresponding to g and h.

We take $X = G$, and let $G' = \{\rho_g : g \in G\}$, where ρ_g is the *right translation* by g:

$$x\rho_g = x \circ g \quad \text{for all } x, g \in G.$$

It isn't clear yet that ρ_g is a permutation; but at least $\rho_g \neq \rho_h$ for $g \neq h$ (consider their effect on the element e), so that we have a one-to-one correspondence. Now we have

$$x\rho_g\rho_h = (x \circ g) \circ h = x \circ (g \circ h) = x\rho_{goh},$$

so the group operation in G corresponds to composition. From this, conditions (P1)–(P3) follow: closure is obvious ($\rho_g\rho_h = \rho_{goh}$); ρ_e is the identity permutation; and $\rho_{g^{-1}}$ is the inverse mapping to ρ_g (from which it follows that ρ_g is indeed a permutation).

It follows of course that every abstract group is an automorphism group, so the three concepts are identical. More is true. Frucht showed that every abstract group is the automorphism group of a graph. (In Section 14.7, we outline a proof of this.) Frucht showed further that in fact this graph can be taken to be trivalent. A sheaf of similar results is known.

From now on, we abbreviate 'abstract group' to 'group', and represent the group operation by juxtaposition gh instead of $g \circ h$. Most accounts now go much further, hiding the origins of the concept by reversing the procedure. A group is *defined* by axioms (A1)–(A3);[1] Cayley's Theorem shows that it makes sense to represent groups by means of permutations in order to study them (nothing is lost by this). Of course, the definition of a permutation group then changes: it is a set of permutations which, equipped with the operation of composition, forms a group!

We need one more concept. This is because the construction in Cayley's Theorem isn't the only way in which a group can be represented by permutations. So we define an *action* of a group G on a set X to be a map θ from G to the set $\mathrm{Sym}(X)$ of permutations of X, satisfying

$$(gh)\theta = (g\theta)(h\theta),$$
$$1\theta = 1,$$
$$g^{-1}\theta = (g\theta)^{-1},$$

where we used the same notation for group operations and permutations (juxtaposition, 1, and $^{-1}$). In fact, the second and third conditions follow from the first, which says that θ is a *homomorphism* from G into $\mathrm{Sym}(X)$.

---

[1] Often 'closure' is given as an axiom. Since a binary operation is defined on all pairs, this is not necessary; it is a historical vestige, or ontogeny repeating phylogeny.

The same group can have many different actions. We need to be able to say when two actions are 'the same'. Let θ, ϕ be actions of G on sets X, Y. We call these actions *equivalent* if there is a bijection $f : X \to Y$ such that

$$(xf)(g\phi) = (x(g\theta))f$$

for all $x \in X$, $g \in G$. In other words, if we use f to identify the sets X and Y, then any element of G induces the same permutation on the two sets.

For an example, let G be the symmetric group S_3, regarded as the automorphism group of a triangle (Fig. 14.1). Then G acts on the vertices and on the edges of the

Fig. 14.1. A triangle

triangle. These actions are equivalent by means of the map f, where $1f = \{2, 3\}$, $2f = \{3, 1\}$, $3f = \{1, 2\}$. (For example, if a permutation g carries 1 to 2, then it carries $\{2, 3\}$ to $\{3, 1\}$.)

14.2. Examples of groups

Perhaps the most famous groups are the *cyclic groups* C_n. The group C_n can be regarded as the additive group of congruence classes modulo n, or as the multiplicative group of all n^{th} roots of unity in $\mathbb{C}$ (that is, $\{e^{2k\pi i/n} : k = 0, \ldots, n-1\}$), or (for $n > 2$) as the automorphism group of the *cyclic digraph* with vertex set $\{0, 1, \ldots, n-1\}$ and edge set

$$\{(i, i+1) : i = 0, \ldots, n-2\} \cup \{(n-1, 0)\}.$$

Algebraically, an important fact is that it is *generated* by a single element g, that is, all its elements are powers of g. Any finite group with this property is cyclic.

(We say that a group G is generated by a set S of elements if each member of g can be expressed as a product of elements of S and their inverses, in any order and allowing repetitions. This is logically equivalent to saying that S is not contained in any proper subgroup of G, but expresses the concept in a more positive way. More generally, if S is a subset of a group G, the subgroup H *generated by* S consists of all products of elements of S and their inverses; it is also characterised as the smallest subgroup of G containing S, that is, the intersection of all subgroups of G containing S. Since every subset of $\mathrm{Sym}(X)$ generates some permutation group, we have a potentially enormous collection of groups; but it is quite difficult to deduce properties of the group from a generating set. We will consider this problem in Section 14.4.)

A closely related group is the *dihedral group* D_{2n} of order $2n$. For $n \geq 3$, D_{2n} is the automorphism group of the cyclic (undirected) graph with vertex set $\{0, 1, \ldots, n-1\}$ and edge set

$$\{\{i, i+1\} : i = 0, \ldots, n-2\} \cup \{\{n-1, 0\}\},$$

or the group of symmetries of a regular n-gon. It contains the cyclic group C_n as a subgroup (the rotations of the n-gon). The remaining elements are reflections of the n-gon in its n axes of symmetry. (If n is odd, all axes of symmetry are alike; but, if n is even, there are two types, one joining opposite vertices and the other joining midpoints of opposite edges.) The dihedral groups can be defined consistently for smaller n: D_2 is the cyclic group of order 2 (generated by one reflection), and D_4 is the *Klein group* $V_4 = \{1, a, b, c\}$, where $a^2 = b^2 = c^2 = 1$, $ab = c$, $bc = a$, $ca = b$. Note that V_4 is the group of symmetries of a rectangle.

We have already met the *symmetric group* $\mathrm{Sym}(X)$, consisting of all permutations of X. If $|X| = n$, it is also denoted by S_n, and its order is $n!$. We saw in Chapter 5 that, for $n > 1$, S_n has a subgroup of order $n!/2$ consisting of the even permutations of X, called the *alternating group* and denoted by $\mathrm{Alt}(X)$ or A_n. We see that S_2 is the cyclic group C_2, while A_3 and S_3 are isomorphic to C_3 and D_6 respectively.

We met briefly the *general linear group* $\mathrm{GL}(n, q)$ consisting of all invertible $n \times n$ matrices over $\mathrm{GF}(q)$ in Chapter 9, where we calculated its order.

Groups can be built up from smaller ones. Two important constructions are the *direct product* and *wreath product*, which we now define.

Let G and H be permutation groups on sets X and Y respectively. We assume that X and Y are disjoint. The *direct product* $G \times H$ consists of all ordered pairs (g, h) with $g \in G$ and $h \in H$, and acts on the disjoint union $X \cup Y$ in the following way:

$$z(g, h) = \begin{cases} zg & \text{if } z \in X; \\ zh & \text{if } z \in Y. \end{cases}$$

(You should check that this is an action.) The group operation is given by

$$(g_1, h_1)(g_2, h_2) = (g_1 g_2, h_1 h_2).$$

The action of $G \times H$ on $X \cup Y$ is called its *natural action*. Another action is its *product action* on $X \times Y$, defined by

$$(x, y)(g, h) = (xg, yh).$$

The *wreath product* $G \, \mathrm{wr} \, H$ is more difficult to define abstractly; I will describe it as a permutation group. Its *natural action* is on the set $X \times Y$; but we take $Y = \{y_1, \ldots, y_n\}$, and regard $X \times Y$ as the disjoint union of n copies $X_1, \ldots, X_n$ of X, where $X_i = X \times \{y_i\}$. Now we define two permutation groups:

- The *bottom group* B is the direct product of n copies of G, in its natural action on $X_1 \cup \ldots \cup X_n$. In other words, B acts by the rule

$$(x, y_i)(g_1, \ldots, g_n) = (xg_i, y_i).$$

- The *top group* T consists of H acting on the second coordinate:

$$(x, y_i)h = (x, y_i h).$$

In other words, T shifts the sets $X_1, \ldots, X_n$ around bodily.

Now the wreath product G wr H is the group generated by B and T (and consists of all products bt for $b \in B$, $t \in T$).

There is another action of the wreath product, the *product action*, on the set X^Y of all functions from Y to X. We can regard a function $f \in X^Y$ as an n-tuple $(f(y_1), \ldots, f(y_n))$ of elements of X. Now the base group acts by

$$f(g_1, \ldots, g_n) = (f(y_1)g_1, \ldots, f(y_n)g_n),$$

(in other words, the image of f under $(g_1, \ldots, g_n)$ is the function f', where $f'(y_i) = f(y_i)g_i$); and the top group acts by the rule that fh is the function f', where $f'(y_i) = f(y_i h^{-1})$.

Puzzles like Rubik's cube give rise to groups, which are most easily described by giving sets of generators. As an example, easier than Rubik's cube, I will describe *Rubik's domino*. This puzzle appears from the outside as a $3 \times 3 \times 2$ rectangular parallelepiped, divided into 18 unit cubes. In the starting position, the nine cubes in one 3×3 face are coloured white, and those in the other square face are black; each cube carries a number of spots of the other colour between 1 and 9, so that on the white face the arrangement is as shown in Fig. 14.2, and each black cube has the

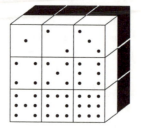

Fig. 14.2. Rubik's domino

same number as the white cube with which it shares a face (giving the mirror image of the above pattern). I will label the white cubes with capital letters from A to I, and the black cubes with the corresponding lower-case letters a to i.

A move consists of a rotation of a face of the parallelepiped. The square faces can be rotated through $90°$, $180°$ or $270°$, while the rectangular faces can only be rotated through $180°$. Thus, moves correspond to powers of the six permutations

$$(A\ C\ I\ G)(B\ F\ H\ D)$$
$$(a\ c\ i\ g)(b\ f\ h\ d)$$
$$(A\ c)(C\ a)(B\ b)$$
$$(C\ i)(I\ c)(F\ f)$$
$$(I\ g)(G\ i)(H\ h)$$
$$(G\ a)(A\ g)(D\ d)$$

The *domino group* is the group of all permutations of the cubes which can be produced by applying a sequence of moves. It is the group generated by the above permutations; but, to see this, we must resolve one difficulty.

The permutations listed above correspond to applying basic moves to the domino in its ordered state. However, if it is disordered, different permutations result because the cubes which are moved have different letters! A move can be regarded as a fixed *place-permutation*, or permutation of the positions; but we have represented states of the domino as *entry-permutations*, or permutations of the cubes. We must examine the distinction.

Let g be a permutation of $\{1, \ldots, n\}$. In two-line form, it is

$$g = \begin{pmatrix} 1 & 2 & \cdots & n \\ 1g & 2g & \cdots & ng \end{pmatrix}.$$

If we compose g with the entry-permutation h, then the entry in position i, which is ig, is replaced by its image under h, which is igh; the result is

$$\begin{pmatrix} 1 & 2 & \cdots & n \\ 1gh & 2gh & \cdots & ngh \end{pmatrix},$$

which is our usual composition of permutations. But if we compose g with the place-permutation h, then the entry ig in position i is carried to position ih; the result is

$$\begin{pmatrix} 1h & 2h & \cdots & nh \\ 1g & 2g & \cdots & ng \end{pmatrix} = \begin{pmatrix} 1 & 2 & \cdots & n \\ 1h^{-1}g & 2h^{-1}g & \cdots & nh^{-1}g \end{pmatrix},$$

so the effect is to compose the inverse of h with g. In particular, choosing g to be the identity, we see that the place-permutation h corresponds to the entry-permutation h^{-1}. So the rule for composing place-permutations is: compose the corresponding entry-permutations from right to left.

In particular, the group generated by a set of permutations is the same, whether they are place-permutations or entry-permutations. Thus, the domino group is indeed generated by the six permutations displayed earlier.

14.3. Orbits and transitivity

If a group G acts on a set X, then as combinatorialists we are mainly interested in X rather than G; we want to know what structures on X are left invariant by G, for example. The action θ is a homomorphism from G to the symmetric group on X, and its image is a permutation group. So we lose little by considering permutation groups rather than abstract groups. (An algebraist, on the other hand, is more concerned with G, and observes that the homomorphism has a kernel N, a normal subgroup of G which measures exactly what is lost in passing to the permutation group $G\theta$.)

In any case, from now on, G will be either a permutation group on X or a group acting on X; I will suppress the map θ in the notation, and write xg for the image of x under (the permutation corresponding to) g.

Our first target is a generalisation of the cycle decomposition of a single permutation (Chapter 3). Let G act on X. Define a relation $\equiv$ on X by the rule

$$x \equiv y \quad \text{if and only if} \quad xg = y \text{ for some } g \in G.$$

(14.3.1) Proposition. $\equiv$ *is an equivalence relation.*

PROOF. There is a kind of historical inevitability about this result; most naturally-occurring equivalence relations in mathematics arise from group actions. The three axioms for an equivalence relation (reflexivity, symmetry and transitivity) are immediate consequences of the three axioms for a permutation group (identity, inverses, and closure under composition). To take the second as an example: suppose that $x \equiv y$. Then $xg = y$ for some $g \in G$; so $yg^{-1} = x$, and $y \equiv x$.

The equivalence classes of the relation $\equiv$ are known as the *orbits* of the group G. So we have, uniquely, a partition of X into orbits. G is said to be *transitive* if there is only one orbit, *intransitive* otherwise.[2] Note that, for intransitive G, we have an action of G on each orbit, and these actions are transitive. So, if we want to describe all the ways in which a group can act on a set, it suffices to describe the transitive actions.[3]

EXAMPLE. The orbits of the domino group are
$\{A, C, I, G, a, c, i, g\}$ (corner cubes);
$\{B, F, H, D, b, f, h, d\}$ (edge cubes);
$\{E\}$ (white centre cube);
$\{e\}$ (black centre cube).

To describe all the transitive actions, we introduce first a special class of these, the *coset actions*. We show that any transitive action is equivalent to a coset action, and we decide when two coset actions are themselves equivalent.

Let H be a subgroup of the group G. A *right coset* of H in G is a set of the form $Hg = \{hg : h \in H\}$ for some fixed $g \in G$. We need the fact that any two cosets are equal or disjoint. (This is the core of *Lagrange's Theorem*.) For this we first show
if $g' \in Hg$, then $Hg' = Hg$.
For, if $g' \in Hg$, then $g' = h_0 g$ for some $h_0 \in H$; then any element $hg' \in Hg'$ lies in hg because $hg' = (hh_0)g$ and $hh_0 \in H$. Similarly, every element of Hg' is in Hg.
Now suppose that cosets Hg, Hg' are not disjoint; let $g' \in Hg \cap Hg'$. Then $Hg = Hg' = Hg'$, as required.

Lagrange's Theorem says that the order of a subgroup H of G divides the order of G. This now follows from the fact that a coset of H has the same number of elements as H itself. (The map $h \mapsto hg$ is a bijection from H to Hg.) We see that the number of cosets of H is equal to $|G|/|H|$. (This number is called the *index* of H in G.)

The *coset space* $(G : H)$ is the set of right cosets of H in G. (It is often denoted by $H\backslash G$, but this is easily confused with the set difference $H \setminus G$.) Now the *coset action* of G on $(G : H)$ is given by the rule

$$(Hk)g = H(kg).$$

---

[2] This is not the same as the distinction between transitive and intransitive motion made so eloquently by Russell Hoban in the quote at the head of this chapter. Hoban's dichotomy is closer to the difference between active and passive forms of a permutation.

[3] The algebraist's job is harder. An intransitive permutation group is contained in the direct product of the transitive permutation groups induced on the orbits, but need not be the whole direct product.

In other words, the permutation corresponding to g maps Hk to Hkg for all $k \in G$. It is easily verified that this is indeed an action.

As promised, we have the following two results.

(14.3.2) Proposition. *Any transitive action of G is equivalent to a coset action.*

PROOF. Let G act transitively on the set X. Choose a point $x \in X$, and let $H = \{g \in G : xg = x\}$. Then H is called the *stabiliser* of x, and is written G_x or $\mathrm{Stab}_G(x)$. We have, by an easy check,
- H *is a subgroup of G.*

Also (and this is the heart of the matter),
- *there is a natural bijection between X and $(G : H)$.*

The bijection is defined as follows. To each point $y \in X$ corresponds the subset $S(y) = \{g \in G : xg = y\}$. The set $S(y)$ is non-empty, by transitivity of G. The sets $S(y)$ (for $y \in X$) form a partition of G, and it is straightforward to identify it with the partition into cosets of H. Finally,
- *this bijection defines an equivalence of the actions of G.*

In other words, if $yg = z$, then $S(y)g = S(z)$; this follows from the definitions.

(14.3.3) Proposition. *Two coset actions on $(G : H)$ and $(G : K)$ are equivalent if and only if the subgroups H and K are conjugate.*

PROOF. H and K are *conjugate* if $K = g_1^{-1}Hg_1$ for some $g_1 \in G$. If this holds, then the map $Kg \mapsto Hg_1g$ is an equivalence. Conversely, suppose that actions on the coset spaces of subgroups H and K are equivalent. Let K correspond to the coset Hg_1 under the equivalence. Then the stabilisers of K and Hg_1 are equal. The first is just K; the second is

$$\{g \in G : Hg_1g = Hg_1\} = \{g \in G : g_1gg_1^{-1} \in H\} = g_1^{-1}Hg_1.$$

So $K = g_1^{-1}Hg_1$ is conjugate to H.

EXAMPLE. How many inequivalent actions of the symmetric group S_3 on $\{1,\ldots,n\}$?

We first describe the transitive actions. S_3 is a group of order 6, containing an identity, three elements of order 2, and two elements of order 3. By Lagrange's Theorem, the possible orders of subgroups are 6, 3, 2 and 1. There is a unique subgroup of each of the orders 6 and 1. Further, the identity and the two elements of order 3 form the unique subgroup of order 3; and there are three subgroups of order 2, each consisting of the identity and an element of order 2. These three subgroups are all conjugate.[4] So, up to equivalence, there is a unique transitive action on a set of size 1, 2, 3 or 6, and no others.

Now an arbitrary action is made up of a disjoint union of these; so the number f_n of different actions on $\{1,\ldots,n\}$ is equal to the number of ways of expresssing

---

[4] Their generators all have the same cycle structure; compare (13.1.2).

n as a sum of ones, twos, threes and sixes. I claim that the generating function is given by

$$\sum_{n=0}^{\infty} f_n t^n = 1/(1-t)(1-t^2)(1-t^3)(1-t^6).$$

This is because the right-hand side is

$$(1 + t + t^2 + \dots)(1 + t^2 + t^4 + \dots)(1 + t^3 + t^6 + \dots)(1 + t^6 + t^{12} + \dots),$$

and the coefficient of t^n is the number of ways of getting a term t^n by multiplying t^a, t^{2b}, t^{3c}, and t^{6d} for some a, b, c, d; that is, the number of expressions $n = a + 2b + 3c + 6d$.

It is possible to find an explicit expression for f_n from this formula. One way is to use analytic tools. Cauchy's integral formula expresses f_n as a contour integral, which can be evaluated by calculating residues at poles, which occur at the sixth roots of unity. But the digression would take us too far afield!

Group actions clarify the distinction between *labelled* and *unlabelled* structures introduced in Section 2.5. Let $\mathcal{C}$ be a class of structures on a set $\{1, \dots, n\}$. ($\mathcal{C}$ might consist of graphs, families of sets, etc.) Two labelled structures C and C' are counted as the same unlabelled structure if and only if they are isomorphic, that is, there is an element of the symmetric group S_n which maps C to C'. We consider the action of S_n on the class $\mathcal{C}$ of labelled structures. In this action, unlabelled structures correspond to orbits; and the stabiliser of a structure C is its *automorphism group* $\mathrm{Aut}(C)$, the set of all permutations fixing it.

(14.3.4) Theorem. *(a) The number of different labellings of a structure C is equal to $n!/|\mathrm{Aut}(C)|$.*

(b) If there are M labelled structures and m unlabelled structures $C_1, \dots, C_m$, then

$$\sum_{i=1}^{m} \frac{1}{|\mathrm{Aut}(C_i)|} = \frac{M}{n!}.$$

Consider, for example, Steiner triple systems on 9 points. Up to isomorphism, there is only one (Chapter 8, Exercise 3), and its automorphism group has order 432 (Chapter 8, Exercise 4); so it can be labelled in $9!/432 = 840$ ways. (This justifies the claim made in Chapter 8, Exercise 15.)

We have more to say about counting unlabelled structures in the next chapter.

14.4. The Schreier–Sims algorithm

What is the order of the domino group?

According to Lagrange's Theorem, if a group G acts on a set X, then the size of the orbit of X is equal to the number of cosets of the stabiliser G_x in G. We can calculate this; and G_x is a smaller group than G, so we could hope to calculate its order, perhaps by a recursive procedure, and then find $|G|$ by multiplying these numbers. We see that what is really needed for this is a generating set for G_x. This simple idea is formalised in the *Schreier–Sims algorithm*; as we'll see, it gives a lot more information too.

First, we review how to compute orbits. Let S be a generating set for the group G acting on X.

(14.4.1) Algorithm: Orbit of x

Start with $Y = \emptyset$. Add the point x to Y.

While any point was added to Y in the previous step, apply all elements of S to the recently added points; whenever a point not in Y is obtained, add it to Y.

At the conclusion, Y is the orbit of x.

While we do this calculation, we can record a witness for each point in the orbit, a permutation carrying x to that point. If $S = \{g_1, \ldots, g_m\}$, this is conveniently done by labelling a new point y with the number i_1 if it is the image of an earlier point under g_{i_1}. Then the earlier point must be $yg_{i_1}^{-1}$ Either it is x, or it has a label i_2, and in the latter case it is obtained by applying g_{i_2} to $yg_{i_1}^{-1}g_{i_2}^{-1}$. Eventually we have $yg_{i_1}^{-1} \ldots g_{i_k}^{-1} = x$, and so $y = xg_{i_k} \ldots g_{i_1}$. Note that we have not only an explicit element carrying x to y, but even an expression for this element as a product of generators.

In fact, all the orbits can be described in this way. We give x a negative label, say -1, to distinguish it as an orbit representative. If $Y = X$, there is a single orbit; otherwise, select an unlabelled point, give it the label -2, and proceed as before. Eventually, every point is labelled, and the labels (together with the generators) give a complete (and compact) description of the orbits and witnesses. The n-tuple of labels is called a *Schreier vector* for G.

Let $Y = \{x = x_1, x_2, \ldots, x_s\}$ be the orbit of x, and let k_i map x to x_i as above, for $i = 1, \ldots, s$ (with $k_1 = 1$). If $H = G_x$, then $Hk_1, \ldots, Hk_s$ are all the cosets of H in G; in other words, $k_1, \ldots, k_s$ are *coset representatives* for H in G.

To find generators for the stabiliser, we use:

(14.4.2) Schreier's Lemma. *Let $\{g_1, \ldots, g_m\}$ generate a group G; let $k_1, \ldots, k_s$ be coset representatives for a subgroup H of G. Let $\overline{g}$ denote the coset representative of the element g; in other words, $\overline{g} = k_i$ if $Hg = Hk_i$. Assume that $k_1 = 1$. Then H is generated by the set*

$$S_H = \{k_i g_j \overline{(k_i g_j)}^{-1} : i = 1, \ldots, s; j = 1, \ldots, m\}.$$

PROOF. All these elements lie in H, since each is the product of an element of G and the inverse of its coset representative. Now suppose that $h = g_{i_1} g_{i_2} \ldots g_{i_r} \in H$. For $j = 0, \ldots, r$, let $t_j = g_{i_1} \ldots g_{i_j}$, and let $u_j = \overline{t_j}$. Then, with $u_0 = 1$, we have

$$h = u_0 g_{i_1} u_1^{-1}.u_1 g_{i_2} u_2^{-1} \ldots u_{r-1} g_{i_r} u_r^{-1},$$

since $u_0 = u_r = 1$ and all the other u_i cancel with their inverses. But $u_{j-1} g_{i_j}$ lies in the same coset as u_j; thus $u_{j-1} g_{i_j} u_j^{-1} \in S_H$, and we have expressed h as a product of elements of S_H.

Now we can apply recursion, to get:

(14.4.3) Group order: Schreier–Sims algorithm

Let S be a set of generators for G.

If $S = \emptyset$ or $S = \{1\}$, then $|G| = 1$.

Otherwise, let x be a point not fixed by all elements of S; calculate the orbit Y of x and Schreier generators for G_x. Apply the algorithm recursively to find $|G_x|$. Then $|G_x| \cdot |Y| = |G|$.

But let's see what is really produced by this algorithm. We end up with a sequence of points $x_1, x_2, \ldots, x_d$ and information about subgroups $G(0), G(1), \ldots, G(d)$, where $G(0) = G$, $G(i) = G(i-1)_{x_i}$ for $i = 1, \ldots, d$, and $G(d) = \{1\}$. In fact, for $i = 1, \ldots, d$, we calculate a set T_i of coset representatives for $G(i)$ in $G(i-1)$. Let $T = T_1 \cup \ldots \cup T_d$. Then $(x_1, \ldots, x_d)$ is called a *base* for G — a base is a sequence of points such that the stabiliser of all these points is the identity — and T is a *strong generating set*. (We'll see soon that it really is a generating set.) Now T_i is the index of $G(i)$ in $G(i-1)$; the order of G is the product of these indices:

$$|G| = |T_1| \cdot \ldots \cdot |T_d|.$$

The recursive nature of the construction is reflected by the fact that $(x_2, \ldots, x_d)$ is a base, and $T_2 \cup \ldots \cup T_d$ a strong generating set, for $G(1)$.

We also have a *membership test* for G. This is a procedure which, given an arbitrary permutation G, decides whether or not $g \in G$, and if so, expresses g in terms of the generators.

(14.4.4) Membership test for G

GIVEN a permutation g of X.

If $G = \{1\}$, then $g \in G$ if and only if $g = 1$. Otherwise, is $x_1 g = x_1 t_1$ for some $t_1 \in T_1$?
- *If not, then $g \notin G$.*
- *If so, then apply the membership test for $G(1) = G_{x_1}$ to gt_1^{-1}; and $g \in G$ if and only if $gt_1^{-1} \in G(1)$.*

Note that this test is also recursive. If g passes the test, we will find unique elements $t_1, t_2, \ldots, t_d$, with $t_i \in T_i$ for $i = 1, \ldots, d$, such that $gt_1^{-1} \ldots t_i^{-1} \in G(i)$ for all i. Then we have $gt_1^{-1} \ldots t_d^{-1} = 1$, so $g = t_d \ldots t_1$. In other words, if $g \in G$, then we find a unique expression for it as a product of elements of $T_d, \ldots, T_1$. This confirms our formula for $|G|$. It also shows that T is indeed a generating set for G, as the name 'strong generating set' suggested. Finally, the Schreier–Sims algorithm enables us to express each element of T, and hence the arbitrary element g of G, in terms of the original set S of generators.

This is just what is needed to solve a puzzle like Rubik's cube or Rubik's domino. We are presented with the puzzle in a disordered state, which is some well-defined permutation g of the initial state. We have to ascertain, first, if g is in the group generated by the moves (so that the given state could indeed have been obtained legally); and, if so, how to express g in terms of the generating permutations (so that, by reversing the sequence, we can return the puzzle to its initial state).

There is one impractical feature of the algorithm as presented here. If the original group has s generators and acts on a set of size n, Schreier's Lemma gives us a set of perhaps as many as sn generators for the stabiliser of a point. Then, the group $G(i)$ fixing i base points might have up to sn^i generators. Of course, $G(d)$ is the trivial group, so all its potential sn^d generators collapse to the identity; and, if we are lucky, the collapse may begin earlier. But, to make the algorithm efficient, it is necessary to have a 'filter' which reduces the number of generators to within a practical bound, without changing the group they generate. This can indeed be done; but we won't pursue this here.

THE DOMINO GROUP. Since we know that the domino group has orbits of sizes 8, 8, 1, 1, it must be a subgroup of the direct product $S_8 \times S_8$. (We can neglect the two fixed points; now $S_8 \times S_8$ is the group of permutations which leave the other two orbits fixed setwise.) Now it turns out that the group is in fact $S_8 \times S_8$. One way to show this is to use the Schreier–Sims algorithm to calculate the order of the group, which turns out to be $(8!)^2$. But a little hand calculation can be used to make the job easier. It we compose the first and third displayed generator, we obtain the permutation

$$(A\,a\,C\,I\,G\,c)(B\,F\,H\,D\,b).$$

The sixth power of this permutation is $(B\,F\,H\,D\,b)$, which fixes all the corner cubes and moves only the edge cubes. Now it can be shown that this and similar permutations generate the alternating group A_8 of permutations of the edge cubes. Similarly, the fifth power of the permutation above fixes all the edge cubes; it and similar permutations generate the symmetric group S_8 on the corner cubes. Thus the group contains at least $S_8 \times A_8$. But the first generator acts as an odd permutation of the edge-cubes. So the group is not $S_8 \times A_8$; and the only larger group it could possibly be is $S_8 \times S_8$.

14.5. Primitivity and multiple transitivity

Just as we've reduced the study of arbitrary group actions to transitive ones, it is possible to make further reductions. We now consider this, in rather less detail.

Let G act on X. Remember that a relation on X is a set of ordered pairs of elements of X, that is, a subset of $X^2 = X \times X$. We say that the relation R is *preserved by* G, or is *G-invariant*, if $x\,R\,y$ implies $xg\,R\,yg$ and conversely. (The converse follows, by applying the inverse of g.) Now G acts on the set X^2, by the rule

$$(x, y)g = (xg, yg);$$

and we have the following:

(14.5.1) Proposition. *The relation R is preserved by G if and only if it is a union of orbits of G on X^2.*

PROOF. G-invariance means that $(x, y) \in R$ if and only if $(xg, yg) \in R$ for any $g \in G$; so the whole of the G-orbit of (x, y) is contained in R. Hence R is a union of orbits. The converse is similar.

A *G-congruence* on X is an equivalence relation R on X which is preserved by G. (We don't require that G fixes the equivalence classes of R.) There are always two trivial G-congruences (if $|X| > 1$): the relation of equality, and the 'all' relation R defined by the rule that $x \mathbin{R} y$ for all $x, y \in X$. The group G is called *imprimitive* if there is a G-congruence other than these two, and *primitive* otherwise.

Let G be a transitive permutation group. If R is a non-trivial G-congruence, let $X_1, \ldots, X_m$ be the congruence classes, and $Y = \{X_1, \ldots, X_m\}$ the set of classes. Now we define two new permutation groups:

- G acts on the set Y; let G_0 be the permutation group on Y induced by G.
- Let H be the subgroup of G which fixes the set X_1 (*not* its pointwise stabiliser), and H_0 the permutation group induced on X_1 by H.

(14.5.2) Theorem. *G is isomorphic to a subgroup of the wreath product H_0 wr G_0; and the given action is equivalent to the restriction to G of the natural action of the wreath product.*

Thus, G can be regarded as being built out of the smaller groups H_0 and G_0. Both these groups are transitive. If either is imprimitive, we can continue the reduction further. We end up with a collection of primitive groups, the *primitive components* of G. (But note that G may have several different congruences, which may give rise to different collections of primitive components.)

Let t be a positive integer not exceeding $|X|$. A permutation group G on X is said to be *t-transitive* if, given any two t-tuples $(x_1, \ldots, x_t)$ and $(y_1, \ldots, y_t)$ of *distinct* points of X, there is a permutation $g \in G$ with $x_i g = y_i$ for $i = 1, \ldots, t$. (In other words, G acts transitively on the set of t-tuples of distinct points.) Now 1-transitivity is the same as transitivity (as defined in Section 14.3).

(14.5.3) Proposition. *Let G be t-transitive on X, with $t \geq 2$. Then*
(a) G is $(t - 1)$-transitive;
(b) G is primitive.

PROOF. (a) Take two $(t - 1)$-tuples $(x_1, \ldots, x_{t-1})$ and $(y_1, \ldots, y_{t-1})$ of distinct elements. Extend them to t-tuples by appending elements x_t and y_t respectively, which are not among the elements in the tuples already. Then choose g with $x_i g = y_i$ for $i = 1, \ldots, t$.

(b) We may assume that G is 2-transitive. Now any G-congruence R is a union of orbits of G acting on X^2 (Proposition 14.5.1), necessarily containing the *diagonal* $\Delta = \{(x, x) : x \in X\}$, since R is reflexive. But, if G is 2-transitive, it has just two orbits on X^2, namely Δ and $X^2 \setminus \Delta$; so there are only two possible congruences.

If $|X| = n$, then the symmetric group on X is n-transitive. Also, the alternating group is $(n-2)$-transitive if $n > 2$. (Given $(n-2)$-tuples $(x_1, \ldots, x_{n-2})$, $(y_1, \ldots, y_{n-2})$ of points, there are just two permutations which carry the first to the second; they differ by a transposition of the remaining points, so they have opposite parity, and one of them is in the alternating group.)

It is known that no other finite permutation group can be more than 5-transitive. This remarkable fact is a consequence of the classification of the finite simple groups, perhaps the greatest collective achievement of mathematicians; but the proof is more than ten thousand pages long, so I must ask you to take it on trust.

14.6. Examples

EXAMPLE: STS(7). We showed in Chapter 8 that there is a unique STS(7), up to isomorphism (see Fig. 14.3). In fact, the argument shows the following:

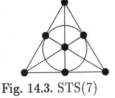

Fig. 14.3. STS(7)

> Let $(X, \mathcal{B})$ and $(Y, \mathcal{C})$ be Steiner triple systems of order 7. Let (x_1, x_2, x_3) be a triangle in the first system, and (y_1, y_2, y_3) a triangle in the second. Then there is a unique isomorphism from the first system to the second which maps x_i to y_i for $i = 1, 2, 3$.

For the isomorphism must map the third point on the block through x_1 and x_2 to the third point on the block through y_1 and y_2, and similarly for the other two sides of the triangle; then it maps the seventh point of X to the seventh point of Y. This map really is an automorphism: three of the remaining blocks consist of a vertex, the 'third point' of the opposite side, and the 'seventh point' of the design; the last block consists of the 'third points' of the three sides.

From this, we can calculate the order of the automorphism group of the Steiner system. By choosing the two systems to be equal (so that the isomorphisms are automorphisms), the number of automorphisms is equal to the number of (ordered) triangles, which is $7 \cdot 6 \cdot 4 = 168$. We also see that a triangle is a base for the automorphism group.

Now the automorphism group is 2-transitive. (The proof is a modification of the proof of Proposition 14.5.3(a). Let (x_1, x_2) and (y_1, y_2) be two pairs of distinct elements. Now choose x_3 so that (x_1, x_2, x_3) is a triangle; and choose y_3 similarly. Then choose an automorphism carrying the first triangle to the second.) In particular, it is primitive.

We can put a name to this automorphism group. In Section 8.5, we saw that the points of the STS(7) can be labelled by the non-zero vectors of a 3-dimensional vector space V over $GF(2)$, so that the blocks are the triples of points with sum zero. Now the group $GL(3,2)$ of invertible 3×3 matrices over $GF(2)$ acts on the non-zero vectors in V, and obviously maps any block to a block; so it is a group of automorphisms. But

$$|GL(3,2)| = (2^3 - 1)(2^3 - 2)(2^3 - 2^2) = 168,$$

so this is the full automorphism group.

EXAMPLE: THE PETERSEN GRAPH. Recall the Petersen graph from Chapter 11 (see Fig. 14.4). (Ignore the labels for the moment.)

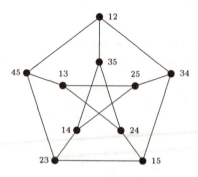

Fig. 14.4. The Petersen graph

We saw in Section 11.12 that any subgraph of shape ⟩—⟨ can be completed in a unique way to a graph on 10 vertices with valency 3, diameter 2 and girth 5. This means, by the same kind of argument as we gave for the Steiner triple system, that the number of automorphisms of the Petersen graph is equal to the number of subgraphs of this type, which is $10 \cdot 3 \cdot 2 \cdot 1 \cdot 2 \cdot 1 = 120$.

Now consider the labels in Fig. 14.4. We have labelled each vertex with a 2-element subset of $\{1, \ldots, 5\}$, so that all $\binom{5}{2} = 10$ 2-subsets are used. A little checking shows that two vertices are adjacent if and only if their labels are disjoint. It follows that any permutation of $\{1, \ldots, 5\}$, in its induced action on the 2-subsets, is an automorphism; and we find a group of automorphisms isomorphic to S_5, with order 120. So the full automorphism group is S_5.

Now the automorphism group is clearly transitive on vertices. It is not 2-transitive, since no automorphism can map two adjacent vertices to two non-adjacent vertices. However, we see that the orbits of S_5 on X^2 are three in number:

• the diagonal $\{(x, x) : x \in X\}$;
• the set $\{(x, y) : x \sim y\}$;
• the set $\{(x, y) : x \neq y, x \nsim y\}$.

The automorphism group is transitive on (ordered) edges and on (ordered) non-edges.

From this information, we can show that S_5 is primitive on X. For a congruence R must be a union of some of these three orbits, and must include the diagonal.

Suppose that it contains the second orbit (the ordered edges). Since we can find vertices x, y, z with $x \sim y \sim z$ and $x \not\sim z$, we have $x \, R \, y$ and $y \, R \, z$, so, by transitivity, $x \, R \, z$. Thus R contains all the ordered non-edges as well, and is the universal relation. A similar argument applies if R contains the orbit of non-edges. So either R is the diagonal, or $R = X^2$. This means that the group is primitive.

14.7. Project: Cayley digraphs and Frucht's Theorem

Let S be a subset of a group G, not containing the identity. The *Cayley digraph* $D(G; S)$ of G with respect to S is defined to have vertex set G, and edges (g, sg) for each $s \in S$ and each $g \in G$. The *Cayley graph* $\Gamma(G; S)$ is the underlying graph of $D(G; S)$; that is, its vertex set is G, and it has edges $\{g, sg\}$ for each $s \in S$ and $g \in G$. We can regard the element s as a 'label' on the edges (g, sg) of $D(G; S)$, or the corresponding edges of $\Gamma(G; S)$. (Note that, if an element s and its inverse both lie in S, they label the same edges of $\Gamma(G; S)$.)

Now the following holds.

(14.7.1) Proposition. (a) $D(G; S)$ *is connected if and only if* S *generates* G.
(b) *For each* $g \in G$, *the map* $\rho_g : x \mapsto xg$ *is an automorphism of* $D(G; S)$.

PROOF. (a) If S generates G, then any $g \in G$ can be written as a product of elements of S and their inverses. This product tells us how to find a path from the identity to g. For example, if $g = s_1 s_2^{-1} s_3$, then we have an edge $(1, s_3)$ labelled s_3, an edge $(s_2^{-1} s_3, s_3)$ labelled s_2 (but going in the wrong direction), and an edge $(s_2^{-1} s_3, s_1 s_2^{-1} s_3)$ labelled s_1.

(This argument shows that the digraph is connected (which means that the underlying graph is connected, see Section 11.8), not that it is strongly connected. In fact, if G is finite, the strong connectedness of $D(G; S)$ follows from the connectedness (see Exercise 12).)

The converse is similar: any path from 1 to g in the underlying graph translates into a product of elements of S and their inverses which is equal to g.

(b) A simple check: if (x, sx) is an edge, then so is $(x\rho_g, sx\rho_g) = (xg, sxg)$, by the associative law.

Note that the permutations ρ_g comprise the permutation group in the proof of Cayley's Theorem (14.1.1), isomorphic to the abstract group G. So we have an action of G on the vertices of the Cayley digraph or graph, as a group of automorphisms. Note that this action is transitive; for $\rho_{g^{-1}h}$ maps g to h. We denote the permutation group by $\rho(G)$, to distinguish it from G (the set of points being permuted): we are thinking here of ρ as the action of G.

More is true:

(14.7.2) Proposition. *Suppose that* S *generates* G. *Then any automorphism of* $D(G; S)$ *which preserves the labels on the edges belongs to* $\rho(G)$.

PROOF. Let f be an automorphism which preserves the labels. Since all elements of $\rho(G)$ also preserve labels, we can compose f with the element $\rho_{1f^{-1}}$ to obtain an automorphism fixing 1; and this automorphism lies in $\rho(G)$ if and only if f does. So we may assume that f fixes 1. Now, for each $s \in S$, there is a unique edge with label s and initial vertex 1 (namely $(1, s)$), and a unique edge with label s and terminal vertex 1 (namely $(s^{-1}, 1)$). So f must fix all elements s or s^{-1} for $s \in S$. In this way we can work out through the digraph, and find that f fixes every element which is a product of elements of S and their inverses. But, by assumption, these elements comprise all of G; so $f = 1 \in \rho(G)$.

Now we can prove Frucht's Theorem:

(14.7.3) Theorem. *Every finite group is the automorphism group of a finite graph.*

PROOF. Let G be a finite group. We assume that G has at least 5 elements. (For smaller groups, it's not difficult to write down suitable graphs.) Now take $S = G \setminus \{1\}$, and construct the Cayley digraph $D(G; S)$. Since S generates G, we know from (14.7.2) that the group of automorphisms of this digraph preserving the edge labels is isomorphic to G. The trick is to replace the labelled directed edges with subgraph 'gadgets' to ensure that the automorphism group remains the same.

Let γ_n denote the graph with $n + 4$ vertices $a, b, c, d, e_1, \ldots, e_n$, having the following edges: $\{a, b\}, \{b, c\}, \{c, d\}, \{b, e_1\}, \{e_i, e_{i+1}\}$ $(i = 1, \ldots, n - 1)$ (see Fig. 14.5). Now let $S = \{s_1, \ldots, s_{m-1}\}$,

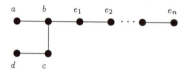

Fig. 14.5. A gadget

where $m = |G|$. Replace each edge (u, v) of $D(G; S)$ with label s_n with a copy of the gadget γ_n, where the vertices a and d of the gadget are identified with u and v. (All the added gadgets are disjoint apart from these identifications.) Let Γ be the resulting graph. Thus, some vertices of Γ are elements of G (coming from $D(G; S)$), while any other vertex belongs to a unique gadget. Now observe:

- *We can recognise the elements of G in Γ,*

since they have valency $m \geq 4$ while any vertex of a gadget has valency at most 3. Moreover, edges with the same label are replaced by isomorphic gadgets, so the label-preserving automorphisms of $D(G; S)$ extend to automorphisms of Γ; but we can recover the label and the orientation of the edge joining any two elements of G from the gadget in Γ, so any automorphism of Γ induces a label-preserving automorphism of $D(G; S)$. Thus, $\mathrm{Aut}(\Gamma) \cong G$, as required.

14.8. Exercises

1. Let G be a permutation group on $X = \{x_1, \ldots, x_n\}$. Regard each permutation g in 'passive' form, that is, as an n-tuple $(x_1 g, \ldots, x_n g)$; then G is a set of n-tuples, that is, an n-ary relation on X. (For example, if $n = 2$ and G is the trivial group, then (X, G) is a digraph with one edge (x_1, x_2).) Show that the result of applying the permutation h to the n-tuple g is the composition gh. Deduce that $\mathrm{Aut}(X, G) = G$.

2. Show that the symmetry group of the regular octahedron is the wreath product $S_2 \,\mathrm{wr}\, S_3$, having its natural action on the six vertices, and its product action on the eight faces. Show that this group is also isomorphic to $S_2 \times S_4$.

3. '*Most naturally-occurring equivalence relations in mathematics arise from group actions.*' Discuss. [HINT: You will find some useful examples in elementary linear algebra.]

4. Consider the STS(7) in cyclic form: the point set is $\mathbb{Z}/(7)$, the blocks are $\{013, 124, 235, 346, 450, 561, 602\}$. Clearly the permutation $x \mapsto x + 1$ (the permutation $(0\ 1\ 2\ 3\ 4\ 5\ 6)$) is an automorphism. Show that the permutation $(2\ 6)(4\ 5)$ is also an automorphism. Now show that these two automorphisms generate the full automorphism group.

5. Show that any subgroup of the symmetric group of degree n can be generated by at most $n(n - 1)/2$ elements.

6. Let G be the symmetric group S_n, acting on the set X of ordered pairs of distinct elements from $\{1, \ldots, n\}$. Show that, for $n \geq 5$, there are three non-trivial G-congruences, defined as follows:

$$(x_1, x_2)\, R_1(y_1, y_2) \text{ if and only if } x_1 = y_1;$$
$$(x_1, x_2)\, R_2(y_1, y_2) \text{ if and only if } x_2 = y_2;$$
$$(x_1, x_2)\, R_3(y_1, y_2) \text{ if and only if } \{x_1, x_2\} = \{y_1, y_2\}.$$

What happens if $n = 4$?

7. A *left coset* of a subgroup H of a group G is a set $gH = \{gh : h \in H\}$. Prove that
(a) the numbers of left and right cosets are equal;
(b) there is a set of elements which are both right coset representatives and left coset representatives.
[HINT FOR (b): Let $\mathcal{L}$ and $\mathcal{R}$ be the sets of left and right cosets. For each $R \in \mathcal{R}$, let $A_R = \{L \in \mathcal{L} : L \cap R \neq \emptyset\}$. Show that the family $(A_R : R \in \mathcal{R})$ satisfies Hall's Condition (6.2.2). (This was essentially Hall's original application of his theorem.)]

If G acts on X, and $H = G_x$, describe the left cosets of H in terms of the action (analogous to the proof of (14.3.2)).

8. Show that all congruence classes of a congruence for a transitive group have the same size. Deduce that a transitive group acting on a prime number of points is primitive.

9. (a) Prove that a graph with 2-transitive automorphism group must be complete or null.
    (b) Find all graphs whose automorphism group is transitive on vertices, ordered edges, and ordered non-edges, but is not primitive on vertices.

10. Let G be t-transitive on X. Prove that the number of orbits of G on the Cartesian power X^t is the Bell number $B(t)$. [HINT: For $t = 3$, the five orbits consist of triples (x, x, x), (x, x, y), (x, y, x), (y, x, x), and (x, y, z), where x, y, z are all distinct.]

11. Show that the two graphs of Fig. 14.6 are isomorphic. Hence write down automorphisms of order 3 and 5 of the Petersen graph. What is the group generated by these two automorphisms?

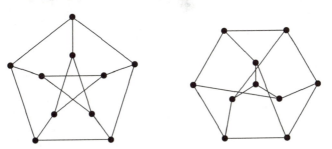

Fig. 14.6. Two isomorphic graphs

12. Let D be a finite digraph whose automorphism group acts transitively on its vertices. Show that, if D is connected, then it is strongly connected. (In other words,

if you can walk from A to B, then you can drive.) Show that this conclusion is false for infinite digraphs.

13. For each group G of order less than 7, find a graph whose automorphism group is G.

14. (a) The *centraliser* of a permutation $g \in S_n$ is the group of all permutations $h \in S_n$ which commute with g, that is, which satisfy $gh = hg$. Prove that the centraliser of g is a subgroup of S_n, and is isomorphic to

$$(C_1 \text{ wr } S_{a_1}) \times (C_2 \text{ wr } S_{a_2}) \times \ldots \times (C_n \text{ wr } S_{a_n}),$$

where g has cycle structure $1^{a_1} 2^{a_2} \ldots n^{a_n}$ (Section 13.1).
   (b) Let Γ be a disconnected graph. Let $\Gamma_1, \ldots, \Gamma_k$ be representatives of the isomorphism types of the connected components of Γ, and suppose that a_i components are isomorphic to Γ_i for $i = 1, \ldots, k$. Prove that

$$\text{Aut}(\Gamma) = (\text{Aut}(\Gamma_1) \text{ wr } S_{a_1}) \times (\text{Aut}(\Gamma_2) \text{ wr } S_{a_2}) \times \ldots \times (\text{Aut}(\Gamma_k) \text{ wr } S_{a_k}).$$

15. (a) Prove that the cyclic group of order n contains $\phi(n)$ elements which generate the group, where $\phi(n)$ is the number of residue classes mod n which are coprime to n. (ϕ is *Euler's function* or the *totient function*.)
   (b) Prove that

$$\frac{\phi(n)}{n} = \sum_{d \mid n} \frac{\mu(d)}{d},$$

where μ is the classical Möbius function.

16. Let G be a permutation group on X. For each subgroup H of G, let $\text{fix}(H)$ be the number of points of X which are fixed by every element of H. Prove that the number of orbits of G on which no non-identity element of G fixes a point is

$$\frac{1}{|G|} \sum_{H \leq G} \text{fix}(H) \mu(H, G),$$

where μ is the Möbius function of the lattice of subgroups of G (ordered by inclusion).

15. Enumeration under group action

'I count a lot of things that there's no need to count,' Cameron said. 'Just because that's the way I am. But I count all the things that need to be counted.'

Richard Brautigan, *The Hawkline Monster* (1974)

TOPICS: Orbit-counting Lemma; cycle index; enumeration of functions by weight

TECHNIQUES: Calculation of cycle index

ALGORITHMS:

CROSS-REFERENCES: Direct and wreath products (Chapter 14); Stirling numbers (Chapter 5); unlabelled structures (Chapters 2, 14); symmetric polynomials (Chapter 13)

In this chapter, we develop a theory of counting which is associated with the names of Redfield and Pólya. Typical of these problems is that the configurations we count 'live' on some basic object, and two of them should not be counted as different whenever one can be transformed into the other by a symmetry of the underlying object. One example of this setup is the counting of unlabelled graphs — review the remarks on this in Chapter 2 — where a graph 'lives' on a vertex set, and isomorphism of graphs is defined by means of permutations of the vertex set. For another example, we will count the number of necklaces that can be made using two colours of beads, two necklaces being counted as the same if one can be transformed into the other by a rotation of the necklace, or by picking it up and turning it over.

1. The Orbit-counting Lemma

I said in the last chapter that naturally occurring equivalences usually come from group actions; that is, the equivalence classes are orbits of a group. The next result gives a formula for the number of orbits.[1]

Let G be a permutation group on a set X. For each element $g \in G$, we let fix(g) denote the number of points $x \in X$ *fixed* by g (that is, satisfying $xg = x$).

---

[1] This result is commonly referred to as 'Burnside's Lemma'. It was given without attribution by Burnside in his book *Theory of Groups of Finite Order*, which introduced the French and German developments in the subject in the second half of the nineteenth century to English mathematicians; but it has been traced back to earlier work of Cauchy and Frobenius. I prefer the impersonal term given here.

> **(15.1.1) Orbit-counting Lemma**
> *The number of orbits of a permutation group G is equal to the average number of fixed points of its elements, viz.*
>
> $$\frac{1}{|G|} \sum_{g \in G} \operatorname{fix}(g).$$

PROOF. We suppose first that G is transitive, and show that the expression in the lemma equals 1, by double-counting pairs (x, g) with $xg = x$. On one hand, the number of pairs is $\sum_{g \in G} \operatorname{fix}(g)$. On the other, it is $\sum_{x \in X} |G_x|$. Now, since G is transitive, G_x has n cosets, where $n = |X|$; so the sum is $\sum_{x \in X} |G|/n = |G|$. Equating the two expressions and dividing by $|G|$ gives the result.

Now let G have t orbits $X_1, \ldots, X_t$, and let $\operatorname{fix}_i(g)$ be the number of fixed points of g in X_i. Since G acts transitively on X_i, we have

$$\frac{1}{|G|} \sum_{g \in G} \operatorname{fix}_i(g) = 1.$$

Also, we have $\operatorname{fix}(g) = \sum_{i=1}^{t} \operatorname{fix}_i(g)$. So

$$\frac{1}{|G|} \sum_{g \in G} \operatorname{fix}(g) = \sum_{i=1}^{t} \frac{1}{|G|} \sum_{g \in G} \operatorname{fix}_i(g) = \sum_{i=1}^{t} 1 = t,$$

as required.

It is immaterial whether we have a permutation group or (more generally) an action of a group in the Orbit-counting Lemma. For let θ be an action of G, with kernel N. Then each permutation in the image of θ is the image of precisely $|N|$ elements of G (comprising a coset of N); also the order of G is $|N|$ times as large as that of its image, so the factors $|N|$ cancel and the average number of fixed points is the same for G and its image.

15.2. An application

> *In how many ways can the faces of a cube be coloured with two colours? Assume that two coloured cubes which differ by a rotation are identical.*

The group in question is the group of rotations of the cube, which has order 24 (and happens to be isomorphic to S_4). We can list its elements as follows. Here, a face-axis, edge-axis, or vertex-axis is an axis of symmetry joining centres of opposite faces, midpoints of opposite edges, or opposite vertices, respectively.

Type	Axis	Order of rotation	No. of elements
1 (identity)	—	—	1
2	Face	2	3
3	Face	4	6
4	Edge	2	6
5	Vertex	3	8
			24

Now we let X be the set of $2^6 = 64$ colourings of the cube. We must calculate fix(g) for each $g \in G$. A colouring is fixed by g if and only if all faces in the same cycle of g have the same colour, so fix$(g) = 2^{c(g)}$, where $c(g)$ is the number of cycles of g on the faces of the cube.

Type	$c(g)$	fix(g)	Contribution
1	6	64	64
2	4	16	48
3	3	8	48
4	3	8	48
5	2	4	32
			240

So the number of different cubes is $240/24 = 10$. Can you describe them?

It is clear that the same method would work for any number r of colours, giving the answer as a polynomial in r (Exercise 1). This observation motivates the introduction of generating functions and enumeration by cycle index, a topic we now consider.

15.3. Cycle index

Given a permutation g on X, there is a *cycle decomposition* of g, an expression for g as a product of disjoint cycles, unique up to the starting points of the cycles and the order of the factors (see Section 13.1). Let there be c_1 cycles of length 1, c_2 of length 2, ..., c_n of length n, where $n = |X|$. (In the cycle notation for permutations, we commonly suppress cycles of length 1, but it is important to count them here.) We define the *cycle index* of g to be the monomial

$$z(g; s_1, \ldots, s_n) = s_1^{c_1} s_2^{c_2} \ldots s_n^{c_n}$$

in indeterminates $s_1, \ldots, s_n$. Now, if G is a permutation group on X, the *cycle index* of G is the average of the cycle indices of its elements:

$$Z(G; s_1, \ldots, s_n) = \frac{1}{|G|} \sum_{g \in G} \prod_{i=1}^{n} s_i^{c_i(g)},$$

where $c_i(g)$ is the number of cycles of g of length i. Just as in Section 15.1, if a group G acts on a set X, the cycle index of G is the same as the cycle index of the induced permutation group.

Before we prove the main result connecting cycle index with enumeration, we give a couple of examples of its use. The group G has an induced action on the set of all k-element subsets of X, and another on the set of k-tuples of distinct elements of X, for each k with $1 \le k \le n$. We let f_k and F_k be the numbers of orbits of G in these actions. By convention, $f_0 = F_0 = 1$. So $f_1 = F_1$ is the number of orbits of G on X, and $F_k = 1$ if and only if G is k-transitive on X (Section 14.5). We show that the ordinary generating function for the numbers f_k, and the exponential generating function for the F_k, can be calculated from the cycle index of G by simple substitutions.

(15.3.1) Proposition. (a) $\displaystyle\sum_{k=0}^{n} f_k t^k = Z(G; 1+t, 1+t^2, \dots, 1+t^n)$.

(b) $\displaystyle\sum_{k=0}^{n} F_k t^k / k! = Z(G; 1+t, 1, \dots, 1)$.

PROOF. We let $\mathrm{fix}_k(g)$ and $\mathrm{Fix}_k(g)$ denote the number of k-subsets, or k-tuples of distinct points, respectively, fixed by g.

(a)
$$\sum_{k=0}^{n} f_k t^k = \frac{1}{|G|} \sum_{g \in G} \sum_{k=0}^{n} \mathrm{fix}_k(g) t^k.$$

Now consider a permutation g, with $c_i(g)$ cycles of length i. For each choice of numbers $b_i \le c_i(g)$ with $\sum_{i=1}^{n} i b_i = k$, we can find k-sets fixed by g which consist of b_i cycles of length i for $i = 1, \dots, n$. Moreover, any fixed k-set is a union of cycles. So
$$\sum_{k=0}^{n} f_k t^k = \frac{1}{|G|} \sum_{g \in G} \sum_{k=0}^{n} \left(\sum{}^{*} \prod_{i=1}^{n} \binom{c_i(g)}{b_i} \right) t^k,$$
where $\sum^{*}$ is over all $(b_1, \dots, b_n)$ with $0 \le b_i \le c_i(g)$ and $\sum i b_i = k$. But since we then sum over k, this is just
$$\frac{1}{|G|} \sum_{g \in G} \prod_{i=1}^{n} \sum_{b=0}^{c_i(g)} \binom{c_i(g)}{b} t^{ib} = \frac{1}{|G|} \sum_{g \in G} \prod_{i=1}^{n} (1 + t^i)^{c_i(g)}$$
$$= Z(G; 1+t, 1+t^2, \dots, 1+t^n).$$
(The manipulations here are similar to those explained in more detail in Section 4.2.)

(b) This one is easier. A k-tuple is fixed if and only if all of its points are fixed; so
$$\mathrm{Fix}_k(g) = c_1(g)(c_1(g) - 1) \dots (c_1(g) - k + 1).$$
Thus,
$$\sum_{k=0}^{n} F_k t^k / k! = \frac{1}{|G|} \sum_{g \in G} \sum_{k=0}^{n} \frac{c_1(g)(c_1(g) - 1) \dots (c_1(g) - k + 1)}{k!} t^k$$
$$= \frac{1}{|G|} \sum_{g \in G} \sum_{k=0}^{n} \binom{c_1(g)}{k} t^k$$
$$= \frac{1}{|G|} \sum_{g \in G} (1 + t)^{c_1(g)}$$
$$= Z(G; 1+t, 1, \dots, 1).$$

Now we turn to the general situation. We take a collection of 'figures' $\phi_1, \phi_2, \ldots$, each of which has a non-negative integral 'weight' $w(\phi_i)$. The set of figures needn't be finite (though it is in many applications), but we do assume that there are *only finitely many figures of any given weight*. We can summarise this information in a *figure-counting series*

$$a(t) = \sum_{n \geq 0} a_n t^n,$$

where a_n is the number of figures of weight n.

Now we are given a permutation group G on a set X — typically G is the automorphism group of some object — and we want to count the number of ways of associating a figure with each point of X, two such 'configurations' being regarded as identical for the purpose of the count if some element of G takes one to the other. (Typically the 'figures' are colours, and we want to count the number of inequivalent colourings, as in the example of the coloured cubes in the last section.) An attachment of figures to points of X is defined by a function $f : X \rightarrow \Phi$, where Φ is the set of figures; it has a total weight

$$w(f) = \sum_{x \in X} w(f(x)).$$

Now G acts on the set of functions, by the rule

$$(fg)(x) = f(xg^{-1}).$$

(The inverse is technically required to make this a valid action; but, informally, it arises because we are regarding the elements of G as place-permutations here — compare the discussion in Section 14.2.) We want to count the orbits of G on functions, which we do by means of the *function-counting series*

$$b(t) = \sum_{n \geq 0} b_n t^n,$$

where b_n is the number of orbits of G on functions of total weight n. (The action of G doesn't change weights of functions.)

(15.3.2) Cycle Index Theorem

$$b(t) = Z(G; a(t), a(t^2), \ldots, a(t^n)).$$

Before proving the theorem, we show that part (a) of (15.3.1) is a consequence of it. We take two figures, with weights 0 and 1: we might as well call the figures themselves 0 and 1. The figure-counting series is just $1 + t$. Now a function from X to the set $\{0, 1\}$ is nothing but the characteristic function of a subset Y of X; and the action of G on functions is equivalent to its natural action on subsets. (If f is the characteristic function of Y, then

$$(fg)(x) = 1 \quad \Leftrightarrow \quad xg^{-1} \in Y \quad \Leftrightarrow \quad x \in Yg,$$

so fg is the characteristic function of Yg.) So $f_k t^k$ is the function-counting series, and the formula for it follows from the Theorem.

We take the proof in four steps.

STEP 1. Let Φ and Φ' be sets of figures, with figure-counting series $a(t)$ and $a'(t)$ respectively. Then the generating function for counting pairs $(\phi, \phi') \in \Phi \times \Phi'$, enumerated by the sum of the weights, is $a(t)a'(t)$.

PROOF. If $a(t) = \sum a_i t^i$ and $a'(t) = \sum a'_i t^i$, then the number of pairs (ϕ, ϕ') with $w(\phi) = j$, $w(\phi') = i - j$, is $a_j a'_{i-j}$. Summing over j gives the total number of pairs with $w(\phi) + w(\phi') = i$, and also the coefficient of t^i in $a(t)a'(t)$.

STEP 2. Let Φ have figure-counting series $a(t)$, and let X be an n-set. Then the generating function for counting functions from X to Φ, enumerated by total weight (that is, the function-counting series for the trivial group on X) is $a(t)^n$.

PROOF. If $X = \{x_1, \ldots, x_n\}$, then functions from X to Φ are represented by n-tuples $(f(x_1), \ldots, f(x_n))$ of elements of Φ. The result now follows from Step 1 by induction. Note that this is the special case of the Cycle Index Theorem for the trivial group on X (whose cycle index is s_1^n).

STEP 3. The series enumerating functions from X to Φ fixed by a permutation g of X, by total weight, is

$$z(g; a(t), a(t^2), \ldots, a(t^n)).$$

PROOF. A function is fixed by g if and only if it is constant on the cycles of g. So a fixed function is specified by giving, for each i, a function from a set of representatives ($c_i(g)$ in number) of the i-cycles of g, to Φ. For fixed i, these functions are enumerated by $a(t)^{c_i(g)}$, by Step 2. However, since such a function has each value repeated i times on X, its contribution to the total weight is multiplied by i, so this contribution is enumerated by $a(t^i)^{c_i(g)}$. Now, by Step 1 and induction, the overall generating function is obtained by multiplying these contributions for all values of i; in other words, it is

$$a(t)^{c_1(g)} a(t^2)^{c_2(g)} \ldots a(t^n)^{c_n(g)} = z(g; a(t), a(t^2), \ldots, a(t^n)),$$

as required.

STEP 4: COMPLETION OF THE PROOF. The number of orbits of G on functions of weight k is the average number of fixed functions of weight k of its elements. By Step 3, this is the coefficient of t^k in

$$\frac{1}{|G|} \sum_{g \in G} z(g; a(t), a(t^2), \ldots, a(t^n)) = Z(G; a(t), a(t^2), \ldots, a(t^n)),$$

by definition of cycle index.

15.4. Examples

We consider some applications of the Cycle Index Theorem.

EXAMPLE 1: COLOURED CUBES. Consider the group of rotations of the cube, acting on its faces. From the table in Section 15.2, the cycle index of this group is

$$\tfrac{1}{24}\left(s_1^6 + 3s_1^2 s_2^2 + 6s_1^2 s_4 + 6s_2^3 + 8s_3^2\right).$$

We can refine our earlier count of ten red-and-blue cubes by enumerating them by number of red faces. Thus, we let red have weight 1, and blue have weight 0. The figure-counting series is $1 + t$, so the function-counting series is

$$\tfrac{1}{24}\left((1+t)^6 + 3(1+t)^2(1+t^2)^2 + 6(1+t)^2(1+t^4) + 6(1+t^2)^3 + 8(1+t^3)^2\right)$$
$$=1 + t + 2t^2 + 2t^3 + 2t^4 + t^5 + t^6.$$

This should check with the listing of cubes you gave in Section 15.2.

Again, to count the number of ways of colouring with a given number r of colours, take all colours to have weight 0. Then the figure-counting series is r, and the number of colourings is

$$\tfrac{1}{24}(r^6 + 3r^4 + 12r^3 + 8r^2) = \tfrac{1}{24}r^2(r+1)(r^3 - r^2 + 4r + 8).$$

For example, when $r = 3$, there are 57 different coloured cubes.

EXAMPLE 2: NECKLACES. To count necklaces, we need to find the cycle indices of the cyclic group (if we allow only rotations) and the dihedral group (if inversions are allowed).

Euler's function $\phi(n)$ (sometimes called the totient function) is the number of congruence classes mod n which are coprime to n. For example, $\phi(12) = 4$, and $\phi(p) = p - 1$ if p is prime. By convention, $\phi(1) = 1$.

(15.4.1) Lemma. The cyclic group of order n contains, for each divisor d of n, $\phi(d)$ elements of order d. Each has n/d cycles of length d.

PROOF. We can identify C_n with $\mathbb{Z}/(n)$. The order of a congruence class m is the smallest positive x such that $mx = ny$ for some y. If $mx = ny$, then $mx/(m,n) = ny/(m,n)$. Since $m/(m,n)$ and $n/(m,n)$ are coprime (we have divided out the common factors of m and n), the least positive solution is $x = n/(m,n)$, $y = m/(m,n)$. So the order of m is $n/(m,n)$.

Now we reverse the argument and ask: how many classes m satisfy $n/(m,n) = d$ for a given divisor d of n? For such an m, we have $m = (m,n)y$, where $(y,d) = 1$; there are $\phi(d)$ choices of y, each giving rise to a unique $m = ny/d$.

Note in particular that the number of elements of C_n which generate the group is $\phi(n)$.

So the cycle index for C_n is

$$\frac{1}{n}\sum_{d|n}\phi(d)s_d^{n/d}.$$

The dihedral group D_{2n} contains the cyclic group C_n, together with n reflections. If n is odd, each reflection has one fixed point and $(n-1)/2$ cycles of length 2; while, if n is even, then half of them have no fixed points and $n/2$ 2-cycles, and the rest have two fixed points and $(n-2)/2$ 2-cycles. Thus

$$Z(D_{2n}) = \frac{1}{2}(Z(C_n) + R_n),$$

where

$$R_n = \begin{cases} s_1 s_2^{(n-1)/2}, & n \text{ odd}, \\ (s_2^{n/2} + s_1^2 s_2^{(n-2)/2})/2, & n \text{ even}. \end{cases}$$

For example, when $n = 10$, we have

$$Z(D_{20}) = \tfrac{1}{20}(s_1^{10} + 4s_5^2 + 4s_{10} + 6s_2^5 + 5s_1^2 s_2^4),$$

so the generating function for black and white necklaces by number of black beads is

$$1 + t + 5t^2 + 8t^3 + 16t^4 + 16t^5 + 16t^6 + 8t^7 + 5t^8 + t^9 + t^{10},$$

while the number of different necklaces with u colours of beads is

$$\tfrac{1}{20}u(u+1)(u^8 - u^7 + u^6 - u^5 + 6u^4 + 4).$$

EXAMPLE 3: GRAPHS. A graph on the vertex set X is determined by its set of edges, a subset of the set of all 2-subsets of X. So, if $|X| = n$, the number of labelled graphs is $2^{\binom{n}{2}} = 2^{n(n-1)/2}$, of which $\binom{n(n-1)/2}{m}$ have m edges. How many unlabelled graphs are there? We identify two graphs if some permutation of X maps one to the other, so the group in question is the symmetric group S_n; but we have to calculate the cycle index for its action on the 2-subsets of X.

The cycle index for the usual action of S_n is implicit in (13.1.5), where we calculated, for each partition of n, the number of permutations with that partition as cycle structure: that is, if $\lambda = 1^{a_1} 2^{a_2} \dots n^{a_n}$,

$$Z(S_n) = \sum_{\lambda \vdash n} \left(\frac{n!}{\prod_{i=1}^{n} i^{a_i} a_i!} \right) s_1^{a_1} \dots s_n^{a_n}.$$

For the action we require, the conjugacy class sizes are the same, but the cycle structures are different. Rather than give an explicit formula, I will explain how the calculation is done, and work an example.

Consider a permutation g of X. We consider two types of 2-subsets of X; those contained within a cycle of g, and those which straddle two different cycles.

(i) In a cycle C of g of length m, if m is odd, there are $(m-1)/2$ cycles of length m on 2-sets; if m is even, there is one cycle of length $m/2$ (consisting of pairs of points which are opposite in C) and $(m-2)/2$ cycles of length m.

(ii) If two cycles have lengths m_1 and m_2, then there are (m_1, m_2) cycles of length $m_1 m_2 / (m_1, m_2)$ on pairs consisting of one point from each cycle.

EXAMPLE: $n = 4$. Using the above rule, we find:

Cycle structure on points	Cycle structure on pairs	Number of permutations
1^4	1^6	1
$1^2 2^1$	$1^2 2^2$	6
2^2	$1^2 2^2$	3
$1^1 3^1$	3^2	8
4^1	$2^1 4^1$	6

So the cycle index is

$$\tfrac{1}{24}(s_1^6 + 9s_1^2 s_2^2 + 8s_3^2 + 6s_2 s^4),$$

and the generating function for graphs by number of edges is

$$1 + t + 2t^2 + 3t^3 + 2t^4 + t^5 + t^6,$$

which is easily checked by listing the graphs.

Note that the generating function gives us confidence that we haven't overlooked any possibilities!

15.5. Direct and wreath products

There are simple formulae for calculating the cycle index of the direct or wreath product of two permutation groups (in its natural action) from those of the factors.

(15.5.1) Proposition. $Z(G \times H) = Z(G)Z(H)$.

PROOF. We have

$$Z(G \times H) = \frac{1}{|G \times H|} \sum_{(g,h) \in G \times H} z((g,h))$$

$$= \frac{1}{|G|} \sum_{g \in G} \frac{1}{|H|} \sum_{h \in H} z((g,h)),$$

and so we have to show that $z((g,h)) = z(g)z(h)$ for any permutations g, h of disjoint sets. But this is immediate from the fact that $c_i((g,h)) = c_i(g) + c_i(h)$. (Recall that the natural action is on the disjoint union of the sets.)

(15.5.2) Proposition. $Z(G \text{ wr } H) = Z(H; Z(G; s_1, s_2, \ldots), Z(G; s_2, s_4, \ldots), \ldots)$.

In other words, $Z(G \text{wr} H)$ is obtained from $Z(H)$ by substituting $Z(G; s_i, s_{2i}, \ldots)$ for s_i, for each i.

PROOF. Rather than direct calculation (which gives little insight), we will show that the 'recipe' given by the right-hand side for calculating the function-counting series is correct. Then we appeal to the principle that, for any permutation group K, $Z(K)$ is the unique polynomial in $s_1, s_2, \ldots$ such that, for *any* figure-counting series $a(t)$, the function-counting series is obtained by substituting $a(t^i)$ for s_i for each i. However, I won't give a proof of this principle.

So let G act on X and H on Y, where $|X| = n$, $|Y| = m$, say $Y = \{y_1, \ldots, y_m\}$. Take a set Φ of figures with figure-counting series $a(t)$. Recall that $G \text{ wr } H$ acts on $X \times Y = \bigcup_{i=1}^{m} X_i$, where $X_i = X \times \{y_i\}$. Elements of the base group act as independent m-tuples from G on the sets $X_1, \ldots, X_m$, while elements of H permute these sets.

The counting series for functions on X fixed by G is $c(t) = Z(G; a(t), a(t^2), \ldots)$. Now we can regard these functions as forming a new set Ψ of 'figures'. Functions f from $X \times Y$ to Φ fixed by the base group can be identified with functions $\overline{f}$ from Y

to Ψ; and f is fixed by G wr H if and only if $\overline{f}$ is fixed by the top group, tht is, by H acting on Y. So, finally, the function-counting series for G wr H is

$$Z(H; c(t), c(t^2), \ldots) = Z(H; Z(G; a(t), a(t^2), \ldots), Z(G; a(t^2), a(t^4), \ldots), \ldots).$$

This is exactly what is obtained from the right-hand side of the proposition by substituting $a(t^i)$ for s_i for each i.

EXAMPLE. S_2 and S_3 have cycle indices $\frac{1}{2}(s_1^2 + s_2)$ and $\frac{1}{6}(s_1^3 + 3s_1s_2 + 2s_3)$. So the cycle index for S_2 wr S_3 is

$$\frac{1}{6}\left(\frac{1}{8}(s_1^2 + s_2)^3 + \frac{3}{4}(s_1^2 + s_2)(s_2^2 + s_4) + (s_3^2 + s_6)\right)$$
$$= \frac{1}{48}(s_1^6 + 3s_1^4 s_2 + 9s_1^2 s_2^2 + 7s_2^3 + 6s_1^2 s_4 + 6s_2 s_4 + 8s_3^2 + 8s^6).$$

Check this by using the fact that this group is isomorphic to the group of symmetries of the cube acting on its faces.

15.6. Stirling numbers revisited

In the preamble to (15.3.2), we introduced the notation F_n for the number of orbits of a permutation group G on the set of n-tuples of distinct points of X. Now let F_n^* be the number of orbits on all n-tuples from X (that is, on the set X^n). The next result gives the relationship between these sequences.

(15.6.1) Proposition. $F_n^* = \sum_{k=1}^{n} S(n, k) F_k,$

where $S(n, k)$ is the Stirling number of the second kind.

PROOF. Given an n-tuple $(x_1, \ldots, x_n)$, we construct from it a partition of $\{1, \ldots, n\}$, corresponding to the equivalence relation in which $i \equiv j$ if and only if $x_i = x_j$. If the partition has k parts, then the n-tuple has k distinct entries; let these be $(y_1, \ldots, y_k)$ (in order of appearance). Now two n-tuples lie in the same orbit of G if and only if both

(a) the partitions of $\{1, \ldots, n\}$ they define are the same; and

(b) the corresponding tuples $(y_1, \ldots, y_k)$ and $(y_1', \ldots, y_k')$ of distinct elements lie in the same orbit of G.

Now there are $S(n, k)$ partitions with k parts, and for each partition there are F_k orbits of G on k-tuples of distinct elements. Multiplying, and summing over k, gives the result.

We examine two extreme cases of this result.

1. If G is n-transitive, then $f_k = 1$ for $k \le n$, and so $F_n^* = \sum_{k=1}^{n} S(n, k) = B(n)$, the Bell number (see Exercise 10 of Chapter 14).

2. Take G to be the trivial group on a set of size t. We have $F_n = t(t-1)\ldots(t - n+1) = (t)_n$, and $F_n^* = t^n$. So

$$t^n = \sum_{k=1}^{n} S(n, k)(t)_k.$$

Since this is true for all positive integers t, it is a polynomial identity. (Compare (5.3.3(b)).)

Using the Orbit-counting Lemma, we can give the combinatorial proof of the 'reverse' of this last formula, promised in Chapter 5, namely

$$(t)_n = \sum_{k=1}^{n} s(n, k) t^k.$$

For this, we consider the set of functions from X to $\{1, \ldots, t\}$, where $|X| = n$, and count orbits of S_n on this set. A function is fixed by a permutation g if and only if it is constant on the cycles of g. Since there are, by definition, $(-1)^{n-k} s(n, k)$ permutations in S_n with k cycles, the number of orbits is

$$\frac{1}{n!} \sum_{k=1}^{n} (-1)^{n-k} s(n, k) t^k = \frac{(-1)^n}{n!} \sum_{k=1}^{n} s(n, k)(-t)^k.$$

But the number of orbits on such functions is just the number of choices of n things from a set of size t, with order unimportant and repetitions allowed. By (3.7.1), this number is

$$\binom{n+t-1}{n} = (-1)^n \binom{-t}{n} = \frac{(-1)^n}{n!} (-t)_n.$$

Thus

$$(-t)_n = \sum_{k=1}^{n} s(n, k)(-t)^k.$$

This holds for all positive integers t, and so it is a polynomial identity. Now substituting $-t$ for t gives the required result.

15.7. Project: Cycle index and symmetric functions

Recall from Section 13.5 the notion of a symmetric polynomial in the indeterminates $x_1, \ldots, x_n$, and some special symmetric polynomials: the *elementary symmetric function* e_r, the sum of all products of r distinct indeterminates; the *complete symmetric function* h_r, the sum of all products of r indeterminates (repetitions allowed); and the *power sum function* $p_r = x_1^r + \ldots + x_n^r$. We'll see that the cycle index of the symmetric group is a recipe for expressing h_n in terms of the power sum functions; and the alternating group plays a similar rôle with respect to the function e_n.

Recall also from Section 13.5
- the generating functions
  $E(t) = \sum_{r \geq 0} e_r t^r$,
  $H(t) = \sum_{r \geq 0} h_r t^r$,
  $P(t) = \sum_{r \geq 1} p_r t^{r-1}$;
- the formula $P(t) = \dfrac{d}{dt} \log H(t)$.

In addition, the formula from Section 13.1 for the number c_λ of permutations with cycle structure λ, viz.

$$c_\lambda = \frac{n!}{1^{a_1} a_1! \, 2^{a_2} a_2! \, \ldots},$$

where $\lambda = 1^{a_1} 2^{a_2} \ldots$.

(15.7.1) **Proposition.** $h_n = Z(S_n; p_1, p_2, \ldots, p_n)$.

PROOF. We have

$$H(t) = \exp \int P(t)\, dt$$

$$= \exp \sum_{r \geq 1} p_r t^r / r$$

$$= \prod_{r \geq 1} \exp(p_r t^r / r)$$

$$= \prod_{r \geq 1} \sum_{a_r \geq 0} \frac{(p_r t^r)^{a_r}}{r^{a_r} a_r!}.$$

Now the coefficient of t^n in the right-hand side is made up of a sum of terms, one for each expression $n = \sum r a_r$; that is, one for each partition $\lambda = 1^{a_1} 2^{a_2} \dots \vdash n$. The contribution which comes from the partition λ is

$$\prod_{r \geq 1} \frac{p_r^{a_r}}{r^{a_r} a_r!} = \frac{c_\lambda}{n!} \prod_{r \geq 1} p_r^{a_r},$$

which is precisely the contribution to $Z(S_n; p_1, p_2, \dots)$ from permutations with cycle structure λ. Summing over λ gives the result, since the coefficient of t^n on the left-hand side is just h_n.

Without proof, I will mention the analogous result for the alternating group.

(15.7.2) Proposition. $h_n + e_n = Z(A_n; p_1, p_2, \dots)$ for $n \geq 2$.

15.8. Exercises

1. Use the Orbit-counting Lemma to find a formula for the number of ways of colouring the faces of a cube with r colours, up to rotations. Repeat for colourings of the edges, and of the vertices.

2. Find the cycle index of the group S_5 acting on 2-subsets. Hence enumerate graphs on 5 vertices by number of edges.

3. Prove Proposition 15.5.1 in the spirit of Proposition 15.5.2. (You will probably find Step 1 in the proof of the Cycle Index Theorem useful.)

4. Show that the cycle index of a direct product of permutation groups, in its product action, can in principle be calculated from the cycle indices of the factors. Perform the calculation for $S_3 \times S_3$. Hence enumerate the 3×3 matrices of zeros and ones, up to row and column permutations, by number of ones.

5. Calculate the cycle index for S_4 acting on (a) the ordered pairs of distinct elements of $\{1, \dots, 4\}$, (b) the subsets of $\{1, \dots, 4\}$. Hence enumerate the (a) loopless digraphs, (b) families of sets, on four points up to isomorphism, by number of edges or sets.

6. Let the cyclic group of prime order p generated by g act on the set of all p-tuples of elements from $\{1, \dots, n\}$ by the rule

$$(x_1, \dots, x_p)g = (x_p, x_1, \dots, x_{p-1}).$$

By counting orbits, prove that $n^p \equiv n \pmod{p}$.

7. Let $F(s_1, \dots, s_n)$ be a polynomial in n variables $s_1, \dots, s_n$. For any polynomial $a(t)$, let $F[a]$ denote $F(a(t), a(t^2), \dots, a(t^n))$. Can you show that, if $F[a] = 0$ for all polynomials $a(t)$ with non-negative integer coefficients, then $F = 0$? Deduce the principle used in the proof of (15.5.2) from this assertion.

16. Designs

Though the uncarved block is small
No one in the world dare claim its allegiance.

Lao Tse, Tao Te Ching (ca. 500 BC)

TOPICS: Designs

TECHNIQUES: Matrix and determinant techniques; Cauchy's Inequality (the 'variance trick')

ALGORITHMS:

CROSS-REFERENCES: Steiner triple systems (Chapter 8), finite geometries (Chapter 9), regular families (Chapter 7), PIE (Chapter 5), Latin squares, [SDRs] (Chapter 6)

Designs are a generalisation of Steiner triple systems. There is no hope of deciding the values of the parameters for which designs exist (as we did for STSs). We will develop just enough theory to resolve the question for small designs, and say a little about some general classes.

16.1. Definitions and examples

Let t, k, v, λ be integers with $t < k < v$ and $\lambda > 0$. A t-(v, k, λ) *design*, or t-*design* with parameters (v, k, λ), is a pair $(X, \mathcal{B})$, where X is a set of v *points*, and $\mathcal{B}$ is a collection of k-subsets of X called *blocks*, with the property that any t points are contained in exactly λ blocks.

EXAMPLES.

1. A non-trivial Steiner triple system is a 2-$(v, 3, 1)$ design, by definition.

2. A 2-(6, 3, 2) design is constructed as follows. Take the six points to consist of a pentagon and the point at the centre; the blocks consist of all triangles formed from these points which contain exactly one edge of the pentagon. (So, if the vertices of the pentagon are 1, ..., 5, and 0 is the centre, the blocks are 012, 023, 034, 045, 051, 124, 235, 341, 452, 513.) The 2-design property is easily checked by inspection (Fig. 16.1(a)).

3. Here is a 2-(7, 3, 2) design. The points are the vertices of a regular heptagon; the blocks are all the *scalene* triangles that can be formed from these points (Fig. 16.1(b)). It is easily checked that any edge or diagonal lies in just two scalene triangles, one the mirror image of the other. In fact, there are two shapes of scalene triangles, one the mirror image of the other; if we take the triangles of one shape, we get a 2-(7, 3, 1) design (a Steiner triple system).

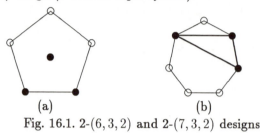

(a)   (b)

Fig. 16.1. 2-$(6, 3, 2)$ and 2-$(7, 3, 2)$ designs

4. Here is a 3-(8, 4, 1) design. The points are the vertices of a cube. There are three types of blocks:

    (i) a face (six of these);
    (ii) two opposite edges (six of these);
    (iii) an inscribed regular tetrahedron (two of these).

Again, the proof is by checking (Fig. 16.2).

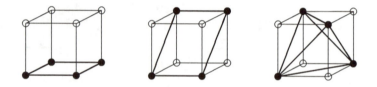

Fig. 16.2. A 3-$(8, 4, 1)$ design

If these examples suggest to you a connection between design theory and geometry, I have not wholly misled you! Now we develop some theory.

(16.1.1) Proposition. *The number b of blocks of a t-(v, k, λ) design is given by*

$$b = \lambda \binom{v}{t} \Big/ \binom{k}{t}.$$

PROOF. A standard double count, of pairs (T, B), where T is a t-set of points, B a block, with $T \subseteq B$: there are $\binom{v}{t}$ choices of T, each contained in λ blocks; and there are b blocks, each containing $\binom{k}{t}$ t-subsets.

We always use b for the number of blocks.

(16.1.2) Proposition. *Let $(X, \mathcal{B})$ be a t-(v, k, λ) design. Given $s \leq t$, let S be a s-subset of X. Let $X' = X \setminus S$, and*

$$\mathcal{B}' = \{B \setminus S : S \subseteq B, B \in \mathcal{B}\}$$

(i.e., take all blocks containing S, and remove S from them). Then $(X', \mathcal{B}')$ is a $(t - s)$-$(v - s, k - s, \lambda)$ design.

PROOF. There are $v - s$ points, and each block contains $k - s$ of them. Let Y be a subset of $X \setminus S$ of size $t - s$. Then $Y \cup S$ is a t-subset of X, so lies in λ blocks in $\mathcal{B}$; removing S, we get λ blocks of $\mathcal{B}'$ containing Y.

The design $(X', \mathcal{B}')$ is called the *derived design* of $(X, \mathcal{B})$ with respect to the set S (or, with respect to the point x, if $S = \{x\}$ is a singleton).

(16.1.3) Corollary. *For $s \leq t$, a t-(v, k, λ) design is also a s-(v, k, λ_s) design, where*

$$\lambda_s = \lambda \binom{v - s}{t - s} \Big/ \binom{k - s}{t - s}.$$

PROOF. λ_s is the number of blocks of a $(t - s)$-$(v - s, k - s, \lambda)$ design.

EXAMPLE. Consider the 3-(8, 4, 1) design we constructed earlier. If we choose one point of this design and remove it from all blocks containing it, we get a 2-(7, 3, 1) design, i.e., a STS of order 7:

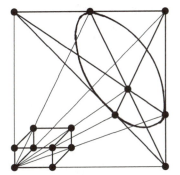

Fig. 16.3. A derived design

16.2. To repeat or not to repeat?

We have defined designs in such a way that the blocks form a set of k-subsets of the point set; that is, each k-set is either a block or not. There is, however, a more general notion, in which a given subset is allowed to occur more than once as a block. If the 'multiplicity' of B is μ, then B contributes μ to the total number of blocks containing each of its t-element subsets.[1] Compare the remarks at the start of Chapter 7.

In statistical design, where this notion first arose, nothing is lost by allowing so-called *repeated blocks*. Imagine that we are testing a number v of varieties of fertiliser. In each experimental trial, we can take k of these varieties and compare them. In order to evaluate the results, it is desirable that each pair of varieties should be compared in the same number (λ, say) of trials. So the experimental design should be a 2-(v, k, λ) design. But the experiment will be just as effective if we can use the same k-set as a block more than once: the trial needs to be repeated the appropriate number of times on each block. The analysis of the results is unaffected.

---

[1] Hughes and Piper call these 'designs' *t-structures*.

What we lose, in fact, is an interesting mathematical problem. Of course, if $\mathcal{B}$ consists of all the k-subsets of X, then $(X, \mathcal{B})$ is a t-design for any $t \leq k$, a so-called *trivial* design. Thus, for a non-trivial design, we require that some k-subset does not occur as a block. Now we can prove the following existence theorem with elementary linear algebra:

(16.2.1) Proposition. *Let* t, k, v *be given with* $t < k < v - t$. *If repeated blocks are allowed, then there exists a non-trivial* t-(v, k, λ) *design, for some* λ.

PROOF. We define a matrix $M = (m_{T,K})$ as follows. M is a $\binom{v}{t} \times \binom{v}{k}$ matrix, whose rows are indexed by the t-subsets of $X = \{1, \ldots, v\}$, and whose columns are indexed by the k-subsets; the (T, K) entry $m_{T,K}$ is equal to 1 if $T \subseteq K$, and 0 otherwise.

Since $t < k < v - t$, we have $\binom{v}{t} < \binom{v}{k}$, so M has more columns than rows. Thus, the columns of M are linearly dependent over $\mathbb{Q}$: there are rational numbers a_K, for K a k-subset of X, such that $\sum a_K c(K) = 0$, where $c(K)$ is the column indexed by K. Multiplying up by the least common multiple of the denominators of these rationals, we can assume that all a_K are integers. Clearly some are positive and some negative; let $-d$ be the least. Now $a_K + d \geq 0$ for all K, and $a_K + d = 0$ for some K.

Consider the 'design' in which the block K is repeated $a_K + d$ times. (Thus, some k-sets do not occur; the others occur a positive integral number of times.) We claim that this is indeed a t-design. Take a t-subset T. To find the number of blocks containing T, we add the multiplicities of the k-subsets K for which $m_{T,K} = 1$. This number is

$$\sum_{|K|=k} m_{T,K}(a_K + d) = \sum_{|K|=k} m_{T,K} d = \binom{v-t}{k-t} d,$$

the first inequality because $\sum m_{T,K} a_K = 0$ (we chose a linear dependence relation between columns), the second because T lies in $\binom{v-t}{k-t}$ subsets of size k.

So we have a 't-design' with $\lambda = \binom{v-t}{k-t} d$.

On the other hand, if we do not allow repeated blocks, the existence question for t-designs is much more difficult. Only in the last few years has it been shown by Luc Teirlinck that non-trivial t-designs exist for all values of t; his designs have $k = t+1$ and $v-t$ divisible by a quite rapidly growing function of t. So the existence question is far from settled!

Note that 1-designs (without repeated blocks) exist for all 'feasible' parameters — this was shown in Section 7.4.[2] So we concentrate on the cases $t \geq 2$. For $t = 2$, a powerful existence theory has been developed by Richard Wilson. From (16.1.2) we see that a necessary condition for the existence of a 2-(v, k, λ) design is that the numbers $r = (v - 1)\lambda/(k - 1)$ and $b = v(v - 1)\lambda/k(k - 1)$ are integers; in other words,

$$(v - 1)\lambda \equiv 0 \pmod{k - 1},$$
$$v(v - 1)\lambda \equiv 0 \pmod{k(k - 1)}.$$

---

[2] These designs were called *regular families* in Chapter 7.

For Steiner triple systems, these necessary conditions assert $v - 1 \equiv 0 \pmod 3$ and $v(v-1) \equiv 0 \pmod 6$; we saw in Chapter 8 that these conditions are also sufficient. Wilson showed:

(16.2.2) Theorem. *There is a function $f(k, \lambda)$ such that, if $v \geq f(k, \lambda)$ and the above necessary conditions are satisfied, then a 2-(v, k, λ) design (possibly having repeated blocks) exists.*

Wilson's proof is a mix of direct and recursive constructions, like the existence proof (8.1.2) for Steiner triple systems but rather more complicated. In the case $\lambda = 1$, of course, the resulting designs have no repeated blocks. Nobody has succeeded in proving a similar theorem for higher values of t.

In the remainder of this chapter, we assume *no repeated blocks.*

16.3. Fisher's Inequality

We are most interested in 2-designs. By (16.1.2), a 2-design is also a 1-design; that is, a point lies in a constant number r of blocks. Now we have $r(k-1) = (v-1)\lambda$ (the formula for $r = \lambda_1$ from (16.1.3)), and $vr = bk$ (applying (16.1.1) to the 1-(v, k, r) design). From these, the result of (16.1.1), viz. $b\binom{k}{2} = \lambda\binom{v}{2}$, follows.

An important result about 2-designs is *Fisher's Inequality:*[3]

(16.3.1) Fisher's Inequality.

In a 2-(v, k, λ) design, $b \geq v$ (i.e. there are at least as many blocks as points).

PROOF. Consider first the case $\lambda = 1$. Take a point x and a block B with $x \notin B$. For each $y \in B$, there is a unique block B_y containing x and y; all these blocks are different. (For if $B_y = B_z$, then B_y and B are blocks containing y and z, and so are equal; but $x \in B_y$, $x \notin B$, a contradiction.) So the number r of blocks through x is at least the number k of points on B, i.e., $r \geq k$. Since $br = vk$, it follows that $b \geq v$.

For the general case, we offer two proofs, the first by linear algebra, the second illustrating a very useful counting argument, variously called *the variance trick* or *Cauchy's Inequality.*[4]

FIRST PROOF. Let $X = \{x_1, \ldots, x_v\}$, $\mathcal{B} = \{B_1, \ldots, B_b\}$, and let M be the $v \times b$ matrix whose (i, j) entry is 1 if $x_i \in B_j$, 0 otherwise. M is called the *incidence matrix* of the design. The use of incidence matrices introduces algebraic methods into design theory, to good effect.

---

[3] R. A. Fisher was one of the most influential statisticians of the twentieth century. In accordance with the remarks in the last section, his inequality is bad news for statisticians: it shows that, to achieve balance between the treatments, at least as many trials are required as the number of varieties being tested.

[4] In geometric language, this inequality asserts that the inner product of two real vectors does not exceed the product of their lengths.

CLAIM 1. MM^T is the $v \times v$ matrix with diagonal entries r and off-diagonal entries λ. For

$$(MM^\mathsf{T})_{ik} = \sum_{j=1}^{b} (M)_{ij}(M^\mathsf{T})_{jk} = \sum_{j=1}^{b} (M)_{ij}(M)_{kj},$$

which is the number of blocks containing x_i and x_k. This number is r if $i = k$, and λ if $i \neq k$, proving the claim.

CLAIM 2. $\det(MM^\mathsf{T}) = rk(r - \lambda)^{v-1}$.

We use the fact that adding a multiple of one row (or column) to another doesn't change the determinant, while multiplying a row (or column) by a constant c multiplies the determinant by c. So we have

$$\det(MM^\mathsf{T}) = \det \begin{pmatrix} r & \lambda & \cdots & \lambda \\ \lambda & r & \cdots & \lambda \\ \vdots & \vdots & \ddots & \vdots \\ \lambda & \lambda & \cdots & r \end{pmatrix}$$

$$= \det \begin{pmatrix} r + (v-1)\lambda & r + (v-1)\lambda & \cdots & r + (v-1)\lambda \\ \lambda & r & \cdots & \lambda \\ \vdots & \vdots & \ddots & \vdots \\ \lambda & \lambda & \cdots & r \end{pmatrix}$$

$$= (r + (v-1)\lambda) \det \begin{pmatrix} 1 & 1 & \cdots & 1 \\ \lambda & r & \cdots & \lambda \\ \vdots & \vdots & \ddots & \vdots \\ \lambda & \lambda & \cdots & r \end{pmatrix}$$

$$= rk \det \begin{pmatrix} 1 & 1 & \cdots & 1 \\ 0 & r - \lambda & \cdots & 0 \\ \vdots & \vdots & \ddots & \vdots \\ 0 & 0 & \cdots & r - \lambda \end{pmatrix}$$

$$= rk(r - \lambda)^{v-1}.$$

(The second equality is obtained by adding all other rows to the first, and the fourth by subtracting λ times the first row from all other rows and using $r + (v-1)\lambda = rk$.) Hence MM^T is non-singular. Since it is $v \times v$, its rank is v. But if $b < v$, then $\mathrm{rank}(M) \leq b < v$, and so $\mathrm{rank}(MM^\mathsf{T}) < v$, a contradiction. So $b \geq v$.

SECOND PROOF. This proof just involves counting, but has the advantage that it more easily gives us information about what happens when the bound is met.

Let B be any block. For $i = 0, \dots, k$, let n_i denote the number of blocks $B' \neq B$ for which $|B \cap B'| = i$. Now we have the following equations:

$$\sum_{i=0}^{k} n_i = b - 1,$$

$$\sum_{i=0}^{k} i n_i = k(r - 1),$$

$$\sum_{i=0}^{k} i(i-1) n_i = k(k-1)(\lambda - 1).$$

(The first equation simply counts blocks different from B. The second counts pairs (x, B') where $x \in B$ and $B' \neq B$ with $x \in B'$: for each of the k points of B, there are $r - 1$ further blocks containing it. The third equation counts triples (x_1, x_2, B'), where x_1, x_2 are two points of B and B' is another block containing both: there are k choices for x_1, $k - 1$ for x_2, and $\lambda - 1$ further blocks containing both.) From these equations, we obtain

$$\sum_{i=0}^{k} i^2 n_i = k(r - 1) + k(k - 1)(\lambda - 1),$$

and hence

$$\sum_{i=0}^{k} (x - i)^2 n_i = (b - 1)x^2 - 2k(r - 1)x + (k(r - 1) + k(k - 1)(\lambda - 1)),$$

where x is an indeterminate. This equation defines a quadratic function of x. From the left-hand side, we see that it is positive semi-definite, that is, its value is at least 0 for all real x. Hence the discriminant of the quadratic form on the right must be negative or zero; that is,

$$k^2(r - 1)^2 - (b - 1)k((r - 1) + (k - 1)(\lambda - 1)) \leq 0.$$

We simplify this by expressing it in terms of the parameters v, k, r, using the equations $bk = vr$ and $r(k - 1) = \lambda(v - 1)$. Multiply by $v - 1$:

$$k^2(r - 1)^2(v - 1) - (vr - k)(r - k)(v - 1) - (vr - k)r(k - 1)^2 \leq 0.$$

After some manipulation, this becomes

$$(k - r)r(v - k)^2 \leq 0.$$

Since $r > 0$ and $(v - k)^2 > 0$, we must have $k \leq r$. Using $vr = bk$, this is equivalent to $b \geq v$, as required.

What happens if equality holds?

(16.3.2) Theorem. For a 2-(v, k, λ) design with $k < v$, the following are equivalent:
    (a) $b = v$;
    (b) $r = k$;
    (c) any two blocks meet in λ points;
    (d) any two blocks meet in a constant number of points.

PROOF. Since $bk = vr$, (a) and (b) are equivalent. Clearly (c) implies (d). We show that (b) implies (c), and that (d) implies (b).

(d) $\Rightarrow$ (b): If any two blocks meet in μ points, then $n_i = 0$ for $i \neq \mu$, and so $\sum (\mu - i)^2 n_i = 0$. This means that $x = \mu$ is a root of the quadratic form above, whose discriminant is thus equal to zero. Thus $(k - r)r(v - k)^2 = 0$, or $r = k$.

(b) $\Rightarrow$ (c): If $r = k$, then $b = v$, and the quadratic form becomes $(v - 1)x^2 - 2k(k - 1)x + k(k - 1)\lambda$; using $k(k - 1) = (v - 1)\lambda$, this becomes $(v - 1)(x^2 - 2\lambda x + \lambda^2)$. Thus $x = \lambda$ is a root. Reversing the previous argument, we see that $n_i = 0$ for $i \neq \lambda$, and so every block meets B in exactly λ points. Since B was arbitrary, (c) holds.

A 2-design satisfying the equivalent conditions of (16.2.2) is called a *square* 2-design.[5] Its parameters (v, k, λ) satisfy the equation

$$k(k-1) = (v-1)\lambda.$$

The *Bruck–Ryser–Chowla Theorem* gives a necessary condition for the existence of a square design.

(16.3.3) Theorem. *Suppose that a square 2-(v, k, λ) design exists. Then:*
(a) if v is even, $k - \lambda$ is a square;
(b) if v is odd, the equation

$$z^2 = (k - \lambda)x^2 + (-1)^{(v-1)/2}\lambda y^2$$

has a solution in integers x, y, z, not all zero.

PROOF. (a) From the first proof of Fisher's Inequality (16.3.1), we see that the incidence matrix M of the design satisfies

$$\det(M)^2 = \det(MM^\top) = k^2(k - \lambda)^{v-1}.$$

So $|\det(M)| = k(k-\lambda)^{(v-1)/2}$. This is an integer; so, if v is even, then $k - \lambda$ is a square.

(b) The second part is a generalisation of the Bruck–Ryser Theorem (9.5.2).[6] The proof is almost identical, and I will not give it here. Instead, I show that, for projective planes, the conclusions of (9.5.2) and (16.3.3)(b) are identical. Let $\lambda = 1$, and set $k = n + 1$, $v = n^2 + n + 1$. The diophantine equation is

$$z^2 = nx^2 + (-1)^{n(n+1)/2}y^2.$$

If $n \equiv 0$ or $3 \pmod 4$, then $n(n + 1)/2$ is even, and the equation has the trivial solution $x = 0$, $y = z$; so the necessary condition is empty. If $n \equiv 1$ or $2 \pmod 4$, then $n(n + 1)/2$ is odd, and the equation is $y^2 + z^2 = nx^2$. As explained in Section 9.8, this has a solution if and only if n is the sum of two squares.

The *complement* of a design $(X, \mathcal{B})$ is $(X, \overline{\mathcal{B}})$, where

$$\overline{\mathcal{B}} = \{X \setminus B : B \in \mathcal{B}\};$$

its blocks are the complements of all the blocks in $\mathcal{B}$.

(16.3.4) Proposition. *The complement of a t-(v, k, λ) design is a t-$(v, v - k, \overline{\lambda})$ design, where*

$$\overline{\lambda} = \sum_{s=0}^{t}(-1)^s \binom{t}{s}\lambda_s.$$

---

[5] Other terms used are *symmetric design* or *projective design*.

[6] It was proved by Chowla and Ryser a year after the Bruck–Ryser Theorem; hence the name.

PROOF. Clearly every block of the complement has $v - k$ points. Let $x_1, \ldots, x_t$ be points. Then the number of blocks of $(X, \overline{\mathcal{B}})$ containing $x_1, \ldots, x_t$, is equal to the number of blocks of $(X, \mathcal{B})$ containing *none* of $x_1, \ldots, x_t$. We find this number by using PIE (Section 5.1). Let $\mathcal{B}_i$ be the set of blocks containing x_i, and for $I \subseteq \{1, \ldots, t\}$, $\mathcal{B}_I = \bigcap_{i \in I} \mathcal{B}_i$. Then $\mathcal{B}_I$ is the set of blocks containing x_i for all $i \in I$, so that $|\mathcal{B}_I| = \lambda_s$ if $|I| = s$. By PIE, the number of blocks containing none of $x_1, \ldots, x_t$ is

$$\sum_{I \subseteq \{1, \ldots, t\}} (-1)^{|I|} |\mathcal{B}_I| = \sum_{s=0}^{t} (-1)^s \binom{t}{s} \lambda_s,$$

since there are $\binom{t}{s}$ sets $I \subseteq \{1, \ldots, t\}$ with $|I| = s$.

Note that the complement of a square 2-design is a square 2-design.

EXAMPLE. Consider the complement of a 2-(7, 3, 1) design. This design has $\lambda_0 = b = 7$, $\lambda_1 = r = 3$, and $\lambda_2 = \lambda = 1$. So

$$\overline{\lambda} = 7 - 2 \cdot 3 + 1 = 2,$$

and the complement is a 2-(7, 4, 2) design.

A design is called *trivial* if every k-set of points is a block. (The set of all k-subsets is the block set of a t-$(v, k, \binom{v-t}{k-t})$ design.)

(16.3.5) Corollary. *A t-(v, k, λ) design with $k \geq v - t$ is trivial.*

PROOF. Let $s = v - k$. Then $s \leq t$, so our design is an s-design, by (16.1.3). So its complement is also an s-design, with block size s. Now some s-set is contained in a block of this design, and so this design is trivial, as is its complement.

16.4. Designs from finite geometry

Recall from Chapter 9 the projective geometry $\mathrm{PG}(n, q)$: its points are all the 1-dimensional subspaces of a vector space V of dimension $n + 1$ over $\mathrm{GF}(q)$, and its i-flats are the $(i + 1)$-dimensional subspaces, for $0 \leq i \leq n$. An i-flat can be identified with the set of points it contains.

(16.4.1) Proposition. *For $1 \leq i \leq n - 1$, the points and i-flats in $\mathrm{PG}(n, q)$ form a non-trivial* 2-$\left(\begin{bmatrix} n+1 \\ 1 \end{bmatrix}_q, \begin{bmatrix} i+1 \\ 1 \end{bmatrix}_q, \begin{bmatrix} n-1 \\ i-1 \end{bmatrix}_q \right)$ *design, where $\begin{bmatrix} a \\ b \end{bmatrix}_q$ is the Gaussian coefficient.*

PROOF. The number of points, and the number of points in a block, are clear. Let x, y be points. Then $\langle x, y \rangle$ is a 2-dimensional subspace of V. The $(i+1)$-dimensional subspaces containing it are in one-to-one correspondence with the $(i-1)$-dimensional subspaces of the quotient space $V/\langle x, y \rangle$, by the Third Isomorphism Theorem.

Note some special cases:

(a) If $i = n - 1$, the design is square. Its blocks are called *hyperplanes*.

(b) If $i = 1$ (the blocks are lines), then $\lambda = 1$. In particular, if $i = 1$ and $q = 2$, we have Steiner triple systems; these are the 'projective triple systems' of Chapter 8.

(c) The intersection of these cases, where $i = 1, n = 2$, consists of the projective planes $\mathrm{PG}(2, q)$. More generally, any projective plane of order q (not necessarily Desarguesian) is a 2-$(q^2 + q + 1, q + 1, 1)$ design.

In a similar way, affine spaces give designs, but with a twist:

(16.4.2) Proposition. *For* $1 \leq i \leq n$, *if either* $q = 2$ *or* $i > 1$, *the points and i-flats in* $AG(n,q)$ *form a non-trivial* $2\text{-}\left(q^n, q^i, \begin{bmatrix} n-1 \\ i-1 \end{bmatrix}_q\right)$ *design. If* $q = 2$ *and* $i > 1$, *it is even a* $3\text{-}\left(2^n, 2^i, \begin{bmatrix} n-2 \\ i-2 \end{bmatrix}_2\right)$ *design.*

The proof is similar. The reason for the exclusion is that, if $q = 2$ and $i = 1$, then lines have just two points and the design is trivial. So any three points are independent, and span a plane; this is why 3-designs are obtained when $q = 2$. (For $q > 2$, some triples of points are collinear and others are not.)

Once again, if $i = 1$ and $q > 2$, we obtain 2-designs with $\lambda = 1$ (the 'affine triple systems' of Chapter 8 in the case $q = 3$); and, if $i = 2$ and $q = 2$, we obtain $3\text{-}(2^n, 4, 1)$ designs (Steiner quadruple systems). The case $i = n - 1$ (blocks are *hyperplanes*) is also interesting; we'll meet it again for $q = 2$ in Section 16.6. Also, any affine plane of order n (Desarguesian or not) is a $2\text{-}(n^2, n, 1)$ design. Now (9.5.7) can be re-phrased in the terminology of design theory as follows:

(16.4.3) Proposition. *For any n, there exists a $2\text{-}(n^2 + n + 1, n + 1, 1)$ design if and only if there exists a $2\text{-}(n^2, n, 1)$ design.*

16.5. Small designs

If we are trying to decide for which values of the parameters a design exists, we may clearly ignore trivial designs; so we may assume that $t < k < v - t$. But, if a design exists, then so does its complement; so it is enough to resolve the question for $t < k < \frac{1}{2}v$.

Here is another construction of new designs from old, which doesn't seem to have an official name; it is a sort of complementation but not to be confused with the operation defined above (before (16.3.4)). In this construction, the 'no repeated blocks' condition is crucial. Let $(X, \mathcal{B})$ be a non-trivial $t\text{-}(v, k, \lambda)$ design. Let $\mathcal{B}^*$ be the set of all k-subsets of X which are *not* in $\mathcal{B}$ (not blocks of the original design). Then $(X, \mathcal{B}^*)$ is a $t\text{-}\left(v, k, \binom{v-t}{k-t} - \lambda\right)$ design. For any t-set lies in $\binom{v-t}{k-t}$ sets of size k altogether; and λ of these are blocks of the old design, the remainder blocks of the new design.

In a non-trivial $t\text{-}(v, k, \lambda)$ design, the value of λ is at most equal to the total number $\binom{v-t}{k-t}$ of k-subsets which contain a given t-subset. If $\lambda = \binom{v-t}{k-t}$, the design is trivial. Since the existence questions for λ and $\binom{v-t}{k-t} - \lambda$ are equivalent, we need only settle the question for $0 < \lambda \leq \frac{1}{2}\binom{v-t}{k-t}$.

We will illustrate by finding all parameters of non-trivial 2-designs with $v \leq 8$. Since $2 = t < k < v - t$, we have $k \geq 3$ and $v \geq 6$. So we have to consider the values $v = 6, 7, 8$.

CASE $v = 6$. We have $k = 3$, and $0 < \lambda < \binom{6-2}{3-2} = 4$. The equations $r(k-1) = (v-1)\lambda$ and $vr = bk$ become $2r = 5\lambda$ and $3b = 6r$, or $b = 2r = 5\lambda$. The first equation shows that λ is even. So $\lambda = 2$. We have a design with these parameters (Example 2 in Section 16.1).

CASE $v = 7$. We have $k = 3$ or 4, and by taking complements, it is enough to consider $k = 3$. Then $0 < \lambda < \binom{7-2}{3-2} = 5$, so $\lambda = 1, 2, 3$ or 4. From Example 2, there are designs with $\lambda = 1$ and $\lambda = 2$. By using the complementary sets of blocks, we obtain designs with $\lambda = 3$ and $\lambda = 4$. Now a short calculation shows that the complementary designs are 2-(7, 4, λ) designs with $\lambda = 2, 4, 6, 8$.

CASE $v = 8$. Now $k = 3, 4$ or 5, and it is enough to consider $k = 3$ and $k = 4$.

SUBCASE $k = 3$. We have $2r = 7\lambda$, so λ is even; and $3b = 8r$, so r, and hence λ, is divisible by 3. However, $0 < \lambda < 6$, so no such design exists. It follows that there is none with $k = 5$ either.

SUBCASE $k = 4$. This time we have $3r = 7\lambda$, and $b = 2r$; so λ is a multiple of 3. Also, $0 < \lambda < \binom{8-2}{4-2} = 15$. So $\lambda = 3, 6, 9$ or 12, and the last two values can be deduced from the first two. The 3-(8, 4, 1) design of Section 16.1, Example 4 is also a 2-(8, 4, 3) design, by (16.1.3). The existence of a 2-(8, 4, 6) design is an exercise. The existence of the other two designs follows.

REMARK. So far, whenever a parameter set for a 2-design satisfies the divisibility conditions and the trivial inequalities, a design happens to exist. But this pattern does not continue. Some designs are excluded by Fisher's inequality; some by more sophisticated theoretical results; and some by exhaustive computer search. For other values, great ingenuity has been used to construct designs.

16.6. Project: Hadamard matrices

How large can the determinant of a matrix with entries of bounded size be? This question was considered by Hadamard. In this section, we prove Hadamard's theorem and investigate its somewhat surprising connection with design theory.

(16.6.1) Hadamard's Theorem. *Let $A = (a_{ij})$ be a $n \times n$ real matrix whose entries satisfy $|a_{ij}| \le 1$ for all i, j. Then $|\det(A)| \le n^{n/2}$. Equality holds if and only if $a_{ij} = \pm 1$ for all i, j and $AA^\mathsf{T} = nI$.*

PROOF. Our proof uses a geometric interpretation of the determinant. $|\det(A)|$ is the volume of the parallelepiped (in n-dimensional Euclidean space) whose sides are the rows of A. Now, if $|a_{ij}| \le 1$ for all i, j, then the Euclidean length of any row is at most $\sqrt{n}$. The volume of the parallelepiped is at most the product of the edge lengths, with equality if and only if the edges are mutually perpendicular. The inequality follows; and equality holds if and only if each row has length $\sqrt{n}$ (so all its entries are ± 1 and $\sum_{j=1}^{n} a_{ij}^2 = n$ for all i) and any two rows are perpendicular (so $\sum_{j=1}^{n} a_{ij} a_{kj} = 0$ for $k \ne i$). The two summations are equivalent to $AA^\mathsf{T} = nI$.

A matrix H which attains Hadamard's bound is called a *Hadamard matrix*. Thus such a matrix has entries ± 1 and satisfies $HH^\mathsf{T} = nI$. We first derive a simple necessary condition on n for the existence of a Hadamard matrix.

(16.6.2) Proposition. *If a Hadamard matrix of order n exists, then either $n = 1$ or 2, or $n \equiv 0$ (mod 4).*

PROOF. Observe first that, if $n \ge 2$, then any two rows of H agree in $n/2$ positions and disagree in $n/2$ positions, since their inner product is 0. So n is even. Also, if we change the sign of any column of a Hadamard matrix, the result is still a Hadamard matrix (most easily because $|\det(H)|$ is unchanged). By a series of such changes, we can arrange that all entries in the first row of H are equal to +1.

Now suppose that $n \geq 3$, and consider the first three rows. Let a, b, c, d be the numbers of columns in which the second and third rows have entries $(+1, +1)$, $(+1, -1)$, $(-1, +1)$ and $(-1, -1)$ respectively.

$$
\begin{pmatrix}
\overbrace{+ \ \cdots \ +}^{a} & \overbrace{+ \ \cdots \ +}^{b} & \overbrace{+ \ \cdots \ +}^{c} & \overbrace{+ \ \cdots \ +}^{d} \\
+ \ \cdots \ + & + \ \cdots \ + & - \ \cdots \ - & - \ \cdots \ - \\
+ \ \cdots \ + & - \ \cdots \ - & + \ \cdots \ + & - \ \cdots \ - \\
\vdots \quad \vdots & \vdots \quad \vdots & \vdots \quad \vdots & \vdots \quad \vdots
\end{pmatrix}
$$

Considering inner products of the three pairs of rows, we have

$$a + b = c + d = n/2,$$
$$a + c = b + d = n/2,$$
$$a + d = b + c = n/2,$$

with solution $a = b = c = d = n/4$. So n is a multiple of 4.

REMARK. We saw that the Hadamard property is invariant under changing signs of rows or columns. It is also invariant under permutations of rows or columns, and under transposition. Two Hadamard matrices are called *equivalent* if one can be transformed into the other by a sequence of operations of this type.

Now we turn to constructions. There is a simple recursive construction, the so-called *tensor product* or *Kronecker product*. If $A = (a_{ij})$ and B are matrices, their tensor product is (in block form)

$$
A \otimes B = \begin{pmatrix}
a_{11} B & a_{12} B & \cdots \\
a_{21} B & a_{22} B & \cdots \\
\vdots & \vdots & \ddots
\end{pmatrix}.
$$

It can be checked that $(A \otimes B)(C \otimes D) = AC \otimes BD$. Now, if H_1 and H_2 are Haramard matrices of orders n_1, n_2 respectively, then

$$(H_1 \otimes H_2)(H_1 \otimes H_2)^\mathsf{T} = (H_1 \otimes H_2)(H_1^\mathsf{T} \otimes H_2^\mathsf{T}) = H_1 H_1^\mathsf{T} \otimes H_2 H_2^\mathsf{T} = n_1 I_{n_1} \otimes n_2 I_{n_2} = n_1 n_2 I_{n_1 n_2},$$

so $H_1 \otimes H_2$ is a Hadamard matrix.

In particular, taking $H = \begin{pmatrix} +1 & +1 \\ +1 & -1 \end{pmatrix}$, we obtain by successive tensor products a Hadamard matrix of order 2^n for any $n \geq 0$. These matrices are said to be of *Sylvester type*.

Another class of examples consists of the matrices of *Paley type*. Let q be a prime power congruent to $-1 \bmod 4$. Let $P(q) = (p_{ij})$ be the $(q+1) \times (q+1)$ matrix with rows and columns indexed by the elements of the field $\mathrm{GF}(q)$ and a new symbol ∞, with

$$
p_{ij} = \begin{cases}
+1 & \text{if } i = \infty \text{ or } j = \infty; \\
-1 & \text{if } i = j \neq \infty; \\
+1 & \text{if } i - j \text{ is a non-zero square in } \mathrm{GF}(q); \\
-1 & \text{if } i - j \text{ is a non-square in } \mathrm{GF}(q).
\end{cases}
$$

It can be shown that $P(q)$ is a Hadamard matrix. (See Exercise 6.)

Of course, this construction can be used in conjunction with the tensor product, to construct Hadamard matrices of all orders which are the product of a power of 2 and numbers of the form $q_i + 1$, where q_i are prime powers congruent to $-1 \pmod 4$. In particular, the existence question is settled for all multiples of 4 less than 36. But this is not the end; here is a construction for order 36.

Let L be a Latin square of order 6 (Chapter 6). First we construct a graph Γ, a so-called *Latin square graph*. The vertices of Γ are the ordered pairs (i, j), for $1 \leq i, j \leq 6$, regarded as the cells of the Latin square. Two vertices are adjacent if the cells lie in the same row or column or contain the same entry. Now H is the 36×36 matrix whose rows and columns are indexed by the vertices of Γ; the (x, y) entry is $+1$ if x and y are adjacent in Γ, and -1 otherwise. Then H is a Hadamard matrix (Exercise 7).

It is conjectured that, for every positive integer n divisible by 4, there exists a Hadamard matrix of order n. The conjecture is still open, though many values have been settled by increasingly ingenious constructions.

What has all this to do with designs?

(16.6.3) Proposition. *For $n > 4$, the following are equivalent:*
(a) there exists a Hadamard matrix of order n;
(b) there exists a 3-$(n, \frac{1}{2}n, \frac{1}{4}n - 1)$ design;
(c) there exists a 2-$(n - 1, \frac{1}{2}n - 1, \frac{1}{4}n - 1)$ design.

PROOF. (a) $\Rightarrow$ (b): Let $H = (h_{ij})$ be a Hadamard matrix of order n. As in (16.6.2), by changing signs of some columns, we may assume that all entries in the first row are $+1$. Now, for $i = 2, \ldots, n$, let $B_i^+ = \{j : h_{ij} = +1\}$ and $B_i^- = \{j : h_{ij} = -1\}$. Each of these $2(n - 1)$ sets has size $\frac{1}{2}n$. We claim that $(X, \mathcal{B})$ is a 3-$(n, \frac{1}{2}n, \frac{1}{4}n - 1)$ design, where $X = \{1, \ldots, n\}$ and $\mathcal{B} = \{B_i^+, B_i^- : i = 2, \ldots, n\}$.

The proof of (16.6.2) shows, in effect, that, given any three rows of H, there are exactly $n/4$ columns where they all agree. Dually, given any three columns (numbered j_1, j_2, j_3, say), there are $n/4$ rows where they all agree. One is the first row; so there are $\frac{1}{4}n - 1$ sets B_i^ϵ ($i = 2, \ldots, n; \epsilon = \pm 1$) containing j_1, j_2, j_3.

(b) $\Rightarrow$ (c): Take the derived design with respect to a point.

(c) $\Rightarrow$ (a): Let D be a 2-$(n - 1, \frac{1}{2}n - 1, \frac{1}{4}n - 1)$ design. Note that D is square. Let A be its incidence matrix. Now replace the entries 0 in A by -1, and border A with a row and column of $+1$s (the first, say); let H be the resulting matrix.

Now any row of A has $\frac{1}{2}n - 1$ entries 1, so any row of H agrees with the first in $1 + (\frac{1}{2}n - 1) = \frac{1}{2}n$ positions. Also, any two rows of A have entries $(1, 1)$ in $\frac{1}{4}n - 1$ places, and $(0, 0)$ in $\frac{1}{4}n$ places (since, by (16.3.4), the complement of D is a 2-$(n - 1, \frac{1}{2}n, \frac{1}{4}n)$ design); so the corresponding rows of H agree in $1 + (\frac{1}{4}n - 1) + \frac{1}{4}n = \frac{1}{2}n$ places. Thus $HH^\mathsf{T} = nI$.

The designs arising in this theorem are called *Hadamard designs*. The designs of points and hyperplanes in projective and affine spaces over $\mathrm{GF}(2)$ are Hadamard 2-designs and 3-designs respectively (check the parameters given in Section 16.4 to see this); the corresponding Hadamard matrices are of Sylvester type.

16.7. Exercises

1. An *extension* of a t-(v, k, λ) design $(X, \mathcal{B})$ is a $(t + 1)$-$(v + 1, k + 1, \lambda)$ design $(Y, \mathcal{C})$ with a point y such that its derived design with respect to y is isomorphic to X. Prove that a necessary condition for a t-(v, k, λ) design with b blocks to have an extension is that $v + 1$ divides $b(k + 1)$. Hence show that, if a projective plane of order $n > 1$ has an extension, then $n = 2, 4$ or 10.[7]

REMARK. Each of the (unique) projective planes of orders 2 and 4 has a (unique) extension. We saw in Chapter 9 that the non-existence of a projective plane of order 10 was established by a massive computation. In fact, a relatively small part of this computation showed that no projective plane of order 10 could have an extension, some years before the non-existence was proved.

2. Prove that, up to equivalence, there is a unique Hadamard matrix of each of the orders 4, 8, 12; and prove that the corresponding Hadamard designs are unique up to isomorphism.

---

[7] This result is due to Dan Hughes.

3. Let H be the (unique) Hadamard matrix of order 12. Let X be the set of columns of H. For $1 \leq i < j \leq 12$, let B_{ij}^+ be the set of columns where the ith and j^{th} rows agree, and B_{ij}^- the set of columns where they disagree. Let $\mathcal{B} = \{B_{ij}^+, B_{ij}^- : 1 \leq i < j \leq 12\}$. Prove that $(X, \mathcal{B})$ is a 5-(12, 6, 1) design.

4. Show that the construction of the preceding problem, applied to an arbitrary Hadamard matrix, always gives a 3-design (if the order of the Hadamard matrix is greater than 4), but is a 4-design only if $n = 12$.

5. A *character* χ of an abelian group A is a homomorphism from A to the multiplicative group of non-zero complex numbers. The *character table* of A is the matrix whose (i, j) entry is the value of the ith character on the jth element of A (for some ordering of the group elements and the characters). Show that, it A is a direct product of cyclic groups of order 2, then its character table is a Hadamard matrix of Sylvester type.

6. Let $P(q)$ be a Hadamard matrix of Paley type. It already has a row and column of $+1$s, so we can read off the corresponding Hadamard 2-design: its points are the elements of $\mathrm{GF}(q)$, and its blocks are the translates of the set of non-zero squares. Show directly that this is a Hadamard 2-design, and deduce that $P(q)$ is a Hadamard matrix.

7. Prove that the Latin square construction gives a Hadamard matrix.

8. Let $(X, \mathcal{B})$ be a Steiner triple system of order 15. For each triple $B \in \mathcal{B}$, let $S(B)$ be the set of all triples equal to or disjoint from B. Prove that

$$(\mathcal{B}, \{S(B) : B \in \mathcal{B}\})$$

is a Hadamard 2-design with 35 points.

9. Let $(X, \mathcal{B})$ be a square 2-(v, k, λ) design, where $X = \{1, \ldots, v\}$ and $\mathcal{B} = \{B_1, \ldots, B_v\}$. Prove that there is a Latin square of order v, having the property that the set of entries occurring in the first k rows and the ith column is B_i, for $i = 1, \ldots, v$. [Such a square is called a *Youden square*.]

10. Let $\mathcal{D} = (X, \mathcal{B})$ be a t-(v, k, λ) design, and let $x_1, \ldots, x_{t+1}$ be points of X. Suppose that μ blocks contain all these points. Use PIE to show that the number of blocks containing none of $x_1, \ldots, x_{t+1}$ is $N + (-1)^{t+1}\mu$, where N depends on t, v, k and λ only.

   Deduce that, if t is even and $v = 2k + 1$, then

$$(X \cup \{y\}, \{B \cup \{y\}, X \setminus B : B \in \mathcal{B}\})$$

is an extension of $\mathcal{D}$, where y is a point not in X.

11. Find all possible parameters of non-trivial designs with 9 points.

12. Show that the family of blocks of a square 2-(v, k, λ) design has at least $k(k - \lambda)^{(v-1)/2}$ SDRs.

17. Error-correcting codes

... flame of incandescent terror
Of which the tongues declare
The one discharge from sin and error

T. S. Eliot, 'Little Gidding' (1942)

TOPICS: Error-correction, minimum distance, linearity, bounds

TECHNIQUES: Linear algebra, projective geometry, number theory

ALGORITHMS: Encoding, syndrome decoding

CROSS-REFERENCES: Packing and covering (Chapter 8); projective geometry (Chapter 9); designs (Chapter 16)

This chapter begins with an example involving 'guessing' a number on the basis of information about it, some of which is incorrect. It looks like a party trick, but in fact the ideas have great practical importance. Information of all kinds is sent through channels where it runs the risk of distortion: pictures of the planets in the solar system from space probes via radio links; musical performances via tapes and compact discs; instructions about how to build a living body via DNA molecules in genes; and so on. We can fancifully regard errors and distortion in the message as 'nature lying to us', and it is important to know how to identify and correct the errors. This is done by means of *error-correcting codes*, whose study could be seen as part of information theory but which has a high combinatorial content.

17.1. Finding out a liar

The panel game 'Twenty Questions', which we referred to in Chapter 4, involves one player trying to guess something thought of by the other, being allowed to ask twenty questions (each of which must have a yes-or-no answer) to gather information. It's clear that 2^{20} different objects can be distinguished. Since this number is slightly greater than 10^6, the game can be played with whole numbers, with the familiar opening gambit, 'Think of a number less than a million'.

What if the respondent lies?

There is a simple scheme for guessing correctly a number between 0 and 15 with seven questions, where the respondent is allowed to lie once. The calculations can be done on the back of a small envelope, or in your head with a little practice. Since I can't demonstrate it to you in this medium, I'll explain how it works, and you can try it out on someone else.

Give your respondent the following instructions.

Instructions

Think of a number between 0 and 15.
 Now answer the following questions.
 You are allowed to lie **once**.

 1. Is the number 8 or greater?
 2. Is it in the set $\{4,5,6,7,12,13,14,15\}$?
 3. Is it in the set $\{2,3,6,7,10,11,14,15\}$?
 4. Is it odd?
 5. Is it in the set $\{1,2,4,7,9,10,12,15\}$?
 6. Is it in the set $\{1,2,5,6,8,11,12,15\}$?
 7. Is it in the set $\{1,3,4,6,8,10,13,15\}$?

The response is a sequence of seven 'yes' or 'no' answers. Writing 1 for 'yes' and 0 for 'no', record it as a binary vector v of length 7. Now multiply v by the 7×3 matrix

$$H = \begin{pmatrix} 0 & 0 & 1 \\ 0 & 1 & 0 \\ 0 & 1 & 1 \\ 1 & 0 & 0 \\ 1 & 0 & 1 \\ 1 & 1 & 0 \\ 1 & 1 & 1 \end{pmatrix}$$

whose i^{th} row is the base 2 representation of i, for $1 \le i \le 7$. (The calculation of vH is done in $\mathrm{GF}(2)$.) The result is a binary vector $w = vH$ of length 3. Alternatively, count the numbers of 1's of v which occur in the sets $\{4,5,6,7\}$, $\{2,3,6,7\}$, $\{1,3,5,7\}$ of coordinates respectively, recording 1 for odd and 0 for even in each case. Now either $w = 0$, in which case no lie was told; or w is the base 2 representation of k, where $1 \le k \le 7$, in which case the answer to the kth question was a lie. Thus, the vector v of responses can be corrected. Then the first four entries of v form the base 2 representation of the chosen number.

We note in passing that no fewer than seven questions would suffice, no matter how they were asked. For at the end, we know not only which of the 16 numbers was chosen, but also which question was answered incorrectly (if any). If, say, six questions sufficed, then we'd have identified one of $16 \cdot 7 = 112$ possibilities with only 6 questions, a contradiction since $112 > 2^6$. (The factor 7 is for the 6 possible positions of the lie and the possibility that no lie was told.) Note that, with 7 questions, we have distinguished $16 \cdot 8 = 2^7$ events; so we succeed with nothing to spare, (no wasted information is generated).

Why does it work?
Let $C = \{c_0, c_1, \ldots, c_{15}\}$ be the set of sixteen 7-tuples of zeros and ones which would be generated by truthful responses to the questions for each of the numbers

$0, \ldots, 15$. For example, the number 13 generates

$$c_{13} = (1,1,0,1,0,0,1).$$

The crucial fact, which can be verified by checking all sixteen cases — there are simpler ways — is that, for any $c \in C$, we have $cH = 0$, where H is the matrix defined above. (Note that the matrix H has rank 3, so its null space has dimension 4; thus C must be precisely the null space of H.) Let e_i be the 7-tuple with 1 in the ith place and 0 in all other positions. Then, if the respondent chooses the number m and lies to the kth question, the replies will form the vector $c_m + e_k$. (Adding e_k changes the kth coordinate and leaves the others unaltered.) Now we compute

$$(c_m + e_k)H = c_m H + e_k H = e_k H,$$

which is the kth row of H and so, by definition, the base 2 representation of the number k. So we have located the error. Once we correct it, we know the vector c_m. On the other hand, if no lie was told, the response is just c_m, and we find $c_m H = 0$. Now, the first four questions we asked about m generate its base 2 representation; so we can read this off from the first four digits of c_m, and then calculate m.

Another important fact is that the correct response to the questions can itself be generated by linear algebra. Let v_m be the base 2 representation of the integer m. Then you can check that $c_m = v_m G$, where

$$G = \begin{pmatrix} 1 & 0 & 0 & 0 & 0 & 1 & 1 \\ 0 & 1 & 0 & 0 & 1 & 0 & 1 \\ 0 & 0 & 1 & 0 & 1 & 1 & 0 \\ 0 & 0 & 0 & 1 & 1 & 1 & 1 \end{pmatrix}.$$

(This follows from the form of the questions. The first four questions ask whether the first, ..., fourth digit in v_m is equal to 1. The fifth question asks whether positions 2, 3, 4 contain an odd number of 1's, that is, whether their sum (mod 2) is 1. Similarly for the sixth and seventh.)

The set C is an example of an *error-correcting code*. We observe that:
• Any two members of C differ in at least three positions.
For suppose, for example, that c_m and c_n differ only in positions i and j. Then $c_m + e_i = c_n + e_j$, and we couldn't distinguish between the possibilities 'm chosen, lie to i^{th} question' and 'n chosen, lie to j^{th} question'. Similar reasoning would apply if two members of C differed in only one position.

Furthermore, since C is the null space of H (or the row space of G), we have:
• The sum of two members of C is again in C.
We say that C is *linear*.

17.2. Definitions

In coding theory, it is customary to assume that information is to be sent in 'words' of fixed length n, each word being an n-tuple of 'letters' taken from an alphabet Q of size q. By far the commonest case in applications is that when $q = 2$, and the alphabet is taken to be $\mathrm{GF}(2) = \{0,1\}$; but this is not essential. Throughout this chapter, n and q have these meanings.

274 Error-correcting codes

We define *Hamming space* $H(n, q)$ to be the set of all words of length n over the fixed alphabet Q of size q. The structure on H will be a metric or distance function. The motivation is that, if a word w is transmitted through a noisy channel, some of the letters in the word may become changed; the more letters that are changed, the further the received word is from the transmitted word. So we define the *Hamming distance* between words v, w to be the smallest number of errors which could change v into w; that is, the number of positions in which the entries in v and w differ. Formally,

$$d(v, w) = |\{i : v_i \neq w_i\}|.$$

(17.2.1) Proposition. *(a)* $d(v, w) \geq 0$, *with equality if and only if* $v = w$.
(b) $d(v, w) = d(w, v)$.
(c) $d(u, v) + d(v, w) \geq d(u, w)$.

PROOF. The first two assertions are obvious from the definition. For the third, we can argue informally as follows: it is possible to change u into w by changing it first into v, making altogether $d(u, v) + d(v, w)$ coordinate alterations, so the smallest number of changes required does not exceed this number; or else, a more formal argument like this can be used. Observe that

$$\{i : u_i \neq w_i\} \subseteq \{i : u_i \neq v_i\} \cup \{i : v_i \neq w_i\},$$

since if $u_i \neq w_i$ then certainly either $u_i \neq v_i$ or $v_i \neq w_i$. Now take the cardinality of both sides, using the fact that

$$|A \cup B| = |A| + |B| - |A \cap B| \leq |A| + |B|$$

to get the desired inequality.

REMARK. A function d from $X \times X$ to the non-negative integers is called a *metric* on X if it satisfies conditions (a)–(c) of (17.2.1). The notion of a metric is an important unifying principle in mathematics; it is very likely that you have met it in analysis or topology, and we saw an application in the 'twice-round-the-tree' algorithm for the Travelling Salesman Problem in Section 11.7. The metric defined here on Hamming space is called, naturally enough, the *Hamming metric*; we also refer to the *Hamming distance* between two words.

A *code* of length n over the alphabet Q is just a subset C of Hamming space $H(n, q)$ which contains at least two words. The elements of the code are called *codewords*. The rationale is that we will perform error correction by restricting our transmissions to be members of the code C, rather than arbitrary words; if the members of C are sufficiently distinguishable (i.e., sufficiently far apart) then, assuming that not too many errors occur, the received word still resembles the transmitted word more closely than any other codeword, and so we can recover the transmitted word. The reason for assuming that there are at least two words is that, in a one-word code, we would know in advance which word was transmitted, and so no information could possibly be sent! Now suppose that there are m possible messages that we might want to send, say $M_1, \ldots, M_m$. We encode a message by associating a unique codeword with it; thus, we take a one-to-one function e from the set of messages to C, and encode M_i as the codeword $e(M_i)$, which is then transmitted. Of course, this requires that the number of codewords is at least as great as the number of messages; indeed, we may assume that every codeword corresponds to a message (by ignoring those which don't). How do we decode?

What is suggested in the preceding paragraph is the concept of *nearest-neighbour decoding*. If the word w is received, then we find the codeword $c \in C$ for which $d(w, c)$ is as small as possible, and assume that the transmitted word was c, and the message was $e^{-1}(c)$. In general, this requires a search through all the codewords to find the nearest one to w, a very time-consuming procedure if the code is large! One of the themes of algebraic coding theory is that, for codes with more algebraic structure, the decoding procedure can be simplified a great deal.

What if there is no unique nearest codeword? We should design the code so that this event is very unlikely, if it can occur at all. Then either choose randomly among the nearest neighbours of w, accepting the small chance of making a mistake; or ask for the message to be retransmitted. Which strategy we use depends on the situation. One important use of error-correction is in obtaining data and pictures from interplanetary space probes; here, the length of time taken by a signal means that re-transmission is out of the question. But, for commercial transactions between banks, the importance of correct information outweighs the cost of a small delay.

For a positive integer e, we say that the code C is *e-error-correcting* if, given any word w, there is *at most one* codeword c such that $d(w, c) \leq e$. This means that, if a codeword is transmitted and at most e errors occur, then nearest-neighbour decoding will recover the transmitted word uniquely. A related parameter is the *minimum distance* d of the code: this is the smallest distance between two distinct codewords.

(17.2.2) Proposition. *A code with minimum distance d is e-error-correcting if and only if $d \geq 2e + 1$.*

PROOF. Suppose that $d \geq 2e + 1$. If a word w lies at distance e or less from two different codewords c_1 and c_2, then

$$d(c_1, c_2) \leq d(c_1, w) + d(w, c_2) \leq e + e = 2e,$$

a contradiction; so C is e-error-correcting.

Conversely, suppose that $d \leq 2e$, and let c_1, c_2 be codewords at distance d. Set $f = \lfloor d/2 \rfloor$; then $f \leq e$ and $d - f \leq e$. We can move from c_1 to c_2 by changing d coordinates one at a time. If w is the word obtained after f changes, then we have $d(c_1, w) = f \leq e$ and $d(c_2, w) = d - f \leq e$; so C is not e-error-correcting. $\blacksquare$

For example, the code which was used for the trick in the last section has minimum distance 3 and is 1-error-correcting.

We see that, for good error-correction properties, we require large minimum distance. Also, we want the code to have as many words as possible, since the number of words limits the number of different messages that can be sent, and hence the rate of transmission of information. These two requirements conflict. In practice, there is a third requirement as well: the processes of encoding and decoding should not be too demanding in terms of computational power. (In the case of space probes, the encoding has to be done in real time by a small, low-powered electronic circuit; the incoming message can be stored and decoded by large computers, but we still want the results within hours rather than centuries!)

The tension between the first two requirements can be formulated as a *packing problem* (compare Section 8.4). The *ball* of radius r and centre w in the Hamming space $H(n,q)$ is the set

$$B_r(w) = \{v \in H(n,q) : d(v,w) \le r\}.$$

(17.2.3) Propositon. *The code $C \subset H(n,q)$ is e-error-correcting if and only if the balls of radius e with centres at the codewords are pairwise disjoint.*

PROOF. The conclusion is just another way of saying that no word lies at distance e or less from two codewords.

So we want to know the maximum number of balls of radius e which can be packed into Hamming space.

17.3. Probabilistic considerations

As we have already suggested, combinatorial coding theory starts from the assumption that (with a sufficiently high degree of certainty) at most a fixed number e of errors are made during transmission. This is at base a probabilistic statement. In this section, we take a superficial look at the probability theory involved, and state Shannon's Theorem.

To simplify matters, we consider only the *binary alphabet* $GF(2) = \{0,1\}$. Words of fixed length n are transmitted through a channel. We make the following assumptions about the channel:

- the probability that a 0 is changed to a 1 is equal to the probability that a 1 is changed to a 0;
- this probability p is the same for each digit, and is less than $\frac{1}{2}$;
- the events that alterations occur to different digits are independent.

A channel satisfying these assumptions is called a *binary symmetric channel*. The assumptions simplify the analysis, but are not very realistic. For example, interference often comes in 'bursts',[1] so that if one digit is incorrect then its successor is more likely to be incorrect also; and the error probability may not be constant (e.g., because of synchronisation problems, errors may be more likely at the start of a word). The assumption $p < \frac{1}{2}$ is harmless, and clearly necessary. If $p > \frac{1}{2}$, then we just reverse each digit received and the error probability becomes $1-p$; if $p = \frac{1}{2}$, then the received message is completely random, and no information can be extracted from it.

We also make the assumption that all words of length n have an equal chance of being transmitted. Again, this is often false. Much information is sent by encoding letters, numerals, and punctuation symbols as 7- or 8-bit binary words, using codes

---

[1] A scratch on a compact disc could destroy a run of consecutive bits, for example.

such as ASCII. In plain text, the words representing a space or the letter 'e' are disproportionately likely to occur. Nevertheless, there are encoding schemes which achieve equiprobability and have the beneficial side-effect of data compression.

The *maximum likelihood* decoding method works as follows. Given a received word w, we decode it to that codeword c such that $\text{Prob}(c \text{ transmitted} : w \text{ received})$ is maximised. In other words, we assume that the codeword sent is the one most likely to result in the given received word.

(17.3.1) Proposition. *Assume that all codewords are equally likely to be transmitted, and that the channel is binary symmetric. Then maximum likelihood decoding coincides with nearest-neighbour decoding.*

PROOF. In a binary symmetric channel, if $d(c, w) = d$, then d errors (in specified positions) change c to w; so $\text{Prob}(w \text{ received} : c \text{ transmitted}) = p^d(1 - p)^{n-d}$. Moreover, $\text{Prob}(c \text{ transmitted}) = 1/|C|$ by assumption. So

$$\text{Prob}(c \text{ transmitted} : w \text{ received}) = p^d(1 - p)^{n-d}(1/|C|)/\text{Prob}(w \text{ received}),$$

which is a decreasing function of d. So it is maximised when c is the codeword nearest to w.

The *rate* of a code C of length n over an alphabet of size q is defined to be $\log_q(|C|)/n$. (The motivation for this definition is that if, say, $|C| = q^k$, then k-tuples of information can be encoded in a one-to-one fashion by codewords and transmitted as n-tuples; information is sent k/n times as fast as it would be without encoding, this being the price paid for error correction.) Shannon proved the following remarkable theorem.

(17.3.2) Shannon's Theorem

Given a binary symmetric channel with error probability p for a single digit,

(a) *if $R < 1 + p\log_2 p + (1 - p)\log_2(1 - p)$ and $\epsilon > 0$, there is a code with rate at least R such that the probability of error in decoding a codeword by nearest-neighbour decoding is less than ϵ;*

(b) *this is best possible, that is, if $R > 1 + p\log_2 p + (1-p)\log_2(1-p)$, then the error probability of any code with rate R is bounded away from 0.*

What is even more remarkable is that the code in Shannon's Theorem is constructed by picking the appropriate number of codewords at random! The number $1 + p\log_2 p + (1 - p)\log_2(1 - p)$ is called the *capacity* of the channel; it represents the maximum rate for 'error-free' transmission. Shannon's Theorem extends to a wider range of situations (arbitrary alphabet size and other channel characteristics).

It is important, however, to realise three important limitations in Shannon's Theorem which mean that it is not the end of coding theory!

- It is *non-constructive*; the code is constructed at random. This doesn't help an engineer who wants an explicit example.
- The length n tends to infinity as $\epsilon \to 0$ or the rate tends to the channel capacity. Using nearest-neighbour decoding with an unstructured code, it is necessary to remember the entire received word before decoding can begin; so, in practice, n is bounded by the memory size of the decoder.
- even for moderate lengths, nearest-neighbour decoding involves a search through (exponentially many) codewords, a very time-consuming process.

17.4. Some bounds

As we saw in Section 17.2, there are three desiderata for a good code:
- *high rate* (large number of codewords);
- *good error-correction*, which for us means large minimum distance;
- *ease of implementation*.

The third of these requires concepts from the theory of computational complexity for its proper discussion. Section 20.1 sketches the ideas, but I won't give a full treatment here. Already we see that the first two requirements conflict; and much of the mathematical interest in the subject comes from this tension. It can be expressed in the form of a question:

> What is the size of the largest code of length n and minimum distance d over an alphabet of size q?

This is the *main problem* of coding theory. Needless to say, the exact answer is known only in special cases. In this section I will prove a simple lower bound and several upper bounds.

(17.4.1) Varshamov–Gilbert bound. *Given n, q, d, there is a q-ary code of length n and minimum distance d or larger, having at least*

$$ q^n \bigg/ \left(\sum_{i=0}^{d-1} \binom{n}{i} (q-1)^i \right) $$

codewords.

PROOF. Recall from Section 17.2 the definition of a *ball* $B_r(c)$ of radius r in Hamming space $H(n, q)$: it consists of all words w satisfying $d(c, w) \leq r$ for a fixed word c (the *centre* of the ball). Now

$$ |B_r(c)| = \sum_{i=0}^{r} \binom{n}{i} (q-1)^i. $$

For a word at distance i from c is obtained by choosing a set of i coordinate positions in which to make errors (in $\binom{n}{i}$ ways), and changing the symbols in these positions (each can be changed to any of the other $q-1$ symbols in the alphabet).

So there are $\binom{n}{i}(q-1)^i$ words at distance i from c, and the result is obtained by summing over i.

Now suppose that C is a code with minimum distance d or larger, and suppose that the union of all the balls $B_{d-1}(c)$, for $c \in C$, is not the whole of the Hamming space. Then we can find a word whose distance from each codeword is at least d. Adding it to C, we obtain a larger code, still with minimum distance d or larger.

But the number of words lying in these balls does not exceed $|C| \cdot |B_{d-1}(c)|$. So, if this product is less than q^n, then C may be enlarged. So we can continue this enlargement at least until $|C| \cdot |B_{d-1}(c)| \geq q^n$, the required result.

Note that
- this is a lower bound, that is, it guarantees a code of the appropriate size;
- the proof is not constructive;
- it is unlikely to be close to best possible (we hope that clever methods will produce much larger codes).

Now we turn to upper bounds. We prove three bounds below. In each case, we ask the question: what does it mean if the bound is attained? The first bound bears a striking resemblance, both in statement and proof, to (17.4.1).

(17.4.2) Hamming bound, or sphere-packing bound. *Suppose that $d \geq 2e+1$. A q-ary code with length n and minimum distance at least d has at most*

$$q^n \left/ \left(\sum_{i=0}^{e} \binom{n}{i}(q-1)^i \right) \right.$$

codewords.

PROOF. By (17.2.3), if C has minimum distance at least $2e+1$, then the balls of radius e with centres at the words of C are pairwise disjoint, and so contain $|C| \cdot |B_e(c)|$ words. This number cannot exceed the total number q^n of words. The result follows. (It should be called the *ball-packing bound*, but the word 'sphere' is often used instead of 'ball' in coding theory.)[2]

Note that the proof of (17.4.1) is a covering argument while that of (17.4.2) is a packing argument.

(17.4.2a) Equality in the Hamming bound. *A code C attains the Hamming bound if and only if every word in $H(n,q)$ lies at distance e or smaller from exactly one word in C.*

This follows immediately from the proof. A code satisfying this condition is called a *perfect e-error-correcting code*.

(17.4.3) Singleton bound. *A q-ary code of length n and minimum distance d has at most q^{n-d+1} codewords.*

---

[2] In mathematical usage, a sphere is the surface of a ball.

PROOF. Consider the first $n - d + 1$ coordinate positions. Two words of C cannot agree in all these positions, since they could then differ in at most the remaining $d-1$ positions, and their distance would be at most $d - 1$. So the number of codewords doesn't exceed the number of $(n - d + 1)$-tuples.

(17.4.3a) **Equality in the Singleton bound.** *A code C attains the Singleton bound if and only if, given any $n - d + 1$ coordinate positions and any $n - d + 1$ symbols from the alphabet, there is a unique codeword having those symbols in those positions.*

This is almost immediate from the proof. (To see that such a code does indeed have minimum distance (at least) d, note that, if two codewords have distance $d - 1$ or less, then they must agree on $n - d + 1$ positions, contrary to the uniqueness requirement.) A code satisfying this condition is called *maximum distance separable*, or MDS.

(17.4.4) **Plotkin bound.** *Let $\theta = 1 - \frac{1}{q}$, and suppose that $d > \theta n$. Then a q-ary code with length n and minimum distance d has at most $d/(d - \theta n)$ codewords.*

PROOF. The argument is more elaborate than those for the earlier bounds. Let C be our code, with M codewords, which we imagine as written out in an $M \times n$ array whose rows are the codewords. We bound in two ways the number N of occurrences of an ordered pair of different symbols in the same column.

First, note that any two rows have Hamming distance at least d, there are at least d columns in which the entries in these rows are different. So

$$N \geq M(M - 1)d. \tag{1}$$

On the other hand, let x_{ij} be the number of occurrences of the ith symbol in the jth column. For each such occurrence, there are $M - x_{ij}$ rows where a different symbol occurs. So the contribution from this column is $\sum_{i=1}^{q} x_{ij}(M - x_{ij})$. But we have $\sum_{i=1}^{q} x_{ij} = M$. This implies that $\sum_{i=1}^{q} x_{ij}^2 \geq M^2/q$, with equality if and only if each x_{ij} is equal to M/q. So we have

$$\sum_{i=1}^{q} x_{ij}(M - x_{ij}) = M^2 - \sum_{i=1}^{q} x_{ij}^2$$
$$\leq M^2 - M^2/q = \theta M^2.$$

Summing the contributions of all n columns, we obtain

$$N \leq n\theta M^2. \tag{2}$$

Combining (1) and (2), we see that $M(M - 1)d \leq n\theta M^2$, giving $M(d - \theta n) \leq d$. If $d \leq \theta n$, this gives no information; but, if $d > \theta n$, we obtain Plotkin's bound.

(17.4.4a) **Equality in the Plotkin bound.** *A code C attains the Plotkin bound if and only if*
(a) any two distinct codewords have distance d; and
(b) each symbol occurs in a given position in the same number M/q of codewords.

PROOF. The result was obtained by combining two inequalities; so, to meet the bound, we must meet it in each of these inequalities. From the proof of (1), we see immediately that equality in (1) is equivalent to condition (a) (which defines an *equidistant code*). We noted in the proof of (2) that equality holds if and only if $x_{ij} = M/q$ for all i, j. Note that this condition resembles that for equality in the Singleton bound. We now extract the common features.

An *orthogonal array of strength t and index λ* (with length n, over an alphabet of size q) is a set C of n-tuples of elements from the alphabet of size q, having the property that, given any t distinct coordinate positions (say $i_1, \ldots, i_t$), and any t elements $a_1, \ldots, a_t$ of the alphabet (not necessarily distinct), there are precisely λ members c of C with the property that they have these entries in these positions; that is, $c_{i_j} = a_j$ for $j = 1, \ldots, t$. Notice that this definition has a similar 'flavour' to the definition of a t-design in the last chapter; there is a body of theory which can be developed for both orthogonal arrays and designs (including the divisibility conditions, derived designs, Fisher's Inequality, etc.)

In this language, we have:
- a code attains the Singleton bound if and only if it is an orthogonal array of strength $n - d + 1$ and index 1;
- a code attains the Plotkin bound if and only if
  (a) it is equidistant with distance d;
  (b) it is an orthogonal array of strength 1.

17.5. Linear codes; Hamming codes

In this section, we see the benefits of giving codes more algebraic structure. We get simpler encoding and decoding algorithms, as well as some easy constructions of good codes. We take our alphabet to be the finite field $\mathrm{GF}(q)$. Then the Hamming space $H(n, q)$ of all words of length n is an n-dimensional vector space over $\mathrm{GF}(q)$. We define a *linear code* to be a vector subspace of $H(n, q)$.

The *weight* $\mathrm{wt}(w)$ of a word w is the number of non-zero entries in w (this is just its distance from the all-zero word). The *minimum weight* of a code C is the smallest weight of a non-zero word in C

(17.5.1) Proposition. *(a) For any $v, w \in H(n, q)$, we have $d(v, w) = \mathrm{wt}(v - w)$.*
(b) The minimum distance and minimum weight of a linear code C are equal.

PROOF. (a) is clear from the definition, since $v_i - w_i \neq 0 \Leftrightarrow v_i \neq w_i$. For (b), observe that any weight in C is a distance (since $\mathrm{wt}(w) = d(w, 0)$), and any distance is a weight (by (a); the linearity of C implies that $v - w \in C$ for all $v, w \in C$). Note that, in general, finding the minimum distance of a code involves comparing all $\binom{N}{2}$ pairs of codewords, but finding the minimum weight involves looking only at the N codewords.

(17.5.2) Linear Varshamov–Gilbert bound. *If q is a prime power, there is a linear code attaining the bound of (17.4.1).*

PROOF. Recall the proof of (17.4.1): as long as $|C| < q^n/|B_{d-1}(c)|$, we can find a word w whose distance from any word in C is at least d, and adjoin it to C to form a larger code. In fact, if C is linear, then the subspace spanned by C and w still has minimum weight at least d. For a typical word in this space has the form $c + \alpha w$, where $c \in C$ and $\alpha \in \mathrm{GF}(q)$. If $\alpha = 0$, $c \neq 0$, the weight is at least d, by our assumption about C. If $\alpha \neq 0$, then

$$\mathrm{wt}(c + \alpha w) = \mathrm{wt}(-\alpha^{-1}c - w) = d(-\alpha^{-1}c, w) \geq d,$$

by the assumption on w (and using the linearity of C).

Since the cardinality of a linear code is a power of q, this allows the Varshamov–Gilbert bound to be improved. To take an example, consider the case $q = 2$, $d = 3$, $n = 15$. By (17.4.1), there is a code with these parameters of cardinality at least $2^{15}/(1 + 15 + \binom{15}{2}) = 270.8\ldots$; that is, at least 271. But (17.5.2) gives a linear code with at least this many words, and it must have cardinality at least 512. (In fact, we'll see soon that there is a code with cardinality 2048 but no larger.)

How do we specify a linear code? Since it is a subspace, we can describe it by giving a basis, a set of k linearly independent words. It is convenient to take these words as the rows of a $k \times n$ matrix G, called a *generator matrix* for the code. In other words, G is a generator matrix of C if and only if its rows are linearly independent and its row space is C.

Closely related to the row space of a matrix A is its null space, the set of words w such that $wA^\top = 0$. A code C can also be specified by giving a matrix (with linearly independent rows) whose null space is C. Such a matrix H is called a *check matrix* for C.[3]

Since the rank of a matrix (the dimension of its row space) and its nullity (the dimension of its null space) sum to n, the number of columns, we see that, if a linear code has length n and dimension k, then a generator matrix is $k \times n$ and a check matrix is $(n - k) \times n$.

(17.5.3) Proposition. *Let G and H be matrices with linearly independent rows, having size $k \times n$ and $(n - k) \times n$ respectively. Then G and H are the generator and check matrices of a code if and only if $GH^\top = 0$.*

PROOF. Suppose that $GH^\top = 0$. Then every row of G lies in the null space of H, so the row space of G is contained in the null space of H. But both spaces have dimension k, so they are equal. The converse is shown by reversing the argument.

There is a *dot product* defined on $H(n, q)$, by the rule

$$v \cdot w = \sum_{i=1}^{n} v_i w_i.$$

---

[3] The name comes from the case $q = 2$, $H = (1, 1, \ldots, 1)$. For any word w, we find that $wH^\top$ is equal to the number of 1's in w (mod 2); that is, it is zero if w has even parity and 1 if w has odd parity. This is called a *parity check*. The code consists of all words of even weight; by evaluating the parity check, we can detect (but not correct) a single error. This is often used in serial communication between computers, where error probabilities are very low and the cost of a re-transmission is small.

(Note that, unlike the case for the Euclidean inner product, it can very well happen that $v.v = 0$ for some non-zero vector v. For example, if $q = n = 2$, consider the vector $(1, 1)$.)

Now the *dual code* $C^\perp$ of a linear code C is defined to be

$$C^\perp = \{w \in H(n, q) : v.w = 0 \text{ for all } v \in C\}.$$

It is a linear code, satisfying $\dim(C) + \dim(C^\perp) = n$. Now a vector lies in the null space of a matrix if and only if its dot product with every row of the matrix is zero. So the null space of a matrix is just the dual of its row space. In other words:

(17.5.4) Proposition. *For any linear code C, the check matrix of $C^\perp$ is equal to the generator matrix of C, and vice versa.*

The generator and check matrices are not just of theoretical interest, but are crucial for the encoding and decoding of linear codes. Let C be a linear code of dimension k, with generator and check matrices G and H respectively.

ENCODING. Since $|C| = q^k$, each k-tuple of digits can be encoded as a word of C in a one-to-one way. The simplest way to do this is as follows. Let v be an arbitrary k-tuple. Then vG is an n-tuple, and is a member of C, since it is a linear combination of rows of G (with the elements of v as coefficients). The linear independence of the rows shows that the map $v \mapsto vG$ is a bijection from $\mathrm{GF}(q)^k$ to C. Moreover, in the case $q = 2$, this matrix multiplication can be performed very efficiently by small, low-power circuits (one of our requirements for efficient encoding, especially for space probes).

DECODING. This is a little more difficult. Suppose that C is e-error-correcting (that is, its minimum distance is at least $2e + 1$). Let the codeword c be transmitted, and the word $w = c + u$ be received, where u is the 'error' (and we assume that $\mathrm{wt}(u) \leq e$). The idea of the decoding procedure is that, rather than remove the error u to reveal the transmitted word c, we remove c to reveal u; this is more reasonable, since we know more about c, and it is equivalent since knowing u, we can find c by subtraction.

We calculate the vector $wH^\top \in \mathrm{GF}(q)^{n-k}$ — this is called the *syndrome* of w. Since $cH^\top = 0$, the syndrome is equal to $uH^\top$, that is, it depends only on the error pattern. Moreover, distinct error patterns have distinct syndromes. For, if $\mathrm{wt}(u_1), \mathrm{wt}(u_2) \leq e$, then $\mathrm{wt}(u_1 - u_2) \leq 2e$, by the triangle inequality. But, if $u_1 H^\top = u_2 H^\top$, then $u_1 - u_2$ is in the null space of H, which is C, so $\mathrm{wt}(u_1 - u_2) \geq 2e + 1$, a contradiction.

Hence, in principle, the error pattern u can be recovered from its syndrome $uH^\top$. In practice, this is the difficult part; linear algebra doesn't help, and we might use a look-up table.[4] Once u is found then, as already described, we obtain c by subtracting u from w.

This method is known as *syndrome decoding*.

---

[4] This would be a table of errors and corresponding syndromes, but ordered by syndrome, so that we can quickly find the error producing any syndrome.

Various information about a code can be read off from these matrices. The most important example of this is as follows.

(17.5.5) Proposition. *A linear code has minimum weight d or greater if and only if any $d - 1$ columns of a check matrix for the code are linearly independent.*

PROOF. The vector $cH^\top$ is a linear combination of the columns of H, with coefficients the elements of c. (Strictly, it is the transpose of this, since $cH^\top$ is a row vector.) So any word of weight m in C gives rise to a linear dependence between m columns of H (with all coefficients non-zero), and conversely. So the minimum weight of C is the smallest number of columns which are linearly dependent.

We examine further a special case. A linear code is 1-error-correcting if and only if it has minimum weight at least 3; by (17.5.5), this occurs if and only if any two columns of H are linearly independent. In other words, we require that no column of H is zero, and no column is a multiple of another. For fixed column length m, let us find the largest such matrix possible.

Define an equivalence relation on the set of all non-zero column vectors of length d, where two columns are equivalent if and only if one is a scalar multiple of the other. The columns of our matrix H must belong to different equivalence classes. How many classes are there? There are $q^d - 1$ non-zero vectors; each one has $q - 1$ non-zero multiples, so each equivalence class has size $q - 1$, and the number of classes is $(q^d - 1)/(q - 1)$. [You should recognise this from Chapter 9 as the number of points in the projective space $\mathrm{PG}(d - 1, q)$; why are these numbers equal?]

So let H be a $d \times (q^d - 1)/(q - 1)$ matrix whose columns are representatives of the equivalence classes of non-zero vectors, and let C be the code with check matrix H. Then C is the *Hamming code* of length $n = (q^d - 1)/(q - 1)$. [Note that q and n determine d; the matrix is not unique, but the only ambiguity is in the choice of representatives and their order; so all codes obtained are *equivalent*, in a sense to be defined.]

(17.5.6) Theorem. *Hamming codes are perfect 1-error-correcting.*

PROOF. This means that they attain the Hamming bound! Certainly it follows from (17.5.5) and the subsequent discussion that a Hamming code is 1-error-correcting. Now its length is $n = (q^d - 1)/(q - 1)$, and its dimension is $n - d$, so the number of codewords is

$$q^{n-d} = q^n/q^d = q^n/(1 + n(q - 1)),$$

and the right-hand side is the Hamming bound for $e = 1$.

Syndrome decoding works especially well for Hamming codes. The syndrome is a d-tuple. If it is zero, then no error has occurred. If it is non-zero, then there is a unique column j and scalar α such that the syndrome is αv_j, where v_j is the jth column of H. Then the error occurred in position j, where α was added; and so subtracting α from the jth coordinate of the received word corrects the error.

If this sounds familiar, re-read the discussion in Section 17.1. The code used there was the binary Hamming code of length $7 = (2^3 - 1)/(2 - 1)$.[5] In the binary case, the only non-zero scalar is 1, so the equivalence relation is equality, and the columns of the check matrix are all the non-zero binary triples, that is, the base 2 representations of the numbers $1, \ldots, 7$. If we arrange them in the obvious order, then the syndrome is the base 2 representation of the number of the position where the error occurred! You should now be able to generalise the method, and explain (for example) how to determine any chosen number between 0 and 2047 in 15 guesses, if one lie is allowed.

In general, how do we compute the check matrix from the generator matrix, or vice versa? The key is the following observation. Let A be a $k \times (n - k)$ matrix. Set $G = (I_k \; A)$ and $H = (-A^\top \; I_{n-k})$. Clearly G and H have ranks k and $n - k$ respectively; and $GH^\top = 0$, so if G is the generator matrix of a code then H is the check matrix, and *vice versa*. Now suppose that we are given an arbitrary generator matrix G. By applying elementary row operations to it, we can put it into reduced echelon form (see Chapter 9). Moreover, elementary row operations don't change the row space (i.e., the code with the given matrix as generator). If we are lucky, the reduced echelon form will have the shape $(I \; A)$ for some A — this means that the leading 1's occur in the first k columns. If so, we can read off the check matrix $(-A^\top \; I)$ directly. In general, we have to apply some permutation of the columns to bring the leading ones into the first k columns, write down the check matrix H, and then apply the inverse permutation to H.

17.6. Perfect codes

Perfect codes had great importance early in the history of coding theory, when two of the pioneers, Hamming and Golay, found some interesting examples. As time went on and very few further examples were found, engineers lost interest and turned to larger and more flexible classes of codes. But perfect codes are unexpectedly important to mathematicians.

We begin with perfect 1-error-correcting codes, and consider first codes over prime-power-size alphabets.

(17.6.1) Proposition. *Let q be a prime power.*
(a) A perfect 1-error-correcting code over an alphabet of size q has length $(q^d - 1)/(q - 1)$ for some integer $d > 1$.
(b) A linear perfect 1-error-correcting code is a Hamming code.

PROOF. (a) Such a code C satisfies $|C| = q^n/(1 + n(q - 1))$; so $1 + n(q - 1)$ divides q^n. Suppose that $q = p^a$ where p is prime, and write $1 + n(q - 1) = q^d p^b$, for some integers d and b with $0 \le b < a$. Then $p^b \equiv q^d p^b \equiv 1 \pmod{q - 1}$. Since $1 \le p^b < q$, we must have $p^b = 1$, whence $1 + n(q-1) = q^d$, and $n = (q^d - 1)/(q-1)$, as required.
    (b) Let $n = (q^d - 1)/(q - 1)$. Now $|C| = q^n/(1 + n(q - 1)) = q^{n-d}$, so a check matrix for C is $d \times n$. But C is 1-error-correcting, so the columns of H are pairwise

---

[5] The matrix H used there is actually the transpose of what we defined as the check matrix.

inequivalent, with respect to the equivalence relation defined before (17.5.6). Since n is equal to the number of equivalence classes, we have one column from each class, and H determines a Hamming code, according to our definition.

Some examples of non-linear codes with the same parameters as Hamming codes are known.

For non-prime-power alphabets, almost nothing is known. The next result is the sum total of our knowledge.

(17.6.2) Proposition. *There is no perfect 1-error-correcting code of length 7 over an alphabet of length 6.*

PROOF. Suppose that C is such a code, over the alphabet $\{1,\ldots,6\}$. We have $|C| = 6^7/(1+7.5) = 6^5 = 6^{7-3+1}$. Thus C, as well as being perfect, is also MDS (it meets the Singleton bound). By (17.4.3a), we see:

$$(*) \begin{cases} \text{Given } a_1,\ldots,a_5 \in \{1,\ldots,6\}, \text{ there are unique} \\ \text{elements } a_6, a_7 \text{ such that } (a_1,\ldots,a_7) \in C. \end{cases}$$

Now fix a_1, a_2, a_3, and define two 6×6 matrices $M = (m_{ij}), N = (n_{ij})$ by the rule that $m_{ij} = k$ and $n_{ij} = l$ if and only if $(a_1, a_2, a_3, i, j, k, l) \in C$. It follows easily from $(*)$ that M and N are two orthogonal Latin squares of order 6, contradicting the proof of Euler's conjecture by Tarry (see Chapters 1, 9).

Unfortunately, since the generalised form of Euler's conjecture is false in all other cases, this argument really is a one-off!

Now we consider e-error-correcting codes for $e > 1$. The first case is $q = 2$, $e = 2$. Such a code C satisfies

$$|C| = 2^n \left/ \left(1 + n + \binom{n}{2}\right) \right. = 2^{n+1}/(n^2 + n + 2).$$

So $n^2 + n + 2 = 2^a$ for some a. Multiplying by 4 and setting $x = 2n+1, y = a+2$, we find

$$x^2 + 7 = 2^y.$$

This is *Nagell's equation*, named after the mathematician who found all the solutions of this equation in integers[6] (in 1930, some time earlier than the development of coding theory). The solutions are $(x,y) = (\pm 1, 3), (\pm 3, 4), (\pm 5, 5), (\pm 11, 7)$ and $(\pm 181, 15)$. In our situation, the code C is 2-error-correcting, and so has minimum distance 5; so $n \geq 5$, and $x \geq 11$. Thus, only the lengths 5 and 90 are possible. We'll see that there is a unique such code of length 5, one of an infinite (but trivial) class, and no such code of length 90.

The *repetition code* of length n over the alphabet A is the simplest code imaginable, consisting of all words $(a, a, \ldots a)$ of length n for $a \in A$. If $|A| = 2$ and $n = 2e+1$, then C is perfect e-error-correcting: any word w of length n has either more zeros than ones (and is closer to $(0, 0, \ldots, 0)$), or more ones than zeros (and is closer to $(1, 1, \ldots, 1)$).

---

[6] The solution of Nagell's equation is about at the limit of what can be covered in an undergraduate course in algebraic number theory. See I. Stewart and D. Tall, *Algebraic Number Theory* (1987).

The non-existence for length 90 will be shown later.

Now consider $e = 3$, $q = 2$. The code C satisfies

$$|C| = 2^n \Big/ \left(1 + n + \binom{n}{2} + \binom{n}{3}\right).$$

This gives a Diophantine equation a bit like Nagell's, of the form $f(n) = 3 \cdot 2^a$, where f is a cubic polynomial. We seem to be worse off than before; but in fact this is not so, since the polynomial f happens to factorise. Putting $m = n + 1$, a little manipulation gives

$$m(m^2 - 3m + 8) = 3 \cdot 2^a,$$

where $m \geq 8$ (since $n \geq 7$ for a 3-error-correcting code).

Now there are two cases:

CASE 1. $m = 2^b$ and $m^2 - 3m + 8 = 3 \cdot 2^c$ for some b, c. If $m \geq 16$, then $m^2 - 3m + 8 \equiv 8$ (mod 16), so $m^2 - 3m + 8 = 24$, which is impossible. So $m = 8$, $n = 7$, and we have a repetition code.

CASE 2. $m = 3 \cdot 2^b$ and $m^2 - 3m + 8 = 2^c$ for some b, c. As before, if $m \geq 48$, then $m^2 - 3m + 8 \equiv 8$ (mod 16), so $m^2 - 3m + 8 = 8$, a contradiction. So $m = 12$ or 24. In the first case, $m^2 - 3m + 8 = 116$ is not a power of 2. So the possibility $m = 24$, $n = 23$ remains. Golay discovered a perfect binary code with these parameters, which was later shown to be unique (up to a suitable definition of isomorphism). This is the so-called *binary Golay code*.

Golay also discovered a ternary perfect 2-error-correcting code of length 11, which is also unique. Now, to cut a long story short, Tietäväinen proved the following result.

(17.6.3) Tietäväinen's Theorem. *For $e > 1$, the only perfect e-error-correcting codes of length n over alphabets of prime-power size q are the binary repetition codes (with $q = 2$, $n = 2e+1$) and the binary and ternary Golay codes (with $q = 2, e = 3, n = 23$ and $q = 3, e = 2, n = 11$ respectively).*

The Golay codes, with their related designs, lattices, and groups, are of enormous importance, which can only be hinted at here. The next result gives a connection between codes and designs. For ease of exposition, we consider linear codes only. The *support* of any word is the set of coordinate positions where its entries are non-zero.

(17.6.4) Proposition. *Let C be a linear perfect e-error-correcting code of length n over $\mathrm{GF}(q)$. Then the supports of the codewords of smallest weight $2e + 1$ in C are the blocks of an $(e + 1)$-$(n, 2e + 1, (q - 1)^e)$ design, each block repeated $q - 1$ times.*

PROOF. That the minimum weight is $2e + 1$ is clear. Now choose any set of $e + 1$ coordinates, and let w be any word whose support is this set. (There are $(q - 1)^{e+1}$ such words w.) There is a unique codeword c with $d(c, w) \leq e$. Now $c \neq 0$, so $\mathrm{wt}(c) \geq 2e + 1$. It follows that $\mathrm{wt}(c) = 2e + 1$ and the support of c contains that of

w (and c agrees with w on its support). So C contains $(q-1)^{e+1}$ words of weight $2e+1$ whose support contains the given $e+1$-set. Now a non-zero scalar multiple of such a codeword has the same support; so there are $(q-1)^e$ supports of size $2e+1$ containing the given $(e+1)$-set, each repeated $q-1$ times. So we have a design with the stated parameters.

REMARK. Linearity is only used here to show that the zero word is in C, and that each block is repeated equally often. Now let C be a binary perfect code, not necessarily linear. By translation in $H(n,2)$, we may assume that $0 \in C$. Then the 'design' has $\lambda = 1$, and the question of repeated blocks doesn't arise. Thus, the conclusion of (17.6.4) holds for $q=2$ without the assumption of linearity.

This enables us to complete the discussion of binary perfect 2-error-correcting codes. The possibility of such a code of length 90 was left open; but its existence would imply that of a 3-(90, 5, 1) design, in which (by (16.1.3)) the number of blocks containing two points is 88/3, a contradiction.

According to this result, the binary and ternary Golay codes (which are both linear) give rise to 4-(23, 7, 1) and 3-(11, 5, 4) designs. The latter is actually a 4-(11, 5, 1) design. Moreover, these designs can be extended to 5-designs. This is done by extending the codes by an *overall parity check* (a new coordinate position such that the entry in that position in any word is chosen so that the sum of all its entries is zero). The supports of words of minimum weight in the extended codes form extensions of the designs: they are a 5-(24, 8, 1) design and a 5-(12, 6, 1) design. These were the first 5-designs known; their automorphism groups are the *Mathieu groups* M_{24} and M_{12}, the first of the 'sporadic' simple groups to be discovered, and the only 5-transitive permutation groups apart from symmetric and alternating groups.[7]

17.7. Linear codes and projective spaces

The basic properties of a code can be expressed in terms of Hamming distance. So it is reasonable to call two codes *equivalent* if one can be transformed into the other by an *isometry* (a distance-preserving transformation) of Hamming space $H(n,q)$. It can be shown that any isometry can be built out of two kinds of transformation:
(a) permutation of the symbols appearing in any coordinate position, where the permutations applied to different coordinates are chosen independently;
(b) permutations of the coordinates.
These generate the *wreath product* of the symmetric groups S_q (on symbols) and S_n (on coordinates), in its *product action* as defined in Chapter 14 — see Exercise 4.

However, if we are interested in linear codes, then this definition of equivalence is too wide, since the symbol permutations don't in general preserve the property of linearity. Assuming that the alphabet is a field F, we should only allow in (a) the multiplication of each coordinate by a non-zero scalar, the scalars applied to different coordinates being chosen independently; we can compose this with an

---

[7] The groups were constructed by means of generating permutations by Mathieu, half a century before the designs were found by Skolem and Witt, which in turn predated the discovery of the codes by Golay.

arbitrary coordinate permutation as in (b). Such a transformation is called *monomial*, and the equivalence relation defined on linear codes is called *monomial equivalence*. (Thus, a monomial transformation is one represented by a matrix which has a single non-zero entry in each row or column.) For example, all Hamming codes with the same parameters are monomially equivalent.

How many monomial equivalence classes of codes are there? We won't answer this question with a formula, but will translate it into projective geometry, revealing an unexpected but very important connection between these fields.

(17.7.1) Theorem. *There is a bijection between*
- *monomial equivalence classes of linear 1-error-correcting codes of length n and dimension $n - d$ over $\mathrm{GF}(q)$; and*
- *orbits of the general linear group $\mathrm{GL}(d, q)$ on n-element spanning subsets of the projective space $\mathrm{PG}(d - 1, q)$.*

REMARK. Orbits of $\mathrm{GL}(d, q)$ can be regarded as 'geometric configurations'. For example, all conics in the projective plane $\mathrm{PG}(2, q)$ (see Section 9.7) form an orbit.

PROOF. We show that each set corresponds in a natural way to an equivalence class of matrices under a relation intermediate between row-equivalence and row-and-column-equivalence. To be precise, given a $d \times n$ matrix, we allow ourselves to apply arbitrary row operations, but restrict the allowable column operations to two types, viz., multiplication of a column by a non-zero scalar, and interchange of two columns. (These column operations obviously generate the group of all monomial transformations, while the row operations generate the whole general linear group of invertible linear transformations.) Now we consider equivalence classes (under this equivalence relation) of $d \times n$ matrices A such that
(a) the rows of A are linearly independent;
(b) any two columns of A are linearly independent.

STEP 1. Given a linear code C of length n and dimension $n - d$ with minimum weight at least 3, its check matrix A satisfies the two conditions (a) and (b) above; and C is determined as the null space of A, or equivalently as $C_0^\perp$, where C_0 is the row space of A. Now elementary row operations have the effect of changing the basis of the row space, and so don't alter C; and monomial transformations of the columns replace C by a monomial equivalent code. So equivalence classes of matrices correspond to monomial equivalence classes of codes.

STEP 2. Let S be a spanning set of n points in $\mathrm{PG}(d-1, q)$, say $S = \{p_1, \ldots, p_n\}$. Let v_i be a vector spanning the 1-dimensional subspace p_i, and let A be the matrix with columns $v_1^\top, \ldots, v_n^\top$. Obviously, any two columns of A are linearly independent. Also, since $v_1, \ldots, v_n$ is a spanning set, some d-element subset is a basis; so the rows of A are linearly independent. Multiplying columns by non-zero scalars doesn't change the points of projective space they span, and permuting columns merely affects the order in which the elements of S are written down. So monomial transformations of columns don't affect S. On the other hand, elementary row operations generate the general linear group $GL(d, q)$, and so S can be transformed into any other set in the same orbit by a sequence of such operations.

These two steps establish the required bijections, and prove the theorem. But it is much more than an enumeration result. Structural properties can be translated back and forth between code and geometry. For example:

- Any word of weight 3 in C corresponds to a set of three dependent columns of A, and hence to a set of three collinear points of S.
- An element of the dual code $C^\perp$ corresponds to a linear map from F^d to F, whose kernel defines a hyperplane of the projective space. So the supports of non-zero elements of $C^\perp$ correspond to the complements of hyperplane sections of S.

17.8. Exercises

1. Write down a check matrix for the ternary Hamming code of length 13. Hence
(a) decode the received word $(1, 2, 1, 0, 2, 1, 0, 0, 1, 0, 2, 1, 0)$;
(b) construct a generator matrix for the code.

2. Show that an orthogonal array of strength t and index λ over an alphabet of size q has cardinality $\lambda \cdot q^t$, and is also an orthogonal array of strength i and index $\lambda \cdot q^{t-i}$ for all $i \leq t$.

3. Show that the design whose blocks are the supports of words of minimum weight in the q-ary Hamming code of length $(q^d - 1)/(q - 1)$ is isomorphic to the design whose blocks are the collinear triples of points in the projective space $PG(d - 1, q)$.

4. Show that the group of isometries of the Hamming space $H(n, q)$ is the *wreath product* $S_q \, \mathrm{wr} \, S_n$, in its *product action* (see Chapter 14).

5. (COMPUTER PROJECT). Investigate solutions of the sphere-packing condition for the existence of a perfect code, viz. $\sum_{i=0}^{e} \binom{n}{i}(q - 1)^i$ divides q^n.

6. (a) Prove that, if C is a linear MDS code, then $C^\perp$ is also MDS.
(b) Show that the code C corresponding to a set S of points in $PG(d - 1, q)$ (as in (17.7.1)) is MDS if and only if S has the property that no d of its points are contained in a hyperplane. (Such a set is called an *arc*.) Deduce that conics in $PG(2, q)$ give rise to MDS codes.

7. (a) Prove that the dual of a Hamming code of length $(q^d - 1)/(q - 1)$ has minimum weight q^{d-1} and attains the Plotkin bound.
(b) Let A be a Hadamard matrix of order n (see Section 16.6). Normalise the first column to -1 and delete it; then change -1 to 0 throughout. Show that the code C whose words are the resulting rows attains the Plotkin bound. When is it linear?

8. (a) Show that, for binary codes, the Hamming bound is always at least as strong as the Singleton bound. Hence or otherwise show that any MDS binary code is equivalent to a repetition code or the dual of one.
(b) Prove that a q-ary perfect 1-error-correcting code has length at least $q + 1$.

18. Graph colourings

On the bank of the river he saw a tall tree: from roots to crown one half was
aflame and the other green with leaves.

'Peredur son of Evrawg'
from *The Mabinogion* (earlier than 1325)

TOPICS: Vertex and edge colourings; perfect graphs; graph minors;
embeddings of graphs in surfaces

TECHNIQUES: Use of Max-Flow Min-Cut Theorem for constructions; alternating chain arguments

ALGORITHMS:

CROSS-REFERENCES: Graphs, networks (Chapter 11); Hall's theorem
(Chapter 6); posets (Chapter 12); [symmetric functions (Chapter
13)]

In Chapter 11, we took the point of view that graphs model connectivity. Here,
the viewpoint is that graphs model 'incompatibility'. For example, suppose that
radio frequencies are being allocated to a number of transmitters. Some pairs of
transmitters are so close that their transmissions would interfere, and they must
be allocated different frequencies. How many frequencies are required? A more
classical example is the *map colouring problem*, where countries sharing a common
frontier must be given different colours on a map; how many colours does the
cartographer need?[1]

We define a *vertex colouring* of the graph $G = (V, E)$ to be a function c from
V to a set of colours such that, for any edge $\{x, y\} \in E$, we have $c(x) \neq c(y)$.
In the frequency-assignment problem, the transmitters are the vertices, and the
incompatible pairs edges, of a graph G; a legitimate frequency assignment is a

---

[1] The celebrated *four-colour problem*, asking whether four colours always suffice, was invented by
Francis Guthrie, who communicated it (via his brother Frederick) to his mathematics professor at
University College London, Augustus De Morgan, in 1852. Two common myths about its origin are:
- *It was known to cartographers for centuries.* Unfortunately there is no evidence at all for this!
- *It was posed by Möbius in a lecture in 1840.* The problem Möbius actually asked was whether
  there exists a map with five countries, any two sharing a frontier. Clearly such a map would
  require five colours; but its non-existence (which we prove in (18.6.3)) doesn't guarantee that no
  other map needs five colours.

For further information, see N. L. Biggs, E. K. Lloyd and R. J. Wilson, *Graph Theory 1736–1936*.

vertex-colouring of G. Similar remarks apply to the map-colouring problem. In each case, we are interested in the smallest number of colours for which a colouring exists.

Note that the only graphs which have vertex colourings with a single colour are the null graphs. More generally, a *clique* in a graph G is a set of vertices, any pair joined by an edge (so that the induced subgraph[2] is complete),[3] and a *coclique* is a set of vertices containing no edges (so that the induced subgraph is null). So, in a vertex colouring, each colour class is a coclique.

18.1. More on bipartite graphs

As defined in Chapter 11, a graph $G = (V, E)$ is bipartite if there is a partition $V = X \cup Y$, $X \cap Y = \emptyset$, so that every edge has one end in X and the other in Y. The partition of V is called a *bipartition* of the graph and its parts are *bipartite blocks*. A connected graph has a unique bipartition.

Thus, in our new terminology, a graph has a vertex colouring with two colours if and only if it is bipartite: the colour classes form a bipartition.

The results in this section seem somewhat unconnected with colourings. Some connections will emerge in the rest of the chapter.

(18.1.1) Proposition. *If the largest coclique in a bipartite graph $G = (V, E)$ has size m, then V can be partitioned into m subsets each of which is a vertex or an edge.*

PROOF. Let $\{X, Y\}$ be a bipartition of G. Let $Y = \{y_1, \ldots, y_n\}$ and, for $i = 1, \ldots, n$, let A_i be the set of neighbours of y_i (so that $A_i \subseteq X$). We use a variant of Hall's Marriage Theorem (see Chapter 6, Exercise 7):

> If a family $(A_1, \ldots, A_n)$ of subsets of X satisfies $|A(J)| \geq |J| - r$
> for all $J \subseteq \{1, \ldots, n\}$, then there is a subfamily of size $n - r$ which
> has a SDR.

Take any $J \subseteq \{1, \ldots, n\}$. Then $\{y_i : i \in J\} \cup (X \setminus A(J))$ is a coclique; so $|J| + |X| - |A(J)| \leq m$, or

$$|A(J)| \geq |J| - (m - |X|).$$

So there is a subfamily of size $d = n - (m - |X|) = |X| + |Y| - m$ which has a SDR; that is, a set of this many disjoint edges of G. If we add in the remaining $|X| - d$ uncovered vertices in X and $|Y| - d$ uncovered vertices in Y, we obtain altogether

$$d + |X| - d + |Y| - d = m$$

disjoint vertices and/or edges whose union is V.

A *matching*[4] is a set of pairwise disjoint edges, and an *edge-cover* is a set of vertices meeting every edge.

---

[2] Recall from Chapter 11 that an induced subgraph of G consists of a subset of its vertices, together with all edges contained within that set.

[3] Sometimes the term 'clique' is used in a more restrictive sense: it is required that no further vertex is joined to every vertex in the clique, that is, it is maximal with respect to inclusion. (No outsider can be admitted to a clique.)

[4] Sometimes called a 'partial matching' to distinguish from a 'complete' or 'perfect' matching which covers all vertices.

(18.1.2) Proposition. *In a bipartite graph, the size of the largest matching is equal to the size of the smallest edge-cover.*

This is König's Theorem (11.10.2).

(18.1.3) Proposition. *Let G be a bipartite graph with maximum valency d. Then the edge set of G can be partitioned into d partial matchings.*

PROOF. Before beginning, we remark that the result is true in the case when G is regular of valency d. For (6.2.3) guarantees the existence of a perfect matching — equivalently, a SDR for the neighbour sets of the vertices in one bipartite block — and then an easy induction gives the result.

The general proof is by induction on the number of edges. As usual, starting the induction is trivial. So assume that the theorem holds for graphs with fewer edges than G. Let (X, Y) be a bipartition of G, and $e = \{x, y\}$ an edge of G, with $x \in X$ and $y \in Y$. Then the edges of $G - e$ can be partitioned into d matchings. It is easier to visualise the edges as being coloured with d colours $1, 2, \ldots, d$, so that a vertex lies on at most one edge of each colour.

Since x and y have valency less than d in $G - e$, at least one colour does not occur on the edges at each of them. If the same colour is missing at both x and y, we can use it to colour the edge e. So we may suppose that colour 1 is missing at x, and colour 2 at y.

Set $x = u_1$, and define $v_1, u_2, v_2, \ldots$ by the rule that $\{u_i, v_i\}$ has colour 2 and $\{v_i, u_{i+1}\}$ has colour 1, as long as such edges exist. Note that all vertices u_i belong to X and all v_i to Y. The sequence cannot revisit any vertex, so it must terminate; and, by assumption, it cannot terminate at either x or y (for example $y \neq v_n$ since y lies on no edge of colour 2). Now we can interchange the colours 1 and 2 on the edges of this path without violating the condition that no two edges of the same colour meet at a vertex. As a result, colour 2 is no longer used on an edge through x, and we can give this colour to e.

REMARK. The method of proof is called the *alternating chains argument*.

A closely related result is the *Gale–Ryser Theorem*, which determines the possible valencies of bipartite graphs.

(18.1.4) Gale–Ryser Theorem. *Let $x_1 \geq \ldots \geq x_m$ and $y_1 \geq \ldots \geq y_m$ be positive integers. Then the following are equivalent:*
(a) *there exists a bipartite graph for which the valencies in the two bipartite blocks are $x_1, \ldots, x_m$ and $y_1, \ldots, y_n$;*
(b) $\sum_{i=1}^{m} x_i = \sum_{j=1}^{n} y_j$ *and*

$$\sum_{i=1}^{k} x_i \leq \sum_{j=1}^{n} \min(k, y_j) \text{ for } k = 1, \ldots, m.$$

PROOF. The necessity of the conditions is straightforward. The first equation $\sum x_i = \sum y_j$ counts in two ways the total number of edges in the graph. For the second, consider the k vertices of largest valency in the first block. They lie on $\sum_{i=1}^{k} x_i$ edges. But the jth vertex in the second block lies on at most $\min(k, y_j)$ of these edges.

For the sufficiency, we outline a proof using the Max-Flow Min-Cut Theorem. This demonstrates the use of this theorem in combinatorial constructions, of which there are many more examples. We construct a network with vertices s (source), $a_1, \ldots, a_m, b_1, \ldots, b_n, t$ (target), and edges as follows:

- (s, a_i) with capacity x_i, for $i = 1, \ldots, m$;
- (b_j, t) with capacity y_j for $j = 1, \ldots, n$;
- (a_i, b_j) with capacity 1, for $i = 1, \ldots, m$ and $j = 1, \ldots, n$.

The edges out of s and the edges into t both form cuts with capacity $\sum_{i=1}^{m} x_i = \sum_{j=1}^{n} y_j = M$, say. Suppose that S is any cut; say that S contains $(s, a_{i_1}), \ldots, (s, a_{i_k})$ and $(b_{j_1}, t), \ldots, (b_{j_l}, t)$. Then S must also contain (a_i, b_j) for $i \neq i_1, \ldots, i_k$ and $j \neq j_1, \ldots, j_l$; its capacity is

$$\sum_{p=1}^{k} x_{i_p} + \sum_{q=1}^{l} y_{j_q} + (m - k)(n - l),$$

and a little calculation (using the conditions (b) of the theorem) shows that this is at least M.

So the minimum capacity of a cut is M. From the Max-Flow Min-Cut and Integrity Theorems (Section 11.9), we conclude that there is an integral flow of value M. In such a flow, all edges (s, a_i) and (b_j, t) must carry their full capacity, since they lie in minimum cuts. Edges (a_i, b_j) carry flow 0 or 1. Let $V = \{a_1, \ldots, a_m, b_1, \ldots, b_n\}$, and let E be the set of pairs $\{a_i, b_j\}$ for which (a_i, b_j) carries flow 1. Since the flow out of a_i is equal to x_i, this vertex has valency x_i in the bipartite graph (V, E). Similarly, b_j has valency y_j.

This proof is, in some sense, algorithmic, since the proof of the Max-Flow Min-Cut Theorem is constructive. Gale's original paper gives a much more directly constructive proof.

18.2. Vertex colourings

The *chromatic number* of a graph $G = (V, E)$, written $\chi(G)$, is the least number r of colours such that G has a vertex colouring with r colours. Equivalently, it is the least r such that V can be partitioned into r cocliques. The introduction to this chapter motivated the study of this invariant; but its computation is difficult. We note that, if G has a clique of size c, then all vertices of this clique must receive different colours in any vertex colouring; so the chromatic number is at least c. But this invariant is also hard to calculate! An upper bound for $\chi(G)$, easier to compute, is the maximum valency of G, or one more if G is a complete graph or an odd cycle. (This is the content of *Brooks' Theorem*, to be proved in the next section.)

One formal approach to the chromatic number is via the *chromatic polynomial*, which is the function f_G on the natural numbers defined by

$$f_G(r) = \left\{ \begin{array}{c} \text{number of colourings of } G \text{ with the set} \\ \{1, \ldots, r\} \text{ of colours.} \end{array} \right.$$

Because of the name, you will not be surprised to learn that $f_G(r)$ is a polynomial in r, though this is not obvious; it will emerge from a recursive calculation of this number. Note that $\chi(G)$ is the least r for which $f_G(r) > 0$.

EXAMPLE. If K_n and N_n are the complete and null graphs on n vertices, then the colours used in a colouring of K_n must all be distinct, while those used for N_n are unrestricted. By (3.7.1),

$$f_{K_n}(r) = (r)_n = r(r-1)\ldots(r-n+1),$$
$$f_{N_n}(r) = r^n.$$

Let $e = \{x, y\}$ be an edge of $G = (V, E)$. We define two operations on G:
- *Deletion* of e yields the graph $G - e = (V, E \setminus \{e\})$.
- *Contraction* of e: replace x and y by a new vertex z, with an edge $\{v, z\}$ whenever $\{v, x\} \in E$ or $\{v, y\} \in E$; edges not containing x or y are unaltered. Call the resulting graph G/e.

(18.2.1) Theorem. $f_G(r) = f_{G-e}(r) - f_{G/e}(r)$.

PROOF. We divide the set of colourings of $G - e$ into two disjoint classes:
- Those for which x and y receive different colours. These are the valid colourings of G.
- Those for which x and y receive the same colour. Such a coluring induces a colouring of G/e, and conversely.

So $f_{G-e}(r) = f_G(r) + f_{G/e}(r)$, as required.

Now, if G is given and e is an edge of G, then $G - e$ has fewer edges, and G/e fewer vertices, than G. Assuming inductively that their chromatic polynomials are known, that of G can be calculated. The induction begins with graphs without edges, for which we calculated the number of colourings already. This inductive argument also proves that the chromatic polynomial of G is a polynomial in r of degree n, where n is the number of vertices of G.

18.3. Project: Brooks' Theorem

Brooks' Theorem asserts that, with known exceptions, a connected graph with maximum valency d has a vertex colouring with d colours.

First note that a graph has a vertex colouring with a given number of colours if and only if all its connected components do; so it is enough to consider connected graphs. Also, the fact that a graph with maximum valency d can be coloured with $d + 1$ colours is straightforward to prove. Consider the vertices one at a time. Each vertex v has at most d neighbours, to which at most d colours have been applied; so there is an unused colour available for v.

We cannot expect to reduce $d+1$ to d here without paying a price. The complete graph on $d+1$ vertices has valency d but obviously requires $d + 1$ colours. Also, a circuit of odd length is divalent but not bipartite (that is, not 2-colourable). Brooks' Theorem asserts that these graphs are the only (connected) exceptions.

The proof of Brooks' Theorem repeats the above argument with more care, in the 'general' case (Case 1 in the argument below), ensuring that some colour does not appear among the neighbours of a vertex when we come to colour it. The other two cases are more in the nature of minor irritants.

First, a piece of terminology. Let k be a positive integer. A graph G is said to be k-connected if, for any $k-1$ vertices $v_1, \ldots, v_{k-1}$ of G, the graph $G - v_1 - \ldots - v_{k-1}$ obtained by deleting them is connected.

(18.3.1) Brooks' Theorem

A connected graph with maximum valency d, which is neither a complete graph nor a cycle of odd length, has a vertex colouring with d colours.

PROOF. The proof is by induction; we assume that the theorem is true for all graphs with fewer vertices than G (and, in particular, for all proper induced subgraphs of G). Assume that G is neither complete nor an odd cycle. We divide the proof into three cases.

CASE 1. G is 3-connected. Since it is not complete, there are two vertices u, w of G at distance 2. Let v be a common neighbour of u and w. Let $v_1 = u$, $v_2 = w$. Now $G - u - w$ is connected. We define a partial order on its vertices by the rule that $x \leq y$ if y lies on a shortest path from v to x. Note that v is the unique maximal element in this order. Take a linear extension of the partial order (12.2.1), say $v_3 < v_4 < \ldots < v_n$. We have $v_n = v$. Moreover, by construction, for any i with $3 \leq i < n$, there exists $j > i$ such that v_i is joined to v_j.

Take d colours $1, 2, \ldots, d$. Now give colour 1 to v_1 and v_2 (this is legitimate since they are not joined). Colour the remaining vertices in turn. For $3 \leq i < n$, at most $d - 1$ neighbours of v_i are already coloured (since it has a neighbour later in the sequence), so there is a colour available for v_i. Finally, when we reach v_n, all its neighbours are already coloured, but two of them (v_1 and v_2) have the same colour; so there is a colour available for v_n.

CASE 2. G is not 2-connected. Thus there is a vertex v whose removal disconnects G; and the vertices different from v can be partitioned into non-empty subsets X and Y such that no edge goes from X to Y. By induction, one of two possibilities holds:
(a) Each of the induced subgraphs on $X \cup \{v\}$ and $Y \cup \{v\}$ can be coloured with d colours. We can change the names of the colours so that v has the same colour in each colouring, and we have a colouring of G.
(b) One of $X \cup \{v\}$ and $Y \cup \{v\}$, say $X \cup \{v\}$, carries either a complete graph K_{d+1}, or a cycle of odd length (with $d = 2$). But this is impossible, since v would have d neighbours in X and at least one in Y.

CASE 3. G is 2-connected but not 3-connected. In this case there are two vertices u, v whose removal disconnects G, say into X and Y as above. The argument of Case 2 applies except in one situation:
(c) Each of $X \cup \{u, v\}$ and $Y \cup \{u, v\}$ requires d colours. Moreover, in any colouring of $X \cup \{u, v\}$ with d colours, u and v have the same colour; and in any colouring of $Y \cup \{u, v\}$ with d colours, u and v have different colours. (In particular, u and v are not joined.)
Let x_u and x_v be the valencies of u and v in $X \cup \{u, v\}$, and y_u, y_v their valencies in $Y \cup \{u, v\}$. Now $x_u + y_u \leq d$ and $x_v + y_v \leq d$. Also, at least $d - x_u$ colours are available for u in $X \cup \{u, v\}$, and similarly for v and for $Y \cup \{u, v\}$. Now the colours of u and v in $X \cup \{u, v\}$ must be uniquely determined, or we could change one of them and violate (c); so $x_u = y_u = d - 1$. Similarly, the sets of colours available for u and v in $Y \cup \{u, v\}$ must be disjoint; so $(d - y_u) + (d - x_u) \leq d$, whence $y_u + y_v \geq d$. Thus

$$2(d-1) + d \leq 2d,$$

or $d = 2$. But then G is either an odd cycle or bipartite, and the theorem is proved.

18.4. Perfect graphs

We've seen that, for any graph G, the chromatic number $\chi(G)$ (the smallest number of cocliques into which G can be partitioned) is not less than the *clique number* $\gamma(G)$

(the size of the largest clique in G). Obviously, graphs in which equality holds are interesting. Claude Berge realised that, to obtain a manageable theory, we should require this condition also for all induced subgraphs of G. Thus, a graph G is *perfect* if, for every induced subgraph H of G, the chromatic number and clique number of H are equal.

We'll also look at complements. (The *complement* $\overline{G}$ of G has the same vertex set as G; two distinct vertices are joined in $\overline{G}$ if and only if they are not joined in G.) The cliques of $\overline{G}$ are the cocliques of G, and *vice versa*. So the clique number of $\overline{G}$ is equal to the *coclique number* of G (the size of its largest coclique), and the chromatic number of $\overline{G}$ is the *clique-partition number* of G (the smallest number of cliques into which it can be partitioned).

A number of earlier results can be phrased to say that certain graphs are perfect. Note that, if a class of graphs is closed under taking induced subgraphs, then to prove that every graph in the class is perfect, we have the seemingly easier task of proving that every graph in the class has chromatic number and clique number equal.

(18.4.1) Proposition. *(a) Bipartite graphs are perfect.*
(b) Complements of bipartite graphs are perfect.

PROOF. Both classes are induced-subgraph-closed, so we show that either type has clique number and chromatic number equal. For bipartite graphs, this is trivial: both numbers are 2 unless the graph is null (in which case they are 1). For (b), this is the content of (18.1.1).

The *line graph* of a graph $G = (V, E)$ is defined as follows. The vertex set of $L(G)$ is E, the edge set of E; two vertices e_1, e_2 are joined in $L(G)$ if and only if (as edges of G) they have a common vertex. There are two kinds of cliques in $L(G)$:
(a) a set of edges of G through a common vertex;
(b) the edge set of a triangle (3-cycle).
Case (b) cannot occur in a bipartite graph G. So the clique number of $L(G)$ is the maximum valency of G. Similarly, the coclique number of $L(G)$ is the size of the largest partial matching in G. Thus, (18.1.2) and (18.1.3) can be phrased as follows:

(18.4.2) Proposition. *(a) Line graphs of bipartite graphs are perfect.*
(b) Complements of line graphs of bipartite graphs are perfect.

Another two classes of perfect graphs arise from posets (Chapter 12). Let $P = (X, \leq)$ be a poset. Two distinct points $x, y \in X$ are *comparable* if $x \leq y$ or $y \leq x$, and *incomparable* otherwise. The *comparability graph* and *incomparability graph* of P are the graphs with vertex set X whose edges are the comparable and incomparable pairs of vertices respectively. (Of course, these graphs are complementary.) Now Dilworth's Theorem (12.5.3) and its (much easier) dual (12.5.2) translate as follows.

(18.4.3) Proposition. *(a) Comparability graphs of posets are perfect.*
(b) Incomparability graphs of posets are perfect.

In view of these results (and others: see, for example, Exercise 2), the next theorem is no surprise. It was conjectured by Berge (under the name 'weak perfect graph conjecture'), and proved by Lovász. I will state it here without proof.

(18.4.4) Perfect Graph Theorem. *The complement of a perfect graph is perfect.*

To get some idea of the power concealed in this harmless-looking theorem, note that using it we may deduce Dilworth's Theorem from its 'trivial' dual, and similarly Hall's Theorem from an even more trivial result. So we can't expect to find a three-line proof of the Perfect Graph Theorem.

I conclude with the main open problem on perfect graphs, which was also conjectured by Berge. It is clear that, if n is odd and $n > 3$, then the n-cycle C_n is not perfect: it has clique number 2 and chromatic number 3. It is also not difficult to show that the complement of C_n fails to be perfect. Thus a graph which contains either C_n or $\overline{C_n}$ as an induced subgraph for n odd and $n > 3$ also fails to be perfect. We call such induced subgraphs *odd holes* and *odd antiholes*. Berge conjectured that these are the only obstructions to perfection:

(18.4.5) Strong Perfect Graph Conjecture. *A graph is perfect if and only if it contains no odd hole or odd antihole.*

If true, this would imply the Perfect Graph Theorem, since the class of graphs which satisfy the conclusion of the conjecture (which are nowadays called *Berge graphs*) is obviously closed under complementation. But the conjecture has so far defeated a small army of graph theorists!

18.5. Edge colourings

An *edge colouring* of a graph $G = (V, E)$ is a map c from E to a set of colours with the property that two edges sharing a vertex have different colours.

Using the notion of line graph defined in the preceding section, we see that an edge colouring of a graph is exactly the same thing as a vertex colouring of its line graph. So, in a sense, the theory of edge colourings is a part of the theory of vertex colourings; but it has its own particular style and results. The *chromatic index* of G is defined to be the least number of colours required for an edge colouring of G.

In an edge colouring, all the edges which meet at a vertex must have different colours. So the chromatic index of G cannot be smaller than its maximum valency. The following theorem of Vizing restricts the chromatic index to two possible values:

(18.5.1) Vizing's Theorem. *If a graph has maximum valency d, then it has an edge colouring with $d + 1$ colours.*

So the chromatic index is either d or $d + 1$. Accordingly, the class of all graphs can be divided into two parts. A graph G belongs to *Class 1* if its chromatic index is equal to its maximal valency, and to *Class 2* otherwise. According to (18.1.3), all bipartite graphs belong to Class 1.

We met edge colourings of complete graphs, in rather different language, in Section 8.6:

> An edge colouring of the complete graph K_n with the smallest possible number of colours is the same thing as a tournament schedule for n teams.

(The teams are the vertices of the graph, the rounds of the tournament are the colours of the edges.) In particular, the chromatic index of K_n is n if n is odd, $n-1$ if n is even. In other words:

(18.5.2) Proposition. *The complete graph K_n belongs to Class 1 if n is even, and to Class 2 if n is odd.*

18.6. Topological graph theory

Although graphs are abstract objects, it is safe to assume that most people think of them as 'dots and lines', the way we've drawn them many times already. In other words, we choose some familiar geometric or topological space as a drawing board, and represent the vertices by distinct points of the space; each edge is represented by a line or curve whose endpoints correspond to its vertices.

The question 'What is a curve?' is a difficult one which took mathematicians nearly a century to resolve. Peano, for example, constructed a continuous curve passing through every point of the unit square. But such curves don't aid intuition. We assume that an edge is represented by a continuous, *piecewise smooth* curve (one having a continuously varying tangent everywhere except perhaps a finite number of 'corners').

For applications such as road layouts and map colouring, we impose a further condition:

> The curves representing two edges are disjoint except for the point representing their common vertex (if any).

We call a drawing of G satisfying this condition an *embedding* of G in the space. It isn't clear whether a given graph can be embedded in a given space. In three dimensions, there is no restriction:

(18.6.1) Proposition. *Any graph can be embedded in $\mathbb{R}^3$.*

PROOF. Take a line L, and represent the vertices by points of L. For each edge e, take a plane Π_e through L (all these planes distinct), and join the vertices of e by a semicircle in Π_e.

In two dimensions, the situation is very different. We call a graph *planar* if it is embeddable in the Euclidean plane. Some experimentation should convince you that the complete graph K_5 and the complete bipartite graph $K_{3,3}$ are non-planar. We'll see that this is a consequence of a theorem of Euler.

First note that embedding in the plane and in the surface of the sphere are 'equivalent' concepts. This is because of *stereographic projection*, which establishes

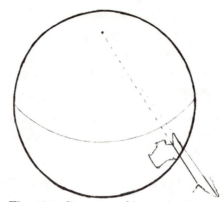

Fig. 18.1. Stereographic projection

a bijection, smooth in both directions, between the plane and the sphere with its north pole removed (Fig. 18.1). Using this bijection, an embedding in the plane is transferred to the sphere. Conversely, given an embedding in the sphere, choose a point lying on none of the curves and use it as the north pole; then projection transfers the embedding to the plane.

Now let G be a connected graph. Given an embedding of G in the plane or sphere, the removal of the image of the embedding leaves a finite number of connected pieces called *faces*. In the case of the plane, just one face — the *infinite face* — is unbounded. The boundary of a face is a closed curve made up of a finite number of vertices and the same number of edges (possibly with repetitions), corresponding to a closed trail in G. (The connectedness of the face boundary depends on that of G: can you see why?) The face itself is topologically equivalent to a disc (the interior of a circle), except in the case of the infinite face in the plane.

(18.6.2) Euler's Theorem. *Let an embedding of the connected graph G in the plane have V vertices, E edges and F faces. Then*

$$V - E + F = 2.$$

PROOF. We use induction on E. A connected graph with one vertex and no edge has one face, and satisfies the theorem.

Suppose that there is an edge e such that $G - e$ is connected. Then $G - e$ has V vertices, $E - 1$ edges and $F - 1$ faces, since, when e is removed, the two faces on either side of it coalesce. (We have to show that these two faces are different. Suppose not; let f be this face. There is a curve in f from one side of e to the other. When e is removed, this becomes a simple closed curve in f. By the Jordan Curve Theorem,[5] this curve divides the plane or sphere into two components, each of which contains a vertex of e; so $G - e$ is not connected.) So $V - (E - 1) + (F - 1) = 2$, and we are done.

---

[5] The *Jordan Curve Theorem* asserts that a simple (non-intersecting) closed plane curve has an 'inside' and an 'outside'; that is, its complement has two connected components, just one of which is unbounded.

So we may assume that there is no such edge e. Then G is a tree. (Choose a spanning tree T of G, by (11.2.2). If $G \neq T$, then the removal of an edge outside T leaves a connected graph.) Thus, $E = V - 1$, by (11.2.1). Moreover, $F = 1$. So $V - E + F = 2$, as required.

If you find that proof a bit unsatisfactory in its (unspoken) appeals to geometrical or physical intuition, you should read Imre Lakatos' *Proofs and Refutations* (1976). Euler's Theorem is used as a test case for an investigation of mathematical rigour, and plausible 'counterexamples' are used to refine and make precise both the statement of the theorem and the arguments used in the proof. (If you are happy with the above proof, then there is even more reason for you to read the book!)

(18.6.3) Corollary. K_5 and $K_{3,3}$ are non-planar.

PROOF. (a) K_5 has 5 vertices and 10 edges, so an embedding would have 7 faces. But each face has at least three edges (a face with 1 or 2 edges can only occur if there are loops or parallel edges in the graph), while each edge bounds at most two faces. Double-counting incident edge-face pairs shows that the number of faces is at most $10 \cdot 2/3 = 6\frac{2}{3}$, a contradiction.

(b) $K_{3,3}$ has 6 vertices and 9 edges, so 5 faces (if embedded in the plane). Now each face has at least four edges: for the graph is bipartite and has no closed trail of odd length. The same argument as before then shows that there are at most $4\frac{1}{2}$ faces, a contradiction.

From this result, we can give further examples of non-planar graphs. A *subdivision* of a graph G is obtained by repeated application of the operation 'insert a vertex into an edge': replace the edge $\{x, y\}$ by two edges $\{x, v\}$ and $\{v, y\}$, where v is a new vertex. It's clear that embeddability in any space is unaffected: choose any point on the path from x to y to represent v, and let the two 'halves' of this path represent $\{x, v\}$ and $\{v, y\}$. So any subdivision of K_5 or $K_{3,3}$ is non-planar, as is any graph containing a subgraph of this form. A still more general construction involves *minors* of a graph.

A graph G_0 is said to be a *minor* of G if it can be obtained from G by a series of deletions and contractions. (See Section 18.2, where it was shown that the chromatic number of G is determined by its proper minors.) Note that a graph can be obtained from any subdivision by contraction; so, if a subdivision of G_0 is a subgraph of G, then G_0 is a minor if G.

The class of planar graphs is closed under taking minors. (It is clear that deleting an edge from a planar graph gives a planar graph. Contraction is a little less obvious. Imagine a continuous deformation in which the curve representing the edge shrinks to a point.) So a planar graph has no K_5 or $K_{3,3}$ minor. Remarkably, the converse is true:

(18.6.4) Kuratowski–Wagner Theorem. *The following conditions on a graph G are equivalent:*
(a) G is planar;
(b) G contains no subdivision of K_5 or $K_{3,3}$;
(c) G has no minor isomorphic to K_5 or $K_{3,3}$.

On the basis of this and other evidence, it was conjectured by Wagner that any minor-closed class of graphs is determined by a finite set of 'forbidden minors'. This has been proved recently in a major piece of work by Robertson and Seymour:

(18.6.5) Robertson–Seymour Theorem. *Let C be a class of graphs which is closed under taking minors. Then there is a finite set S of graphs with the property that $G \in C$ if and only if no member of S is a minor of G.*

What about other 2-dimensional surfaces? Topologists have produced a complete classification of closed surfaces (without boundary points or infinite points), which we now outline. First, such surfaces are divided into *orientable* and *non-orientable* surfaces. A surface is *non-orientable* if it is possible to take a clock on a trip 'round the world' and find, on returning, that its hands turn backwards (its orientation has been reversed). The most famous example is the *Möbius strip*, obtained by taking a strip of paper, giving one end a 180° twist, and joining the ends. It is not closed; but by stitching up the boundary in either of two possible ways, we obtain the *Klein bottle* and the *real projective plane*, both closed and non-orientable. A surface is *orientable* if this phenomenon cannot occur. The sphere is an example. Another is the *torus*, obtained by forming a cylinder (by joining the ends of a strip without a twist) and then bending it round and sewing up the ends without a twist. The classification asserts:

(18.6.6) Classification of closed surfaces

(a) An orientable closed surface is homeomorphic to a 'sphere with g handles', for some $g \geq 0$.

(b) A non-orientable closed surface is homeomorphic to a 'sphere with c cross-caps', for some $c > 0$.

A handle is just like the handle of a teacup, so that a sphere with one handle is a torus.[6] Another metaphor is a bridge. (If a graph drawn in the plane has two edges which cross, then the crossing can be removed by replacing the level-crossing by a bridge. So the class of graphs embeddable on the torus is larger than for the sphere.) A cross-cap is more mysterious; it is like a black hole such that, if you enter the event horizon at one point, you instantly find yourself leaving at the opposite point with your orientation reversed.[7] (This also gives a mechanism for resolving crossings.)

An embedding of a graph in a surface is called *simple* if each face is homeomorphic to a disc. Not all embeddings are simple. For example, take a graph in the

---

[6] 'A topologist is someone who can't distinguish his doughnut from his teacup.'

[7] It is instructive at this point to compare the topologist's 'real projective plane' with the geometer's (Chapter 9). To a geometer, the points are the lines through the origin in $\mathbb{R}^3$. Each affine point (i.e., line not in the equatorial plane) can be represented by the unique point where it meets the southern hemisphere of the unit sphere; points at infinity correspond to antipodal pairs of points on the equator. This can be realised by taking a cross-cap covering the entire northern hemisphere.

plane; draw it inside a small disc, and paste the disc onto a torus. Now the 'infinite face' is not a disc (in topologists' language, it is not simply-connected). For simple embeddings, there is a generalisation of Euler's Theorem:

(18.6.7) Euler's Theorem for surfaces. *Suppose that a simple embedding of a graph in a surface S has V vertices, E edges and F faces.*
(a) If S is a sphere with g handles, then $V - E + F = 2 - 2g$.
(b) If S is a sphere with c cross-caps, then $V - E + F = 2 - c$.

The number on the right-hand side of each of these equations is called the *Euler characteristic* of the relevant surface. Euler's Theorem gives restrictions on graphs embeddable in a surface, by the same argument as in (18.6.3). Sometimes, exact bounds can be obtained.

(18.6.8) Ringel–Youngs Theorem. *K_n can be embedded in a sphere with g handles if and only if*
$$n \le \tfrac{1}{2}\left(7 + \sqrt{48g + 1}\right).$$

PROOF. K_n has n vertices, $n(n-1)/2$ edges, and so at most $n(n-1)/3$ faces (arguing as before). So
$$n - n(n-1)/2 + n(n-1)/3 \ge 2 - 2g.$$

Rearranging as a quadratic in n, we find the inequality of the theorem. Now it is necessary to construct a complete graph with $\lfloor \tfrac{1}{2}(7 + \sqrt{48g + 1}) \rfloor$ vertices, embedded in a sphere with g handles. This is the content of a long project by Ringel and Youngs. For $g = 0, 1$, the formula gives 4 and 7 respectively. Exercise 9 asks you to show that K_7 is embeddable in a torus.

In general, the class of graphs embeddable in a surface S is minor-closed. According to the Robertson–Seymour Theorem (18.6.5), it is characterised by a finite set of excluded minors. But this set can be quite large. For example, 35 excluded minors are required to characterise graphs embeddable in the projective plane, and over 800 for the torus!

One of the main areas of interest in topological graph theory is the connection with colouring problems. Any plane map can be described by a graph whose vertices are the countries, with edges joining countries which share a boundary. (If two countries share several unconnected segments of boundary, use multiple edges.) Now a colouring of the map is the same thing as a vertex colouring of the graph. The famous *four-colour problem* was resolved in 1976 by Appel and Haken, with the help of extensive computation:

(18.6.9) Appel–Haken Theorem, or Four-colour Theorem. *Any planar graph has a vertex colouring with four colours.*

It is impossible to summarise here the techniques used; but Appel and Haken, and others, have written several good accounts. On the other hand, we prove in the next section that five colours suffice; the proof illustrates the basic ideas which grew into the Appel–Haken proof.

In this case, it is trivial that there are maps which require four colours, but very difficult to show that no more than four are needed. For other orientable surfaces, the difficulty is the other way around. It is fairly straightforward to give an upper bound for the number of colours needed. This bound turns out to be precisely the number in the Ringel–Youngs Theorem! This theorem guarantees that a complete graph of the appropriate size is embeddable in the surface, and it requires as many colours as it has vertices. We conclude:

(18.6.10) Map Colouring Theorem. *The minimum number of colours required for a vertex colouring of any graph embeddable in the sphere with g handles is $\lfloor \frac{1}{2}(7 + \sqrt{48g + 1}) \rfloor$.*

18.7. Project: The Five-colour Theorem

In this section, I show that a planar graph can be coloured with five colours. The argument is due to Kempe, who thought (incorrectly) that he had proved the Four-colour Conjecture. The mistake was pointed out by Heawood, who salvaged the Five-colour Theorem (and more). See R. J. Wilson and J. J. Watkins, *Graphs: An Introductory Approach* (1990), for further discussion.

(18.7.1) Five-Colour Theorem. *A planar graph has a vertex colouring with five colours.*

PROOF. The proof is by induction on the number of vertices. We assume the result for graphs with fewer vertices than G. We also assume that G is drawn in the plane, and that G has no repeated edges (since these don't affect the chromatic number).

Let G have V vertices, of which n_i have valency i for each i; let there be E edges and F faces. Now, as in the proof of (18.6.3), we have $2E \geq 3F$. Counting vertices and incident vertex-edge pairs,

$$\sum n_i = V,$$
$$\sum i n_i = 2E.$$

From Euler's Theorem, we conclude that

$$\sum (6 - i) n_i \geq 12.$$

The left-hand side of this inequality must be positive; so $n_i > 0$ for some $i < 6$, whence G *contains a vertex of valency at most* 5.

Let v be such a vertex. By induction, $G - v$ has a colouring with five colours 1, 2, 3, 4, 5. If not all colours are used on the neighbours of v, then there is a free colour which can be applied to v. So we may assume that v has valency 5, and that all its neighbours have different colours. Let the neighbours be $z_1, \ldots, z_5$ in anticlockwise order, where we may assume that z_i has colour i.

Let S be the set of all vertices which can be reached from z_1 by a path using vertices with colours 1 and 3 only. Then we can legitimately interchange colours 1 and 3 throughout the set S, without affecting the property that adjacent vertices have different colours. If $z_3 \notin S$, then after this interchange no neighbour of v has colour 1, and we can use this colour for v. So we may suppose that $z_3 \in S$. Thus, there is a path $z_1, x_1, \ldots, x_k, z_3$ consisting of vertices with colours 1 and 3. Adjoining v to this path, we obtain a simple closed curve C.

By the Jordan Curve Theorem, C divides the plane into two parts, and clearly z_2 and z_4 lie in different parts; suppose that z_2 is inside C. Let T be the set of vertices which can be reached from z_2 by a path using vertices with colours 2 and 4 only. No such path can cross C, so T lies wholly inside C, and $z_4 \notin T$. Then we can interchange the colours 2 and 4 throughout T, freeing colour 2 for use on v.

18.8. Exercises

1. Find the clique number and the chromatic number of (a) the complement of the n-cycle C_n; (b) the Petersen graph.

2. Find the chromatic polynomial of the path P_n and of the cycle C_n with n vertices.

3. (a) Let $x_1 \geq \ldots \geq x_m$ and $y_1 \geq \ldots \geq y_n$ be positive integers. Show that the following are equivalent:
 - there is a bipartite graph with valencies $x_1, \ldots, x_m$ in one bipartite block and $y_1, \ldots, y_n$ in the other;
 - there is a matrix with entries 0 and 1 only, having row sums $x_1, \ldots, x_m$ and column sums $y_1, \ldots, y_n$.

(b) Recall the notion of partition of an integer, conjugate partitions, and natural partial order of partitions from Section 13.1. Use the Gale–Ryser Theorem (18.1.4) to show that, if λ and μ are partitions of the same integer, then there is a zero-one matrix whose row sums are the parts of λ and whose column sums are the parts of μ if and only if $\mu \leq \lambda^*$.

REMARK. In fact, with the notation of Section 13.6, if we express the elementary symmetric polynomial e_λ in terms of the basic polynomials m_μ, by

$$e_\lambda = \sum_{\mu \vdash n} a_{\lambda\mu} m_\mu,$$

then $a_{\lambda\mu}$ is equal to the number of zero-one matrices whose row sums form the partition λ and whose column sums form the partition μ. (This is the content of Exercise 10 of Chapter 13.) We showed in the proof of Newton's Theorem (13.5.1) that $a_{\lambda\mu} = 0$ unless $\mu \leq \lambda^*$; the Gale–Ryser Theorem asserts the converse, viz., if $\lambda \leq \mu^*$ then $a_{\lambda\mu} > 0$.

4. A graph G is called an *interval graph* if its vertices are a collection of non-empty intervals of the real line $\mathbb{R}$, with two vertices adjacent if and only if they have non-empty intersection. Interval graphs are useful in modelling time-dependent phenomena. By slight perturbations of the endpoints of the intervals, we may assume that these endpoints are all distinct, and that the intervals are closed.

(a) Prove that interval graphs are perfect. [HINT: Given a set of intervals, let $n(x)$ be the number of intervals containing the real number x. The clique number is the maximum value of this function. Take the smallest x at which the maximum is attained; then some interval in the collection starts at x. Repeat at the smallest x not yet covered at which the maximum is attained, as long as one exists. In this way, we construct a coclique covering every x at which $n(x)$ is maximum. Now use induction.]

(b) Prove that complements of interval graphs are perfect. [HINT: Let $C_1, \ldots, C_m$ be the cocliques of maximum size in an interval graph. Let x_i be the right-hand end of the leftmost interval in C_i, and let x be the minimum of $x_1, \ldots, x_m$. Show that the leftmost interval of each C_i contains x.]

5. Let g be a permutation of $\{1, \ldots, n\}$. The *permutation graph* defined by g has vertex set $\{1, \ldots, n\}$; its edges are all the pairs whose order is reversed by g (that is, all $\{i, j\}$ with $i < j$ and $ig > jg$).

(a) Prove that the complement of a permutation graph is a permutation graph.

(b) Recall the dimension of a poset (Section 12.6). Prove that a graph is a permutation graph if and only if it is the incomparability graph of a poset of dimension at most 2.

(c) Prove that a graph is a permutation graph if and only if it is both a comparability graph and an incomparability graph.

(d) Prove that a permutation graph is perfect. Can you find a direct argument? [HINT: Use the argument of Erdős and Szekeres used to prove (10.5.1).]

6. A graph is called *N-free* if it doesn't contain the path on 4 vertices as an induced subgraph. (See the Hasse diagram of the poset N in Fig. 12.1.)

(a) Show that the complement of an N-free graph is N-free.

(b) Show that an N-free graph is connected if and only if its complement is disconnected.

(c) Prove that the class of N-free graphs is the smallest class containing the 1-vertex graph and closed under disjoint union and complementation. (In other words, any N-free graph can be built from 1-vertex graphs by these operations.)

(d) Hence show that N-free graphs are perfect.

7. Show that an N-free graph is a comparability graph. [HINT: Exercise 6(c).] Hence show that an N-free graph is a permutation graph.

8. Show that the Petersen graph belongs to Class 2.

9. Find an embedding of K_7 in a torus, and an embedding of the Petersen graph in the real projective plane.

10. Show that any finite graph can be embedded in $\mathbb{R}^3$ so that edges are represented by straight line segments. [HINT: Consider points (t, t^2, t^3).]

11. Show that a plane triangulation with no vertex of valency less than 5 has at least 12 vertices of valency 5. Construct an example with exactly 12 vertices of valency 5, and colour it with four colours.

19. The infinite

In the Middle Ages the problem of infinity was of interest mainly in connection with arguments about whether the set of angels who could sit on the head of a pin was infinite or not.

N. Ya. Vilenkin, *Stories about Sets* (1965)

... the true mathematician and physicist know very well that the realms of the small and the great often obey quite different rules.

Kurt Singer, *Mirror, Sword and Jewel* (1973)

TOPICS: Set theory, cardinal and ordinal numbers; König's Infinity Lemma, Zorn's Lemma and equivalents; infinite Ramsey Theorem; the 'random graph'

TECHNIQUES: Transfinite induction; free constructions; back-and-forth; probabilistic existence proofs

ALGORITHMS:

CROSS-REFERENCES: SDRs (Chapter 6); projective planes (Chapters 7, 9); Steiner triple systems (Chapter 8); posets (Chapter 12); graph colourings (Chapter 18)

Counting is a less precise tool for infinite sets than for finite ones. The shepherdess who can count her flock of a hundred sheep will know if the wolf has taken one; but, if she has an infinite flock, she won't notice until almost all of her sheep have been lost.

Nevertheless, combinatorics depends on counting. So, in the first section, you will find a quick tour through set theory and the two kinds of numbers used for infinite counting.

The remainder of the chapter describes some topics in infinite combinatorics. Most of these could be described as 'climbing up from the finite'; truly infinite reasoning is more recondite and is done mostly by set theorists.

19.1. Counting infinite sets

This section gives a very brief account of set theory and cardinal and ordinal numbers. It is no substitute for a textbook account (such as K. J. Devlin's *Fundamentals of Contemporary Set Theory*), however.

Before beginning infinite combinatorics, we must look at how to count infinite sets. We will see that two kinds of counting (progressing in order from one number to the next, and measuring the 'size' of a finite set), which are essentially the same in the finite case, have to be distinguished. First, though, there are two more fundamental difficulties: What is an infinite set? And, anyway, what is a set?

In Chapter 2, I took the point of view that we understand the natural numbers from our early experience with counting. In much the same way, we have intuition about sets (or 'collections', 'classes', or 'ensembles' of objects) against which to test our conclusions. However, Russell's paradox demonstrates that we cannot uncritically allow any collection of elements to form a set, or we introduce contradictions into the foundations of mathematics.[1]

The basic idea adopted to rectify this problem is that we start with some collection of fundamental objects or 'urelemente' which are not themselves sets, and then construct sets in stages: at each stage, we can gather together objects constructed at earlier stages into sets.[2] Logicians prefer to build the mathematical universe out of nothing, and traditionally start with the empty set of objects. It is not sufficient just to go through stages 1, 2 and so on (indexed by the natural numbers), since the sets we would obtain would all be finite. (Beginning with $\emptyset$, at the first stage, we get $\{\emptyset\}$; at the second, $\{\{\emptyset\}\}$ and $\{\emptyset, \{\emptyset\}\}$; etc.) We must continue the construction into the transfinite, and need infinite sets to describe the stages properly. To avoid circularity, mathematicians adopted an axiomatic approach.

However, logicians know well that axioms[3] can never entirely capture a mathematical structure. Kurt Gödel showed that, if a structure has at least the richness of the natural numbers (with their ordering, addition, and multiplication), then any set of axioms which can be written down (actually or potentially) is 'incomplete': some assertions about the structure can be neither proved nor disproved using these axioms. Subsequent work showed that no infinite structure can be completely specified by axioms; there will always be other structures satisfying the same axioms. So, if we decide to base set theory on axioms, we must be prepared for there to be different 'set theories', and statements which are true in some and false in others.

It is worth sparing a moment to see why the ambiguities come in. In terms of our intuition, the gathering of elements into sets in each stage is not precisely defined, and there is room for manœuvre on what subsets are included. Now everything has a set-theoretic description. An ordered pair is a set (the standard definition is $(x, y) = \{\{x\}, \{x, y\}\}$[4]); a function is a set of ordered pairs. In particular:

- We want to say that two sets have the same number of elements if there is

---

[1] As Bertrand Russell wrote to Gottlob Frege, 'Consider the set of all sets which are not members of themselves. Is it a member of itself?'

[2] This procedure avoids Russell's paradox: the elements of Russell's 'set' continue appearing at every stage in the construction, so there is no stage at which they all exist to be gathered into a set at the next stage.

[3] The logical system used for the discussion here is 'first-order logic', widely accepted as the best logical basis for mathematics.

[4] The important feature of this definition is that $(x, y) = (u, v)$ if and only if $x = u$ and $y = v$. Any set-theoretic construct with this property would do.

a bijection between them. But different set theories have different bijections; so two sets may have the same number of elements in one theory and not in another.

- The notorious *Axiom of Choice* asserts that given any 'family' (that is, set) of non-empty sets, we can choose representatives of the sets. More formally, if A_i is a set for each i in some index set I, then there exist elements a_i for $i \in I$ such that $a_i \in A_i$ for each $i \in I$. Now these elements are described by a function from I to the union of the sets A_i; this function may be present in some models but not others.

Gödel showed that the Axiom of Choice is consistent (it cannot be disproved from the other axioms). He did this by constructing a model or universe in which the collections of elements which can be gathered into a set at any stage are those satisfying some formula of logic (this is called the 'constructible universe'), and showing that the Axiom of Choice holds in this model. Later, Cohen showed by a technique known as 'forcing' that it is independent (it cannot be proved either).

Since there is no way of resolving questions like 'Is the Axiom of Choice true?' on the basis of the standard axioms, the only hope of progress is to try to refine our intuition about what set theory is, until perhaps there is general agreement about the need for a new axiom which would decide some of these questions. In the meantime we explore consequences of these statements and of their negations.[5] Many of these consequences are of a combinatorial nature.

Now what about counting? Corresponding to the 'stages' in the construction of sets, we define a transfinite sequence of numbers, the *ordinal numbers*, as follows. The empty set is an ordinal number (the number 0); if the set n is a number then so is $n \cup \{n\}$ (this number, representing $n+1$, will be constructed at the stage after n); and, to enable us to leap up into the transfinite, a 'transitive set' of ordinal numbers (containing all members of members) is itself an ordinal number. This condition, for example, allows us to gather up all the natural numbers into a single ordinal number, the first infinite ordinal, conventionally called ω. (ω is a transitive set, since by construction the members of any ordinal number are the smaller ordinal numbers.)

In the construction, we distinguish three kinds of ordinals:

- *zero*, or $\emptyset$;
- *successor ordinals*, of the form $n + 1 = n \cup \{n\}$;
- *limit ordinals*, with no immediate predecessor, obtained by the 'gathering up' procedure.

Now we can say that the 'stages' of the intuitive construction of sets are indexed by the ordinal numbers.

Having defined the ordinal numbers, we have (almost by definition) the principle of *transfinite induction*:

---

[5] We do mathematics with the Axiom of Choice or with its negation, in much the same way that we do Euclidean or non-Euclidean geometry.

(19.1.1) Transfinite induction

Suppose that P is a property of ordinal numbers. Assume
- *$P(0)$ holds;*
- *if $P(n)$ holds, then $P(n+1)$ holds;*
- *if n is a limit ordinal and $P(m)$ holds for all $m < n$, then $P(n)$ holds.*

Then $P(n)$ holds for all ordinal numbers n.

Transfinite induction can be used in constructions as well as proofs, just as the more usual induction in Chapter 2.

Ordinal numbers capture the notion of succession. But they don't measure the size of a set. *Hilbert's hotel*[6] illustrates this. Consider a hotel with ω rooms (numbered 0, 1, 2, ...). One day, when all the rooms are full, a new guest arrives. To accommodate him, the manager simply moves each guest into the next room along, freeing room 0 for the newcomer. Next day, infinitely many new guests arrive. Undeterred, the manager shifts the guest from room n into room $2n$ for each n, freeing the odd-numbered rooms for the new arrivals.

As we saw already, two sets *have the same cardinality* if there is a bijection between them.[7] Hilbert's hotel shows that there is a bijection between ω and $\omega + 1$, and also between ω and $\omega + \omega$. So the ordinal numbers are too discriminating. There are two ways to proceed:

We may decide that, having defined what it is for sets to have the same cardinality, we have implicitly defined the cardinality of a set. Roughly speaking, cardinalities are equivalence classes for the relation 'same cardinality'; but care is required, since the equivalence classes are not sets (by the same reasoning as in Russell's paradox; singletons, for example, continue appearing at all stages).

An alternative approach depends on the fact:

Any non-empty set of ordinal numbers has a least element.

(This is proved by transfinite induction in the same way that the same assertion for the natural numbers is proved by induction — see Chapter 2.) Now, given any set X, the set of all those ordinal numbers which are bijective with X has a least element (if there are any such numbers!), and we take this least element to be the cardinality of X. In other words, a *cardinal number* is an ordinal number which is not in one-to-one correspondence with any smaller ordinal number. With this approach, all natural numbers, and ω, are cardinal numbers, but $\omega + 1$ and $\omega + \omega$ are not.

---

[6] As described in Stanislaw Lem's story 'The Interstellar Milkman, Ion the Quiet'. See N. Ya. Vilenkin, *Stories about Sets* (1965).

[7] Thus, a set is *countable* if and only if it is bijective with $\mathbb{N}$.

In this approach, a set is finite if and only if it is bijective with some natural number. This is precisely how natural numbers are used in ordinary counting (as 'standard sets' of each possible size); our approach generalises this to the transfinite.

An alternative notation for cardinal numbers is due to Cantor, the 'aleph notation'. ($\aleph$ (aleph) is the first letter of the Hebrew alphabet.) Using transfinite induction, we define $\aleph_n$ for all *ordinal* numbers n by the rules:

- $\aleph_0 = \omega$;
- $\aleph_{n+1}$ is the next cardinal number after $\aleph_n$;
- if a is a limit ordinal and $\aleph_b$ is defined for all $b < a$, then $\aleph_a$ is the least cardinal number exceeding all these.

Life is simplified by the following theorem of Zermelo:

(19.1.2) Well-ordering Theorem. *The Axiom of Choice is equivalent to the assertion that every set admits a one-to-one function onto some ordinal number.*

Thus, if we assume that our set theory satisfies the Axiom of Choice (as is almost universally done), then every set has a unique cardinal number.

Cardinal numbers, being special ordinal numbers, are totally ordered. We have $a \leq b$ if there is a one-to-one function from a set of cardinality a into one of cardinality b. It follows from (19.1.2) that, assuming the Axiom of Choice, given any two sets, there is a one-to-one function from one to the other (in some order!)

We can do arithmetic with cardinal numbers. If A and B are disjoint sets with cardinalities a and b respectively, then $a + b$, $a \cdot b$ and a^b are the cardinalities of $A \cup B$, $A \times B$ (Cartesian product), and A^B (the set of functions from B to A) respectively. (Representing subsets of B by their characteristic functions, we see that 2^b is the cardinality of the power set of B.) But the rules are a bit different. The next result assumes the Axiom of Choice.

(19.1.3) Proposition. *(a) If a and b are infinite, then $a + b = a \cdot b = \max(a, b)$.*
(b) If $a > 1$, then $a^b > b$ for all b.

In particular, $2^a > a$ for all a. It is known that 2^ω is the cardinality of the set of real numbers. Cantor's *continuum hypothesis* is the assertion that 2^ω is the smallest uncountable cardinal number; in other words, no subset of $\mathbb{R}$ has cardinality strictly between those of $\mathbb{N}$ and $\mathbb{R}$. (In aleph notation, $2^{\aleph_0} = \aleph_1$.) More generally, the *generalised continuum hypothesis* (GCH) asserts that, for any cardinal number b, 2^b is the next cardinal after b (or $2^{\aleph_n} = \aleph_{n+1}$ for all ordinals n). It is known that the GCH is undecidable; it holds in Gödel's constructible universe, but there are models in which it is false.

19.2. König's Infinity Lemma

The Axiom of Choice (which I will abbreviate to AC) is fundamental to most infinite combinatorics, and I will assume that it holds. However, some students, learning about it for the first time, overrate its influence, and worry that it is being invoked in an argument along the lines, 'The set X is non-empty, so choose an element $x \in X$ ...'. This does not require AC; one choice, or indeed finitely many choices, are permitted by the other axioms. Similarly, AC is not required if there is a rule for making the choices.[8] Only when infinitely many genuine free choices must be made is AC required.

Often, we have a situation where later choices depend on earlier ones. The so-called 'Principle of Dependent Choice' allows us to make such choices; it is a consequence of AC. (AC allows us to choose an element from any set which could conceivably arise in the process.) In this form, it is invoked in proving a result which is very useful in applying AC to combinatorics: *König's Infinity Lemma*.

A *one-way infinite path* in a digraph D is a sequence $v_0, v_1, v_2, \ldots$ of distinct vertices such that (v_i, v_{i+1}) is an edge for all $i \geq 0$.

(19.2.1) König's Infinity Lemma. *Let v_0 be a vertex of a digraph D. Suppose that*
(a) every vertex has finite out-valency;
(b) for every positive integer n, there is a path of length n beginning at v_0.
Then there is a one-way infinite path beginning at v_0.

REMARK. The result is false if condition (a) is relaxed. Take a path of length n for every finite n, all starting at the same point, but otherwise disjoint.

PROOF. We call a vertex v of D *good* if, for every n, a path of length n starts at v. We claim:

> If v is good, then there exists v' such that (v, v') is an edge and v'
> is a good vertex of $D - v$.

For let w_n be the next point after v on a path of length n starting at v. Since there are only finitely many vertices x for which (v, x) is an edge, one of them (say v') must occur infinitely often as w_n. This means that there are arbitrarily long finite paths starting at v', and hence paths of all finite lengths starting there, none of which contain v.

Now, by assumption, v_0 is good. For each $i \geq 0$, choose v_{i+1} so that (v_i, v_{i+1}) is an edge and v_{i+1} is a good vertex of $D - \{v_0, \ldots, v_i\}$. Then $v_0, v_1, v_2, \ldots$ is the required one-way infinite path.

Another infinite principle was invoked in the proof of the Claim above: an infinite form of the *Pigeonhole Principle* (cf. (10.1.1)).

(19.2.2) Pigeonhole Principle (infinite form)
If infinitely many objects are divided into finitely many classes, then some class contains infinitely many objects.

The infinite form of Ramsey's Theorem is a generalisation of this; see Section 19.4.

Now we give an application of König's Infinity Lemma, showing how it can be used to transfer information between the finite and the (countably) infinite.

---

[8] Bertrand Russell's example: If a drawer contains infinitely many pairs of shoes and we must choose one shoe from each pair, we can take all the left shoes. But, for infinitely many pairs of socks, AC is required.

(19.2.3) Proposition. *Let Γ be a countably infinite graph. Suppose that any finite induced subgraph of Γ has a vertex colouring with r colours. Then Γ has a vertex colouring with r colours.*

PROOF. Let $v_1, v_2, \ldots$ be the vertices of Γ. For each n, let C_n be the (non-empty) set of all vertex colourings of the induced subgraph on $\{v_1, \ldots, v_n\}$ with the r colours $1, \ldots, r$. Form a digraph D with $\bigcup_{n \geq 0} C_n$ as vertex set (we take C_0 to be a singleton whose only member c_0 is the empty set!) and with edges as follows: for $c_n \in C_n$ and $c_{n+1} \in C_{n+1}$, let (c_n, c_{n+1}) be an edge if and only if c_n is the restriction of the colouring c_{n+1} to the vertices $v_1, \ldots, v_n$. Then each vertex has out-valency at most r (since at most r colours can be applied to v_{n+1} if $v_1, \ldots, v_n$ are already coloured). Moreover, $d(c_0, c_n) = n$ for all $c_n \in C_n$.

So the hypotheses of König's Infinity Lemma are satisfied. We conclude that there is a one-way infinite path $c_0, c_1, c_2, \ldots$. This gives us a rule for colouring all the vertices of Γ; for v_n is assigned a colour in c_n, and by definition it gets the same colour in all c_m for $m > n$. Moreover, it is a legitimate vertex colouring; for, if $\{v_i, v_j\}$ is an edge, then v_i and v_j are assigned different colours in c_n, where $n = \max(i, j)$.

(19.2.4) Corollary. *Any plane map, finite or infinite, can be coloured with four colours.*

PROOF. For finite maps, of course, this is the Four-colour Theorem (18.6.9). A plane map has at most countably many countries, since each country contains a point with rational coordinates, and there are only a countable number of such points. So the infinite case follows from (19.2.3).

In fact, (19.2.3) holds for arbitrary infinite graphs, not just countably infinite ones. To prove this, we need a stronger principle, *Zorn's Lemma*, to be described in the next section.

19.3. Posets and Zorn's Lemma

One of the most striking differences between finite and infinite posets is that the latter need not have maximal elements, as shown by the natural numbers (for example). An important theorem giving conditions under which maximal elements exist is *Zorn's Lemma*:

(19.3.1) Zorn's Lemma. *Let P be a non-empty poset. Suppose that every chain in P has an upper bound. Then P has a maximal element.*

PROOF. Recall how we showed that every finite poset has a maximal element: if not, pick an element, and repeatedly pick a larger element, yielding an infinite ascending chain. The same trick works here. Suppose that $P = (X, \leq)$ has no maximal element. By transfinite induction, define elements x_a, for all ordinal numbers a, such that $x_a < x_b$ for $a < b$. This is done as follows:

- Let x_0 be any element of x.

- If x_a is defined, let x_{a+1} be any strictly greater element (this exists since x_a is not maximal).
- If a is a limit ordinal, then the elements x_b for $b < a$ form a chain; let x_a be an upper bound for this chain.

Obviously, all the elements x_a are distinct. But this leads to a contradiction: take a to be a cardinal number greater than the cardinality of X, and there are not enough elements available in X for such a chain!

Note that we used the Axiom of Choice in this proof: we have to choose each term of the series from a set of 'admissible' elements. This is in fact inevitable: Zorn's Lemma is 'equivalent to' AC; the latter can be proved from the former and the other axioms of set theory. (See Exercise 3.)

Here is a fairly typical application of Zorn's Lemma, to an infinite version of (12.2.1):

(19.3.2) Theorem. *Any poset has a linear extension.*

PROOF. Let (X, R) be a poset. We let $\mathcal{R}$ be the set of relations $R' \supseteq R$ for which (X, R') is a poset, partially ordered by inclusion. We claim:

> *Every chain in $(\mathcal{R}, \subseteq)$ has an upper bound.*

For let C be a chain, and let R' be the union of the members of C (each member of C being a relation on X, that is, a set of ordered pairs). Then (X, R') is a partial order. (This involves checking the axioms. The arguments are all similar: here is the proof of transitivity. Suppose that $(x, y), (y, z) \in R'$. Then, say, $(x, y) \in R_1$ and $(y, z) \in R_2$ for some $R_1, R_2 \in C$. Since C is a chain, one of these relations contains the other; say $R_1 \subseteq R_2$. Then $(x, y), (y, z) \in R_2$; so $(x, z) \in R_2$ (because (X, R_2) is a poset), and $(x, z) \in R'$, as required.) Clearly $R' \supseteq R$, and R' is thus an upper bound for C in $\mathcal{R}$.

By Zorn's Lemma, there is a maximal element of R, say R'. We show that (X, R') is a total order. If it were not, then there would be some pair (a, b) of points which are incomparable in (X, R'). Now exactly the same argument as in the proof of (12.2.1) shows that we could enlarge R' to make a and b comparable, by setting $R' = R' \cup (\downarrow a \times \uparrow b)$. But this would contradict the maximality of R'.

So (X, R') is a linear extension of (X, R), as required.

Zorn's Lemma is often conveniently applied in the form of the *Propositional Compactness Theorem*, which we now develop with an application.

An *ideal* in a lattice is a non-empty down-set which is closed under taking joins. Equivalently, I is an ideal in L if
- $0 \in I$;
- $x, y \in I \Rightarrow x \vee y \in I$;
- $x \in I, a \in L \Rightarrow x \wedge a \in I$.

An ideal is *proper* if it is not the whole of L; equivalently, if it does not contain 1.

(19.3.3) Proposition. *Any lattice contains a maximal proper ideal.*

PROOF. Straightforward application of Zorn's Lemma to the set of proper ideals, partially ordered by inclusion. (If no ideal in a chain contains 1, then the union doesn't contain 1 either.)

Slightly more generally, *any proper ideal I in a lattice is contained in a maximal ideal*. This is proved by modifying the argument to use only the set of proper ideals containing I.

Our application depends on the following observation: in a Boolean lattice L, if I is a maximal ideal, then for each $a \in L$, exactly one of a and a' belongs to I. (They cannot both belong, since their join is 1. If neither lies in I, then the set

$$J = \{y : y \leq a \vee x \text{ for some } x \in I\}$$

is an ideal containing I and a but not a', contradicting maximality.)

Recall the definition of propositional formulae and valuations from Section 12.4. One small piece of terminology: A set Σ of propositional formulae is *satisfiable* if there is a valuation v such that $v(\phi) = \text{TRUE}$ for all $\phi \in \Sigma$.

(19.3.4) Propositional Compactness Theorem. *Let Σ be a set of propositional formulae. Suppose that every finite subset of Σ is satisfiable. Then Σ is satisfiable.*

PROOF. We work in the Boolean lattice L of equivalence classes of formulae, and identify a formula with its equivalence class. Let I be the ideal *generated by* $\Sigma' = \{(\neg\phi) : \phi \in \Sigma\}$: that is, I is the set of elements of L which lie below some finite disjunction of elements of Σ'. The hypothesis implies that $1 \notin I$. For, if $1 \in I$, then 1 would be a (finite) disjunction of elements of Σ'. By assumption, there is a valuation giving all these elements the value FALSE; but then 1 would have the value FALSE, which is impossible.

By the extension of (19.3.3), there is a maximal ideal I^* containing I. Now define a valuation v^* by

$$v^*(\phi) = \begin{cases} \text{TRUE} & \text{if } \phi \notin I^*, \\ \text{FALSE} & \text{if } \phi \in I^*. \end{cases}$$

Check that v really is a valuation; clearly $v(\Sigma) = \text{TRUE}$.

The Propositional Compactness Theorem is a more powerful tool than König's Infinity Lemma, allowing arguments to be extended to arbitrary infinite cardinality, as we'll see shortly. It is in fact less powerful than the Axiom of Choice: there are models of set theory in which AC fails but Propositional Compactness is true.

As an application, we extend (19.2.3) to arbitrary infinite graphs.

(19.3.5) Proposition. *Suppose that every finite subgraph of Γ has a vertex colouring with r colours. Then Γ has a vertex colouring with r colours.*

PROOF. We take the set

$$\{p_{x,i} : x \text{ a vertex of } \Gamma, \ i = 1, \ldots, r\}$$

of propositional variables. Let Σ be the set of formulae of the following types:
- for each vertex x of Γ, a formula asserting that $p_{x,i}$ is true for exactly one value of i;
- for each edge $\{x, y\}$ of Γ, a formula asserting that $p_{x,i}$ and $p_{y,i}$ are not true for the same value of i.

For example, if $r = 3$, these formulae would be

$$(p_{x,1} \vee p_{x,2} \vee p_{x,3}) \wedge (\neg(p_{x,1} \wedge p_{x,2}) \wedge \neg(p_{x,1} \wedge p_{x,3}) \wedge \neg(p_{x,2} \wedge p_{x,3}))$$

and

$$\neg(p_{x,1} \wedge p_{y,1}) \wedge \neg(p_{x,2} \wedge p_{y,2}) \wedge \neg(p_{x,3} \wedge p_{y,3})$$

respectively.

This set of formulae is satisfiable if and only if a vertex colouring with r colours exists. For, if v is a valuation making Σ true, then give vertex x the colour i if $v(p_{x,i}) = \text{TRUE}$; and conversely.

By assumption, any finite subset of Σ is satisfiable. For a finite subset involves the variables $p_{x,i}$ for only finitely many vertices x; these form a finite subgraph which can be coloured with r colours; use this colouring to define $v(p_{x,i})$ for vertices x in the subgraph, and define the other values arbitrarily.

So the Propositional Compactness Theorem gives the desired result.

19.4. Ramsey theory

The infinite form of Ramsey's Theorem can be stated as follows.

(19.4.1) Ramsey's Theorem (infinite form)

Suppose that k and r are positive integers, and let X be an infinite set. Suppose that the set of k-element subsets of X are partitioned into r classes. Then there is an infinite subset Y of X, all of whose k-element subsets belong to the same class.

For example, the case $k = 1$ is the infinite form of the Pigeonhole Principle (19.2.2). I will give a proof for $k = 2$; the general case is an exercise (with hints — Exercise 4).

We may suppose that X is countable, say $X = \{x_1, x_2, \ldots\}$. (Simply choose a sequence of distinct elements of X and use these.) Now we define a subsequence $y_1, y_2, \ldots$ of distinct elements, and a sequence of infinite subsets $Y_0, Y_1, Y_2, \ldots$ such that
(a) $Y_1 \supseteq Y_2 \supseteq \ldots$;
(b) $y_i \notin Y_i$, and all pairs $\{y_i, z\}$ for $z \in Y_i$ have the same colour;
(c) $y_j \in Y_i$ for all $j > i$.
This construction is done by induction, starting with $Y_0 = X$. In the ith step, choose $y_i \in Y_{i-1}$; observe that the infinitely many pairs $\{y_i, x\}$, for $x \in Y_{i-1} \setminus \{y_i\}$, fall into r disjoint 'colour classes'; so there is an infinite subset Y_i of $Y_{i-1} \setminus \{y_i\}$ for which (b) holds, by the Pigeonhole Principle.

At the conclusion of the inductive argument, we have arranged that the colour of a pair $\{y_i, y_j\}$ for $j > i$ depends only on i, not on j. Let c_i be this colour. By the Pigeonhole Principle again, there is an infinite subset M of the natural numbers such that c_i is constant for $i \in M$. Then $\{y_i : i \in M\}$ is the required infinite monochromatic set.

The finite version of Ramsey's Theorem can be deduced from the infinite, using König's Infinity Lemma. The argument is very similar to (19.2.3). We suppose that the finite version is false, for some choice of r, k, l; that is, for every positive integer n there is a colouring of the k-subsets of $\{1, \ldots, n\}$ with no monochromatic l-set. Let C_n be the set of such colourings. Form a digraph with vertex set $\bigcup_{n>0} C_n$, edges from C_n to C_{n+1} being defined by restriction just as before. König's Infinity Lemma gives us an infinite path, which tells us how to colour the k-subsets of the natural numbers without creating a monochromatic l-set, contrary to the infinite Ramsey Theorem.

The remainder of this section concerns possible infinite extensions or quantifications of Ramsey's Theorem. The proofs are sketched or omitted; you should regard it as a Project.

There are three natural ways in which we could try to extend Ramsey's Theorem:
(a) quantify the two infinities in the statement (as infinite cardinals);
(b) allow infinitely many colours;
(c) colour subsets of infinite size.
These three will be considered in turn.

(a) QUANTIFYING THE INFINITIES. For simplicity, we assume that $k = r = 2$. Here are one positive and one negative result.

(19.4.2) Theorem. (a) *Let a be an infinite cardinal. If $|X| > 2^a$, and the 2-subsets of X are coloured with two colours, there must exist a monochromatic set of cardinality greater than a.*
(b) *The 2-subsets of $\mathbb{R}$ can be coloured so that no uncountable set is monochromatic.*

(Since $|\mathbb{R}| = 2^{\aleph_0}$, part (b) says that the result of (a) is best possible for $a = \aleph_0$. In the notation of Chapter 10, $R(2, 2, \aleph_1)$ is the next cardinal after $2^{\aleph_0}$.)

I won't prove (a) — for the proof, which is not difficult, see for example *Ramsey Theory*, by R. L. Graham *et al.* (1990) — but the construction for (b) is quite easy. It depends on the following fact. Let a family (x_a) of real numbers indexed by ordinal numbers be given, and suppose that, if $a < b$, then $x_a < x_b$. Then the family is at most countable. For there is a 'gap' between x_a and the next number in the sequence, and this gap (an interval of $\mathbb{R}$) contains a rational number q_a. All these rationals are distinct. The result follows since there are only countably many rationals.

Now, by the Axiom of Choice, there is a bijection between $\mathbb{R}$ and an ordinal number. Let x_a be the real corresponding to the ordinal a. For $a < b$, colour $\{x_a, x_b\}$ red if $x_a < x_b$, blue if $x_a > x_b$. Now, according to the last paragraph, a monochromatic red set is at most countable; the same holds for a monochromatic blue set, by reversing the order of $\mathbb{R}$ in the argument.

(b) INFINITELY MANY COLOURS. There are two different directions possible here. The first is a simple extension, illustrated by the following negative result:

(19.4.3) Theorem. *The 2-subsets of a set of size 2^a can be coloured with a colours without creating a monochromatic triangle.*

PROOF. We take our set of size 2^a to be the set of all functions from the ordinal number a to $\{0, 1\}$. Now, for each $b \in a$, we colour the pair $\{f, g\}$ with colour b if b is the smallest point at which f and g disagree. Now there cannot be three functions pairwise disagreeing at the same point!

To motivate the other approach, we have to return to the basic philosophy of Ramsey theory, as expressed in the phrase 'complete disorder is impossible'. We expect that, if an infinite set carries an arbitrary colouring, there should be an infinite subset on which the colouring is particularly simple. With only finitely many colours, 'simple' has to mean 'monochromatic'; but in general there are other possibilities, for example, all the colours may be different! This leads to so-called 'canonical' forms of the theorems, first developed by Erdős and Rado. For example:

(19.4.4) Canonical Pigeonhole Principle. *If the elements of an infinite set are coloured with arbitrarily many colours, then there is an infinite subset in which either all the colours used are the same, or all the colours are different.*

This is clear because, if the first alternative fails, then each colour appears only finitely often, so infinitely many colours must be used; and using AC we can choose one point of each colour.

Erdős and Rado proved the canonical Ramsey theorem (sometimes called the *Erdős–Rado Canonisation Theorem*. Here is the formulation for $k = 2$.

(19.4.5) Erdős–Rado Canonisation Theorem, case $k = 2$. *Suppose that the 2-subsets of* $\mathbb{N}$ *are coloured with arbitrarily many colours. Then there is an infinite subset* Y *in which one of the following alternatives holds (where, in each pair* $\{x, y\}$, *we assume that* $x < y$):
- *all colours are equal;*
- $\{x, y\}$ *and* $\{u, v\}$ *have the same colour if and only if* $x = u$;
- $\{x, y\}$ *and* $\{u, v\}$ *have the same colour if and only if* $y = v$;
- *all colours are different.*

(c) COLOURING INFINITE SETS. The result here is wholly negative:

(19.4.6) Theorem. *For any infinite set* X, *there is a colouring of the countable subsets of* X *with no monochromatic subsets.*

PROOF. Let $\mathcal{P}_\omega(X)$ denote the set of countably infinite subsets of X. Define two equivalence relations on $\mathcal{P}_\omega(X)$ by
- $A \simeq B$ if $|A \triangle B|$ is finite;
- $A \sim B$ if $|A \triangle B|$ is finite and even;

where $A \triangle B$ is the symmetric difference of A and B.

Then each $\simeq$-class is the union of two $\sim$-classes, so that Y and $Y \setminus \{y\}$ belong to different $\sim$-classes for each $y \in Y \in \mathcal{P}_\omega(X)$. Choose one $\sim$-class in each $\simeq$-class and colour its members red; colour the other sets blue.

Nevertheless, mathematicians are reluctant to call this the end. Two developments are possible. Recognising that AC is used in that short proof, they look for positive results in set theory without AC; or they allow, not all colourings, but only those which are 'nice' with respect to some structure, such as Borel sets in a topological space.

19.5. Systems of distinct representatives

Hall's Condition is not sufficient for a SDR for a family of sets if finiteness is not assumed. Consider the following example: X_0 is the set of all *positive* integers, and $X_i = \{i\}$ for all positive integers i. Now $X(J) = J$ if $0 \notin J$, and $X(J)$ is infinite if $0 \in J$. But there is no SDR since, whichever number n we choose to represent X_0, there will be no possible representative for X_n.

However, of the two ways we could relax finiteness (allowing infinitely may sets, and allowing infinite sets), it is the second which is crucial to the failure of Hall's Theorem. This was shown by Marshall Hall, who proved the following result.

(19.5.1) Theorem. *Let* $\mathcal{A} = (A_i : i \in I)$ *be a family of finite sets, and suppose that* $|A(J)| \geq |J|$ *for all finite sets* J *of indices. Then the family* $\mathcal{A}$ *has a SDR.*

PROOF. The simplest proof uses the Propositional Compactness Theorem. Take a set of propositional variables $p_{i,x}$, for all choices of $i \in I$ and $x \in A_i$. Let Σ consist of all formulae of the following types:
- for each $i \in I$, a formula asserting that $p_{i,x}$ is true for exactly one $x \in A_i$;
- for each pair i, j of distinct indices, and each $x \in A_i \cap A_j$, the formula $(\neg(p_{i,x} \wedge p_{j,x}))$.

A valuation v satisfying Σ defines a SDR $(x_i : i \in I)$, by the rule that x_i is the unique element $x \in A_i$ for which $v(p_{i,x}) = \text{TRUE}$: the formulae of the second kind guarantee that the representatives are distinct.

Now, if Σ_0 is a finite subset of Σ, and J the set of indices i for which some $p_{i,x}$ is mentioned in Σ_0, then the subfamily $(A_i : i \in J)$ satisfies (HC), and so has a SDR; thus, there is a valuation satisfying Σ_0. Now the result follows by compactness.

REMARK. Hall's proof uses Zorn's Lemma directly and is considerably more complicated; you can read it in his book *Combinatorial Theory* (1989). There is a simpler
proof in the countable case, using König's Infinity Lemma (see Exercise 5).

A great deal of work has been done on necessary and sufficient conditions for
arbitrary families of sets to have SDRs.

19.6. Free constructions

One striking difference between finite and infinite combinatorics is that infinite
objects of some specail kinds are much easier to construct. There is plenty of room
to manœuvre; we just go on until the construction 'closes up'. A couple of examples
wiill illustrate this.

The first concerns projective planes (Chapters 7 and 9). A *projective plane* is an
incidence structure of points and lines, in which any two points are incident with a
unique line and any two lines with a unique point, and satisfying a mild condition
to exclude degenerate cases (there exist four points, no three collinear). All known
finite projective planes have a rich algebraic structure, depending ultimately on finite
fields. Infinite planes are not so restricted:

(19.6.1) Proposition. *Any infinite incidence structure of points and lines, in which
two points lie on at most one line, can be embedded into a projective plane.*

PROOF. We begin by adding some 'isolated' points if necessary, to ensure that there
are four points with no three collinear. Now perform a construction in stages as
follows:
- at odd-numbered stages, for each pair of points which are not collinear in the
  structure so far, add a line incident with just those two points;
- at even-numbered stages, for each pair of lines which are not concurrent in the
  structure so far, add a point incident with just those two lines.

Now, after progressing through the natural numbers, we take the structure
consisting of all points, lines, and incidences constructed. Given any two points,
there is a stage at which both have been added to the structure; not later than
the next stage, a line incident with both of them is added, and no further line
incident with both will ever appear. The dual assertions hold similarly. So we have
a projective plane.

For example, this 'free construction' produces planes which do not satisfy Desargues' Theorem (9.5.3). (Start with a 'broken Desargues configuration', the structure
shown in Fig. 9.2 with one 3-point line replaced by three 2-point lines.)

Obviously, the free construction is very flexible and can be adapted to produce
various other kinds of objects. Sometimes, however, countably many stages are
not enough, and we need the power of transfinite induction. Here is an example.
This concerns Steiner triple systems (Chapter 8). A *Steiner triple system* (STS) has
blocks of size 3 with any two points in a unique block; a *Steiner quadruple system*
(SQS) has blocks of size 4 with any three points in a unique block. Infinite Steiner
triple systems exist; for example, they can be produced by the free construction
(Exercise 7). If $\mathcal{D} = (Y, \mathcal{C})$ is a SQS and $y \in Y$, the *derived system* $\mathcal{D}_y$, with point set

$X = Y \setminus \{y\}$ and as blocks all those 3-sets B for which $B \cup \{y\}$ is a block of $\mathcal{D}$, is a STS (see Chapter 16). Conversely, which STS can be *extended* to a SQS?

(19.6.2) Proposition. *Any infinite STS can be extended to a SQS.*

PROOF. Let $(X, \mathcal{B})$ be a STS, which we propose to extend to a SQS $(Y, \mathcal{C})$ by adding a point $y \notin X$. Then we must have $Y = X \cup \{y\}$, and

$$C \supseteq \{B \cup \{y\} : B \in \mathcal{B}\};$$

indeed, the set on the right consists of all blocks in $\mathcal{C}$ which contain y.

An *n-arc* is a set of n points of X containing no block of $\mathcal{B}$. We see that any block not containing y must be a 4-arc; indeed, the set of all such blocks is a set of 4-arcs with the property that any 3-arc is contained in exactly one of them. So the extension problem is equivalent to the existence of such a set; and we propose now to construct one by transfinite induction. Note first that there are plenty of 4-arcs: given any 3-arc, all but three of the remaining points extend it to a 4-arc.

A short argument with cardinal numbers (Exercise 8) shows that the set of 3-arcs has the same cardinality as the set X of points. Let this cardinal be m (an initial ordinal), and index the 3-arcs as $(T_n : n < m)$. Now we perform the following construction, over stages indexed by the ordinal numbers up to m. We build a set $\mathcal{F}_n$ of 4-arcs for each $n \le m$ as follows:

- *Stage 0*: Set $\mathcal{F}_0 = \emptyset$.
- *Stage $n + 1$*: If T_n is contained in some 4-arc in $\mathcal{F}_n$, then set $\mathcal{F}_{n+1} = \mathcal{F}_n$. Suppose not. Then fewer than m 4-arcs have been put into $\mathcal{F}_n$, and they contain fewer than m points. Three more points fail to extend T_n to a 4-arc. So we can find a point x such that $T_n \cup \{x\} = F$ is a 4-arc and x lies in no member of $\mathcal{F}_n$. Thus, no 3-subset of F is contained in a member of $\mathcal{F}_n$. Set $\mathcal{F}_{n+1} = \mathcal{F}_n \cup \{F\}$.
- *Limit stage n*: let $\mathcal{F}_n = \bigcup_{l<n} \mathcal{F}_l$.

At stage m, we have ensured that every 3-arc lies in a unique member of $\mathcal{F}_m$, and the theorem is proved.

REMARK. Using techniques of logic, it can be deduced from (19.6.2) that only finitely many finite STS fail to be extendable. No exampes of non-extendable STS are known!

19.7. The random graph

I will end this chapter with what I confess is one of my favourite topics in combinatorics.

(19.7.1) Erdős–Rényi Theorem

There is only one countably infinite random graph.

Some explanation is called for. By a *random graph* I mean one produced by the following stochastic process. Fix a set X of vertices. For each 2-element subset

$\{x, y\}$ of X, toss a fair coin;[9] if it comes down heads, then join x and y by an edge, otherwise leave them unjoined. If X is finite, this procedure gives each (labelled) graph the same (non-zero) chance of being picked. Moreover, if we are interested in unlabelled graphs, we can see that the more symmetric a graph is, the less its chance of occurring. [The symmetric group $\mathrm{Sym}(X)$ acts on the set of labelled graphs; its orbits are the isomorphism classes (the unlabelled graphs), and the stabiliser of a graph Γ is its automorphism group $\mathrm{Aut}(\Gamma)$. By (14.3.4), the product of the probability of a given unlabelled graph and the order of its automorphism group is $n!/2^{n(n-1)/2}$, where $n = |X|$.][10] By contrast,

> there is a countably infinite graph R such that, with probability 1, a random countably infinite graph is isomorphic to R. Moreover, R has a very large automorphism group.

It is my contention that this illustrates an important difference between mathematics and virtually all other subjects. In no other field could such an apparently outrageous claim be made completely convincing by a short argument, as I propose to give. The claim also illustrates that our intuition about the infinite is likely to be caught out very often.

Probability theory (or measure theory) for infinite spaces resembles the familiar finite theory, with a few additions. The significant one here is the concept of a *null event* (or *null set*), one with probability zero. If an event E has the property that, for any $\epsilon > 0$, there is an event $E_\epsilon \supseteq E$ with probability $\mathrm{Prob}(E_\epsilon) \le \epsilon$, then E is null ($\mathrm{Prob}(E) = 0$). It is an easy exercise to show that the union of a countable set of null events is null. [Suppose that E_n is null for all $n \ge 1$. Given $\epsilon > 0$, choose $E_{n,\epsilon}$ containing E_n with $\mathrm{Prob}(E_{n,\epsilon}) \le \epsilon/2^n$, and set $E_\epsilon = \bigcup_{n \ge 1} E_{n,\epsilon}$. Then $\mathrm{Prob}(E_\epsilon) \le \epsilon$, and $\bigcup_{n \ge 1} E_n \subseteq E_\epsilon$.]

Now we begin on the proof. It depends on the following property $(*)$, which a graph may or may not have:

> Given any two finite disjoint sets U, V of vertices, there exists a vertex z joined to every vertex in U and to no vertex in V.

The Erdős–Rényi Theorem follows from the following two assertions:

1. *With probability 1, a random countable graph satisfies* $(*)$.

2. *Up to isomorphism, there is a unique countable graph which satisfies* $(*)$.

PROOF OF 1. We have to show that the event that $(*)$ fails is null. Now there are only countably many pairs (U, V) of disjoint finite sets of vertices; so it is enough to prove that, for a *fixed* choice of U and V, the probability that no vertex z exists satisfying the conditions is zero. Call a vertex *good* if it is joined to everything in

---

[9] In the language of probability theory, tosses of a fair coin are independent, and each outcome of a toss has probability $\frac{1}{2}$ of occurrence.

[10] Finite random graphs are not as unstructured as this discussion might suggest; global patterns arise from the local chaos. This will be discussed in the next chapter.

U and nothing in V, and *bad* otherwise. Any vertex z is most likely to be bad; the probability of this is $1 - \frac{1}{2^n}$, where $n = |U \cup V|$. But there are infinitely many vertices, and the events that they are bad are all independent. So the probability that vertices $z_1, \ldots, z_N$ are all bad is $(1 - \frac{1}{2^n})^N$. Since this tends to zero as $N \to \infty$, the assertion is proved.

PROOF OF 2. This illustrates a logical technique called *back-and-forth*. Suppose that Γ and Δ are two countably infinite graphs satisfying $(*)$, with vertex sets $X = \{x_1, x_2, \ldots\}$ and $Y = \{y_1, y_2, \ldots\}$ respectively. We build, in stages, an isomorphism θ between them, as follows. At the beginning of any stage, the value of θ has been determined on finitely many points of X.

At an odd-numbered stage, let x_n be the first point of X on which θ has not been defined (that is, the point with lowest index). Let U' and V' be the (finite) sets of neighbours and non-neighbours of x_n respectively on which θ has been defined. In order to extend θ to x_n, we must find a point $z \in Y$ which is joined to every point of $\theta(U')$ and to no point of $\theta(V')$. Since Δ satisfies $(*)$, such a point z exists; choose one (for definiteness, the one with lowest index), and set $\theta(x_n) = z$.

At an odd-numbered stage, let y_m be the first point of Y not in the range of θ. Argue as above, using the fact that Γ satisfies $(*)$, to find a suitable pre-image of y_m.

After countably many stages, we have ensured that every point of X is in the domain of θ, and every point of Y is in the range. (This is the point of going back-and-forth; if we only went 'forth', we would define a one-to-one map but couldn't guarantee it to be onto.) Moreover, θ is clearly an isomorphism, and we are done.

The name R stands for 'random graph'. The proof we have given is an existence proof; if an event (such as property $(*)$) occurs with probability 1, then it certainly occurs, so there exists a graph with this property; assertion 2 shows its uniqueness.[11] For Erdős and Rényi, an existence proof was enough; but an explicit construction is more satisfactory. R can be produced by a variant of the 'free construction' of the preceding section: at each stage, add vertices fulfilling all instances of $(*)$ where U and V consist of previously constructed vertices. But one can be even more definite. A direct construction was given by Rado, whose name is also commemorated by the letter R.

Rado took the vertex set to be the set of non-negative integers. Given x and y, where $x < y$, to decide whether to join x to y, we express the larger number y to base 2; that is, we write it as a sum $\sum_{z \in X} 2^z$ of distinct powers of 2. If 2^x is one of these powers (that is, if $x \in X$), then join x to y; otherwise, don't join. Property $(*)$ is easy: adding an element to U if necessary, we can assume that $\max(U) > \max(V)$; then $z = \sum_{u \in U}$ has the required property.

The graph R has many nice properties. I will describe two of these, known

---

[11] This shows that, paradoxically, probability theory is an important tool in proving the existence of objects. Exercise 12 gives another instance, and a finite example was given in Chapter 10. A related concept in topology, *Baire category*, has similar uses. See J. C. Oxtoby, *Measure and Category* (1980), for many entertaining illustrations.

as *universality* and *homogeneity*. A graph is said to be *universal* if it satisfies the conclusion of the next result.

(19.7.2) Proposition. *Any finite or countable graph is an induced subgraph of R.*

PROOF. We use the machinery of back-and-forth, but going forth only. In other words, take the graph Δ to be R (i.e., to have property $(*)$), and let Γ be any finite or countable graph. Proceeding only from Γ to Δ (as in the 'odd-numbered steps' before), we construct a one-to-one map from Γ to Δ whose image is an induced subgraph of Δ isomorphic to Γ.

A graph Γ is said to be *homogeneous* if the following condition holds:

> Let ϕ be any isomorphism between finite induced subgraphs of Γ. Then there is an automorphism θ of Γ which extends ϕ.

(19.7.3) Proposition. *R is homogeneous.*

PROOF. This is again proved by back-and-forth. We take the two graphs Γ and Δ to be equal to R, but modify the start of the construction: instead of starting with no information about θ, we take its initial value to be the given map ϕ. Then the argument produces an isomorphism from Γ to Δ (that is, an automorphism of R) which agrees with ϕ on its domain.

It follows that the automorphism group of R is infinite. For let the vertices be $\{x_1, x_2, \ldots\}$. Since all 1-vertex induced subgraphs are isomorphic, there is an automorphism θ_n mapping x_1 to x_n for each n. In fact this group has cardinality $2^{\aleph_0}$, the same as that of the symmetric group on a countable set; see Exercise 12.

It can be shown that any countable homogeneous graph which contains all finite graphs as induced subgraphs is necessarily isomorphic to R. A much more difficult result is a theorem of Lachlan and Woodrow which determines all countable homogeneous graphs.

19.8. Exercises

1. Prove that the set of finite subsets of a countable set is countable, but that the set of all subsets is not.

2. (a) Use König's Infinity Lemma to show that every countable poset has a linear extension.

(b) Use the Propositional Compactness Theorem to show that every poset has a linear extension.

3. Prove the Axiom of Choice, assuming Zorn's Lemma. [HINT: consider the set of partial choice functions for a family of sets, ordered by inclusion.]

4. Prove the infinite Ramsey Theorem for all k. [HINT: The proof is by induction on k. Follow the argument given, but replacing condition (b) by

(b') $y_i \notin Y_i$, and for all $(k-1)$-subsets Z of Y_i, the sets $\{y_i\} \cup Z$ have the same colour.

(Use Ramsey's Theorem with $k-1$ replacing k to construct Y_i.) Now, in the constructed sequence $\{y_1, y_2, \ldots\}$, the colour of a k-set depends only on its element y_i with smallest index i. The final application of the Pigeonhole Principle is essentially the same.]

5. Use König's Infinity Lemma to prove the countable version of Hall's Theorem for families of finite sets (19.5.1).

6. Why do König's Infinity Lemma and the Propositional Compactness Theorem allow us to prove the finite Ramsey Theorem from the infinite, but the infinite Four-colour Theorem from the finite?

7. Modify the free construction of (19.6.1) to produce infinite Steiner triple systems.

8. Prove that the number of 3-arcs in an infinite STS is equal to the number of points. [HINT: $a^3 = a$ for all infinite cardinals a.]

9. Show that, in an infinite projective plane, the (cardinal) number of points on any line is equal to the total number of points. Hence show that any infinite projective plane contains a set S of points such that $|S \cap L| = 2$ for all lines L.

10. Prove that a countably infinite graph Γ is a spanning subgraph of R if and only if Γ satisfies the following condition:

for any finite set V of vertices, there is a vertex z joined to no vertex in V.

11. (a) Prove that R is isomorphic to its complement.

(b) Prove that R is isomorphic to $R - v$ for any vertex v, and to $R - e$ for any edge e. (In other words, R is immune to any finite amount of tampering.)

12. Let S be a set of positive integers. Let $\Gamma(S)$ be the graph with vertex set $\mathbb{Z}$, in which x and y are joined if and only if $|x - y| \in S$.

(a) Prove that the map $x \mapsto x + 1$ is an automorphism of $\Gamma(S)$, permuting all the vertices in a single infinite cycle (a *cyclic automorphism*). (In the language of Section 14.7, $\Gamma(S)$ is the Cayley graph of the additive group of $\mathbb{Z}$ with respect to the set S.)

(b) Choose S at random by tossing a fair coin for each positive integer n, putting $n \in S$ if the result is heads and not otherwise. Prove that, with probability 1, $\Gamma(S)$ is isomorphic to R.

(c) Deduce that R has a cyclic automorphism.

(d) Show that an event with probability 1 is uncountable, and deduce that the automorphism group of R is uncountable. (Read 'cardinality $2^{\aleph_0}$' for 'uncountable' here and try the resulting harder problem.)

20. Where to from here?

> This kind of rather highflown speculation is an essential part of my job. Without some capacity for it I could not have qualified as a Mobile, and I received formal training in it on Hain, where they dignify it with the title of Farfetching.
>
> Ursula K. LeGuin, *The Left Hand of Darkness* (1969)

This final chapter has two purposes. A few topics not considered earlier are discussed briefly; usually there is a central problem which has served as a focus for research. Then there is a list of assorted problems in other areas, and some recommended reading for further investigation of some of the main subdivisions of combinatorics.

20.1. Computational complexity

This topic belongs to theoretical computer science; but many of the problems of greatest importance are of a combinatorial nature. In the first half of this century, it was realised that some well-posed problems cannot be solved by any mechanical procedure. Subsequently, interest turned to those which may be solvable in principle, but for which a solution may be difficult in practice, because of the length of time or amount of resources required. To discuss this, we want a measure of how hard, computationally, it is to solve a problem. The main difficulty here lies in defining the terms!

PROBLEMS.

Problems we may want to solve are of many kinds: anything from factorising a large number to solving a system of differential equations to predict tomorrow's weather. In practice, we usually have one specific problem to solve; but, in order to do mathematics, we must consider a class of problems.

For example, from a mathematician's point of view, finding a winning strategy for chess is trivial, since there are only finitely many configurations to consider. (The laws of chess put an upper bound on the number of moves in a game, and the number of possibilities at each move is clearly finite.) So mathematicians define 'generalised chess' played on an $n \times n$ board.

For illustration, we consider the following class of problems, known by the term HAMILTONIAN CIRCUIT: given a graph Γ, does it have a Hamiltonian circuit? We saw in Section 11.7 that there is a trivial algorithm which solves this problem, but it is extremely inefficient!

Obviously the 'complexity' if the problem depends on the size of the input data — bigger graphs will pose harder problems, in general — so we need first a measure of the size of the data. We use an information-theoretic measure: the number of bits of information needed to present the data. For example, a graph with n vertices can be encoded as n^2 bits, as follows: number the vertices as $v_1, v_2, \ldots, v_n$; then let

$$a_{ij} = \begin{cases} 1 & \text{if } v_i \text{ is joined to } v_j \text{ by an edge;} \\ 0 & \text{otherwise.} \end{cases}$$

(This is essentially the adjacency matrix.) Then represent the graph by the sequence

$$a_{11}a_{12} \ldots a_{1n}a_{21}a_{22} \ldots a_{2n} \ldots a_{n1}a_{n2} \ldots a_{nn}.$$

Note that this method is somewhat redundant: we know that $a_{ii} = 0$ for all i (no vertex is joined to itself), and $a_{ij} = a_{ji}$ for all i, j (joins are not directed), so we could encode the same information in half the space. We will not strive for the most efficient representation! This leads to an important principle: Our complexity measures should be such that (within reason)

> *different representations of the input data don't change the complexity of a problem.*

Rather than defining 'within reason', I'll illustrate with a data representation which is unreasonably wasteful. Consider the problem PRIME: given an integer N, decide whether it is prime. The integer N could be given as a sequence of N ones; but this is ridiculous, since the base 2 representation of N uses only $1 + \lceil \log_2 N \rceil$ binary digits.

We make one further simplification: we consider only problems with a simple yes-no answer, so-called *decision problems*. The HAMILTONIAN CIRCUIT problem is of this form, as is PRIME. A more general problem can be reduced to several decision problems using the 'twenty questions' principle (Chapter 4). For example, the problem 'What is the longest circuit in the graph G?' can be solved by a number of instances of 'Does G have a circuit of length $\geq k$?'; only $\lceil \log_2 n \rceil$ questions are required, where n is the number of vertices.

RESOURCES.

The complexity of a problem should be a measure of the computational resources needed to solve it.

There are various resources which have been considered: time, memory space, number of processors (in a parallel processing system), etc. In practice, time is usually the limiting factor, and I will consider only this one.

Of course, different computers run at different speeds, and we have to allow for this. We standardise by taking the unit of time to be that required for the processor to carry out one operation. The effect of processor speeds is not so significant. For example, if a computation takes 10^{30} processor cycles to perform, it doesn't matter whether the computer runs at 1 or 1000 million cycles per second. (There are fewer than 10^9 seconds in a year; and the universe is fewer than 10^{10} years old, according to current theory.)

Different processors can carry out different amounts of work in a single cycle. Again, this dictates that our complexity measure should not be too precise:

> *processor details should not change the complexity of a problem.*

The theoretical analysis is based on a *Turing machine*, almost the most primitive machine imaginable. It consists of a 'tape' on which information can be written, extending infinitely far in both directions (but having all but a finite amount blank), and a 'head' positioned over the tape so that it can read or write to one location on the tape. (The tape and head correspond to the memory and CPU of a real computer.) The head can also be in one of a finite number of 'internal states', and each tape location can have one of a finite number of 'symbols' (including 'blank') written on it. In one cycle, the machine can write a symbol on the tape, change its internal state, and move one position left or right.

The details are not too important. What is important is that[1]

- any computation possible on any machine (theoretical or practical) can be performed by a Turing machine; and
- any processor ever made can in a *single* cycle perform only the equivalent of a bounded number of Turing machine steps.

Thus we define the *complexity* of a class C of problems to be the function f defined by

$$f_C(n) = \begin{cases} \text{the least } m \text{ such that a Turing machine can solve} \\ \text{any instance of } C \text{ whose data consists of } n \text{ bits in} \\ \text{at most } m \text{ steps.} \end{cases}$$

We call two classes C_1 and C_2 *equivalent* if there is a polynomial $p(x)$ such that $f_{C_2}(n) \le f_{C_1}(p(n))$ and $f_{C_1}(n) \le f_{C_2}(p(n))$. This definition encompasses our principles that different data representations and different processor details should not alter the complexity of a problem (that is, the resulting complexity measures should give equivalent results). So all classes C for which $f_C(n)$ is bounded by a polynomial in n are equivalent, but are not equivalent to problems which take exponentially long to solve.

ALGORITHMS.

We haven't specified *how* the problem is to be solved. The definition of complexity presupposes that the most efficient algorithm is used. This means that upper bounds on complexity are much easier to prove than lower bounds. To show that the complexity of C is *at most* $F(n)$, we just have to exhibit an algorithm which solves any instance of C of size n in at most $F(n)$ steps. But to show that the complexity is *at least* $F(n)$ is much tougher; we have to prove that **no** algorithm can exist which takes fewer than $F(n)$ steps.

Consider the problem PRIME, for instance. Remember that an integer N is to be input in base 2 representation, so that if the input has size n, then N may be as large as $2^n - 1$. If we use 'trial division', checking all numbers up to $\sqrt{N}$ to see if they divide N, we will take at least $2^{n/2}$ steps: exponentially many! Using much more elaborate number theory, it has been shown that the complexity of PRIME doesn't exceed $n^{c \log \log n}$. Conceivably, it is polynomial in n.

---

[1] Alan Turing would have argued that these statements hold true for the human brain as well as any artificial machine.

THE CLASS **P**.

We say that the class C of problems is *polynomial-time solvable*, or *belongs to* **P**, if its complexity is not greater than a polynomial function of n. As we saw earlier, this property does not depend on details about data representation or processors.

To show that a class C belongs to **P**, we have to describe an algorithm which solves problems in C in polynomial time. This has been done in a number of instances, some of which are not at all obvious. For example:

(a) Given integers M and N, does M divide N? The primary-school long division algorithm decides this in polynomial time.

(b) Given a graph G, is it connected? We saw an algorithm for this problem in Section 11.11; it runs in polynomial time.

(c) The greedy algorithm for a minimal connector (Section 11.3) is polynomial.

(d) LINEAR PROGRAMMING. This is traditionally solved by the 'simplex method'. Though this is efficient in practice, there are some contrived problems which it takes exponentially long to solve. In the last decade, Khachiyan found a different algorithm (the 'ellipsoid method') which runs in polynomial time. Subsequently Karmarkar found another polynomial-time algorithm.

Current 'received wisdom' is that the problems in C are 'tractable' if C belongs to **P**, and are 'intractable' otherwise. (In fact, properties which are not quite polynomial, such as PRIME, are regarded as 'tractable' as well. Large numbers are routinely tested for primality by known algorithms of complexity $n^{c \log \log n}$. For $n = 1000$, that is, numbers with about 300 decimal digits, $\log \log n$ is only 7.742.)

THE CLASS **NP**.

There is an important type of problem for which no polynomial-time algorithms are known; but, if a solution is proposed, then its validity can be checked quickly.

Imagine that you are a travelling salesman with a briefcase full of Hamiltonian graphs. Your customers don't have a quick way of deciding the HAMILTONIAN CIRCUIT problem — if they did, they wouldn't be your customers — but they want to buy graphs with Hamiltonian circuits. You show them a graph, and tell them a Hamiltonian circuit in the graph; they can easily (meaning 'in polynomial time') check that your claim is correct.

A class C of decision problems is said to *belong to* **NP** if, for any problem in the class for which the answer is 'yes', there is a 'certificate', a piece of information using which it is possible to verify the correctness of the answer in polynomial time. Thus, an explicit Hamiltonian circuit is a certificate for HAMILTONIAN CIRCUIT, showing that it belongs to **NP**.

The letters **NP** stand for 'non-deterministic polynomial', deriving from another way of viewing this concept. A class C belongs to **NP** if a problem in C of size n with answer 'yes' can be solved in time which is polynomial in n by a program which is allowed to make some lucky guesses. (You can find a Hamiltonian circuit in polynomial time by guesswork, if there is one and you are lucky!)

The class **P** is contained in **NP**: problems in **P** can be solved quickly without recourse to certificates or guesswork. Many other classes of problems are in **NP**: HAMILTONIAN CIRCUIT (as we've seen), GRAPH ISOMORPHISM (deciding whether two graphs are 'the same'), SATISFIABILITY (does a Boolean formula take the value

TRUE for some choice of values for its variables?), DECODING A LINEAR CODE. (Encoding a linear code is in **P**, since it is simple linear algebra.) It is true, though not obvious, that PRIME belongs to **NP**. But the opposite problem, COMPOSITE (i.e., is N composite?) is clearly in **NP**. (A certificate for the compositeness of N is a proper factor M of N: we saw that divisibility can be checked in polynomial time.) The main open problem is:

Is **P** $\neq$ **NP**?

NP-COMPLETENESS.
We say that a class C_1 is *reducible* to a class C_2 if, given a problem in C_1 (with data of size n), we can compute the data for a problem in C_2 with the same answer, in a time which is polynomial in n. Thus, if C_1 is reducible to C_2, and if C_2 belongs to **P** (or **NP**, respectively), then so does C_1. Intuitively, it means that problems in C_1 are no harder than those in C_2.

Stephen Cook proved in 1969 that **NP** contains a class C of problems such that any class in **NP** is reducible to C. The class he gave was SATISFIABILITY of Boolean formulae. A class with Cook's property consists of 'the hardest problems in **NP**', in the sense that if a polynomial-time algorithm for such a class were ever found, then it would follow that every problem in **NP** would have a polynomial-time solution, that is, that **P** = **NP**. Such a class is called **NP**-*complete*.

Since Cook's work, hundreds of classes have been shown to be **NP**-complete, including HAMILTONIAN CIRCUIT and DECODING A LINEAR CODE.

In summary, then, we regard a class of problems as 'easy' if it lies in **P**, or nearly so; and as 'hard' if it is at least as hard as an **NP**-complete class. (There are problems which are much harder than anything in **NP**; typical examples are finding winning strategies in positional games, where we have to consider each possible response of our opponent, each response we could make to it, each response of our opponent to our move, and so on — the branching tree of possibilities require exponential time to analyse.) Other problems, notably GRAPH ISOMORPHISM, are in **NP** but not known to be either in **P** or **NP**-complete; it is thought that they may lie strictly between these two classes.

The standard reference on **P** and **NP** is *Computers and Intractability: A Guide to the Theory of NP-Completeness* by M. R. Garey and D. S. Johnson (1979), which lists hundreds of problems (many of them combinatorial) together with their classification as in **P**, **NP**-complete, neither, or 'don't know'.

20.2. Some graph-theoretic topics

This section presents thumbnail sketches of three topics in graph theory (reconstruction, higher regularity conditions, and random graphs), which haven't been mentioned yet.

GRAPH RECONSTRUCTION.
The *reconstruction problem* for graphs (in two versions, one for vertices and one for edges) has the fascination of a long-standing open problem, and also has unexpected links with other topics. In its original, vertex form, the problem is as follows. Given

a graph G with n vertices, construct a 'deck of cards'; the ith card carries a drawing or specification of the subgraph of G obtained by deleting the ith vertex of G, for $i = 1, \ldots, n$. Now we ask: can G be reconstructed, up to isomorphism, from the information provided by the deck of cards? Graphs on two vertices cannot be reconstructed in this way, since each of them has a deck of two cards, each with a 1-vertex graph on it. However, the *vertex-reconstruction conjecture* asserts that any graph with more than two vertices is reconstructible.

More formally, call graphs G and H *hypomorphic* if there is a bijection ϕ from the vertex set of G to that of G such that $G - v$ and $H - \phi(v)$ are isomorphic, for each vertex v of G. The conjecture asserts that hypomorphic graphs with more than two vertices are isomorphic.

The problem is open, but many partial results exist. On one hand, it is known to be true for many particular classes of graphs (disconnected graphs, trees, regular graphs, etc.). On the other, many properties are known such that, if two graphs are hypomorphic and one has the property in question, then so does the other (number of induced subgraphs of a particular kind, existence of a Hamiltonian circuit, etc.).

It is known that the vertex-reconstruction conjecture fails for directed graphs; infinitely many pairs of digraphs are known which are hypomorphic but not isomorphic.[2]

There is also an *edge-reconstruction conjecture*, in which the information given is the deck of edge-deleted subgraphs. It is also open. The largest known counterexample is the pair of graphs G, H on four vertices, where G consists of a triangle and an isolated vertex, and H is the 'star' $K_{1,3}$. There are many partial results, of which the strongest is the theorem of Lovász and Müller, which shows that the conjecture is true for the vast majority of graphs:

(20.2.1) Theorem. *A graph with n vertices and more than $n \log_2 n$ edges is edge-reconstructible.*

The edge-reconstruction conjecture can be formulated in the language of permutation groups; the result of (20.2.1) extends to a general theorem about permutation groups.

HIGHER REGULARITY CONDITIONS.
A graph G is said to be *strongly regular*, or SR, with parameters (n, k, λ, μ), if the following conditions hold:
- G has n vertices;
- G is regular with valency k;
- any two adjacent vertices have exactly λ common neighbours;
- any two non-adjacent vertices have exactly μ common neighbours.

We have seen some examples already. The Petersen graph is SR, with parameters $(10, 3, 0, 1)$; indeed, any Moore graph of diameter 2 is SR (Section 11.12). The complete bipartite graph $K_{k,k}$ is SR with parameters $(2k, k, 0, k)$. Many other examples exist. On the other hand, various necessary conditions are known for the

---

[2] The existence of a finite number of counterexamples is not regarded as invalidating a conjecture of this kind, merely as modifying its statement.

quadruple (n, k, λ, μ) to be parameters of a SR graph. A simple counting argument shows that $k(k - \lambda - 1) = (n - k - 1)\mu$. (Double-count edges joining a neighbour of v to a non-neighbour.) Another powerful condition comes from linear algebra, by the argument we used for Moore graphs in Section 11.12. However, there is still a wide gap between the known necessary conditions and the sufficient conditions (arising from explicit constructions). As a sample question, it is not known whether or not a SR graph with parameters $(99, 14, 1, 2)$ exists.

Strongly regular graphs are closely connected with topics in finite geometry (nets, partial geometries), design theory, Euclidean geometry, permutation groups, and a number of other areas.[3]

The definition can be unified and strengthened. For a positive integer t, we say that the graph G is *t-tuple regular* if, for any set S of at most t vertices, the number of common neighbours of the vertices in S depends only on the subgraph induced on S. For $t = 1$ and for $t = 2$, this condition reduces precisely to regularity and strong regularity, respectively; and the condition becomes stronger as t increases. For large t, we have the following:

(20.2.2) Theorem. *Let G be 5-tuple regular. Then G is one of the following: a disjoint union of complete graphs of the same size; a regular complete multipartite graph (i.e., the complement of the preceding); a pentagon; or the line graph of $K_{3,3}$. All these graphs are t-tuple regular for all t.*

Another variant is a weakening of the condition of strong regularity, to *distance-regularity*. A connected graph G is distance regular if, given integers j and k, the number of vertices at distance j from vertex v and k from vertex w depends only on the distance i between v and w. (In fact, only a small subset of these conditions are required to guarantee the whole set.) A connected graph is strongly regular if and only if it is distance-regular of diameter 2. However, it seems that distance-regular graphs of large diameter are not so common, and there is some hope of a classification of these. See the book *Distance-Regular Graphs* by A. E. Brouwer *et al.* (1989) for further information.

RANDOM GRAPHS.

Twice already (in Chapters 10 and 19) we've met the notion of a random graph, whose edges are selected independently with probability $\frac{1}{2}$ (so that all labelled graphs are equally likely, if the number of vertices is finite). To develop a theory, we need more flexibility! Two models are commonly used. Let n denote the number of vertices.

FIRST MODEL. We choose edges independently with probability p, for some fixed p with $0 < p < 1$.

SECOND MODEL. We specify the number m of edges of the graph, and choose the set of edges from the $\binom{n(n-1)/2}{m}$ possible m-sets (all such sets equally likely).

We examine the behaviour of a 'typical' graph as $n \to \infty$, where p or m is equal to a prescribed function of n. Of course, the two models are not the same; the

---

[3] A surprising recent connection is with the theory of knot polynomials.

second always has exactly m edges, whereas the number of edges in the first has a binomial distribution with mean $pn(n-1)/2$. Nonetheless, it is not too surprising that these models behave quite similarly, if m and p are related by $m = pn(n-1)/2$. For the sake of exposition, I'll use the second model.

One feature of the theory is that, for various properties P of graphs, there are sharp thresholds. In other words, there is a function f such that, if m is a bit less than $f(n)$, then *almost no* graphs have property P (that is, the proportion of graphs on n vertices which satisfy P tends to 0 as $n \to \infty$), and if m is a little greater than $f(n)$, then *almost all* graphs (a proportion tending to 1) satisfy P. We say that P holds *almost surely* if almost all graphs have P.

Two basic results illustrate these ideas. In both results, we consider random graphs with n vertices and m edges, according to the second model.

(20.2.3) Proposition. *Suppose that* $m \sim cn$.

(a) *If* $0 < c < \frac{1}{2}$, *then almost all graphs have the property that almost all components are trees or unicyclic, the largest component having approximately* $\log n$ *vertices.*

(b) *If* $c = \frac{1}{2}$, *then almost surely the largest component has about* $n^{2/3}$ *vertices.*

(c) *If* $c > \frac{1}{2}$, *then almost surely the largest component has size about* $c'n$, *for some constant* c' *(depending on* c*).*

(20.2.4) Proposition. *Suppose that* $m \sim cn \log n$.

(a) *If* $0 < c < \frac{1}{2}$, *then almost all graphs are disconnected.*

(b) *If* $c > \frac{1}{2}$, *then almost all graphs are Hamiltonian.*

Many similar results are known, and the 'sharpness' of the thresholds has been greatly improved. One helpful way of describing the results is in terms of the 'evolution' of a random graph, as the number of edges is gradually increased. The existence of thresholds shows that the process in some way resembles the 'punctuated equilibrium' model of biological evolution,[4] where short periods of rapid change are interspersed with long stretches of relative uniformity.

Bollobás' book *Random Graphs* (1985) gives a detailed account.

20.3. Computer software

An essential part of the training of a statistician or numerical analyst involves the use of 'standard' computer software packages. A few years ago, I was asked what the equivalent packages in combinatorics were. I answered, 'C or Pascal'.

Today I would still give that answer, though now more by taste than by necessity. In a general-purpose programming language, you can do anything; and you do not pay the price in overheads associated with translation from one language into another, or with 'user-friendliness'.

[If you do combinatorial computing in a general-purpose language, it is very important to remember Wirth's dictum, 'Algorithms + Data Structures = Programs'. The data structures you will use (perhaps large integers, partitions, permutations,

---

[4] Stephen Jay Gould, *Ever Since Darwin* (1977).

trees, graphs, families of sets) are not usually well represented by the built-in data structures of the language (perhaps small integers, floating-point reals, characters, and strings of characters). Time spent on designing well-adapted data structures will not be wasted.]

Now it is increasingly common to find specialised systems which are useful in combinatorics. This is particularly true for the material of Chapter 14. The Schreier–Sims algorithm for getting information about the group generated by a set of permutations is quite sophisticated, and can be integrated in a system where it might use the output from an algorithm for finding generators for the automorphism group of a graph, or from the *Todd–Coxeter algorithm* (which takes as input generators and relations for a group G, and generators for a subgroup H, and returns permutations generating the action of G on the coset space $G : H$). The output from the Schreier–Sims algorithm might itself be subjected to further group-theoretic analysis. Two integrated systems (both of which have much wider capabilities) are the long-established CAYLEY (and its successor MAGMA), and the newer and smaller GAP. These algorithms are also making their way into more conventional 'computer algebra' systems.

Various packages for combinatorial optimisation are available. Often, these are centred around linear programming. However, dramatic new algorithms for finding approximate solutions to 'hard' (e.g. **NP**-complete) optimisation problems, such as *simulated annealing* and *genetic algorithms*, are becoming available.

Finally, I should mention the language ISETL, designed for the purpose of teaching discrete mathematics: see Baxter, Dubinsky and Levin, *Learning Discrete Mathematics with ISETL* (1989). This language handles large integers, sets, sequences, functions, etc., with a syntax almost identical to that used by a mathematician.[5] It is very easy to learn (no type declarations are required), and is freely available on a wide range of personal computers and operating systems.

20.4. Unsolved problems

This section presents some further problems which have guided the direction of research in the past. Unlike earlier chapters, you are not expected to solve all of these.

GRAPH COLOURING.
Despite the work of Appel and Haken, and of Robertson and Seymour, this area still abounds with hard problems. Here are three:

- The *Strong Perfect Graph Conjecture*: a graph Γ and all its induced subgraphs have clique number and chromatic number equal if and only if neither G nor its complement contains an induced cycle of odd length greater than 3 (see Section 18.4).

---

[5] For example, in ISETL, one can define the Cartesian product of sets A and B to be

```
{ [x,y] : x in A, y in B }.
```

- *Hadwiger's Conjecture*: a graph with chromatic number n has K_n as a minor. [This would imply the Four-Colour Theorem, since the class of planar graphs is minor-closed and K_5 is not planar.]
- The *List Colouring Conjecture*: Let G have edge-chromatic number n. Suppose that S is any set of 'colours' and, for each edge e of G, a list $L(e)$ of n elements of S is given. Then G can be edge-coloured using S, so that the colour of any edge e belongs to $L(e)$.

EXTREMAL SET THEORY.

Rather than mention specific problems, I will describe how a number of questions can be put into this framework.

If we identify a binary word of length n with the subset of $\{1,\ldots,n\}$ of which it is the characteristic function, then the 'main problem' of coding theory over the binary alphabet (Section 17.4) takes the form: *Given n and d, what is the largest family $\mathcal{F}$ of subsets of $\{1,\ldots,n\}$ such that $|F_1 \triangle F_2| \geq d$ for all $F_1, F_2 \in \mathcal{F}$ with $F_1 \neq F_2$?*

A permutation π of $\{1,\ldots,n\} = N$ can be regarded as a subset $S(\pi)$ of the square array $N \times N$ containing exactly one element from each row or column. The number of fixed points of $\pi_1 \pi_2^{-1}$ is then $|S(\pi_1) \cap S(\pi_2)|$. So 'metric' questions about permutations can be phrased in terms of families of sets of this special form.

DESIGN THEORY.

Asking for less than a complete list of parameters of t-designs, we could pose the problems:

- Do Steiner systems $S(t, k, v)$ (or t-$(v, k, 1)$ designs) exist for all t?
- Is there a projective plane whose order is not a prime power?
- Is there a Hadamard matrix of every order divisible by 4?

One can define q-analogues of t-(v, k, λ) designs: the blocks are k-dimensional subspaces of a v-dimensional vector space over $\mathrm{GF}(q)$, and any t-dimensional subspace lies in exactly λ blocks. Do non-trivial q-ary t-designs (without repeated blocks) exist for all t? Or even for $t = 4$?

POSETS.

A question that continues to tantalise is the $\frac{1}{3}$-$\frac{2}{3}$ *problem*. If x and y are incomparable elements of the poset P, let $\pi(x, y)$ be the proportion of linear extensions of P in which $x < y$. Is it true that every poset which is not a chain contains elements x and y with $\frac{1}{3} \leq \pi(x, y) \leq \frac{2}{3}$?

The problem of finding the cardinality of the free distributive lattice on n generators was mentioned in Section 12.3.

ENUMERATION.

A glance at the *Encyclopedia of Integer Sequences*, shows a number of combinatorial counting sequences of which only a few terms are known. Each poses the problem of finding either a general formula or more terms. The authors, N. J. A. Sloane and S. Plouffe, mention several, including various kinds of 'polyominoes', non-attacking queens, polytopes, Latin squares, linear spaces (families of sets with any two points in a unique set of the family), and knots.

MISCELLANEA.

- Is there a perfect 1-error-correcting code over an alphabet not of prime-power size (Section 17.6)?
- Is there a Moore graph of diameter 2 and valency 57 (Section 11.12)?
- A conjecture of Isbell: Let $n = 2^a b$ with b odd. If a is sufficiently large compared to b, then an intersecting family of subsets of $\{1, \ldots, n\}$ with cardinality 2^{n-1} cannot be invariant under a transitive group of permutations of $\{1, \ldots, n\}$.

20.5. Further reading

I expect that, if you've read this far, you are feeling that on some topics I stopped just as things were getting interesting, while on others I said more than anyone would reasonably want to know. But I can't predict which topics will fall into which class. This section should help you explore further.

The authors of a book have to make some compromise between coverage and exposition. The result will lie at some point on the scale between light bedtime reading and an encyclopædia. Where I list two books, I have tried to put the textbook before the reference book.

GENERAL.

There are a number of general combinatorics books. Those which go beyond the introductory material tend to reflect their authors' interests. For example, M. Hall's *Combinatorial Theory* (1986) is strong on codes and designs, as well as the asymptotics of the partition function and the proof of the Van der Waerden Conjecture. Other books include L. Comtet, *Advanced Combinatorics* (1974), J. Riordan, *An Introduction to Combinatorial Analysis* (1958), H. J. Ryser, *Combinatorial Mathematics* (1963), and the recent book by J. H. van Lint and R. M. Wilson, *A Course on Combinatorics*.

The *Handbook of Combinatorics*, due out soon, will contain commissioned surveys on all parts of combinatorics. Another good source of more specialised surveys is the Proceedings of the biennial British Combinatorial Conference. Speakers are invited to survey their subject areas, and the papers are published in advance of the meetings. These have appeared in the London Mathematical Society Lecture Note Series since 1981, and most topics have been covered.

Another very useful book is L. Lovász's *Combinatorial Problems and Exercises* (1979), a vast collection of problems ranging from routine exercises to major theorems. The book is in three parts: Problems, Hints, and Solutions; the third part is the longest!

ENUMERATION.

F. Harary and E. M. Palmer, *Graphical Enumeration* (1973) is a good exposition; I. P. Goulden and D. M. Jackson, *Combinatorial Enumeration* (1983) contains everything you need about manipulating generating functions.

One book which is indispensible to enumerators is *The Encyclopedia of Integer Sequences* (1995), by N. J. A. Sloane and S. Plouffe. This is a list of over 5000 sequences in lexicographic order, with detailed bibliographic information about each one. If the sequence you've just discovered counting ultra-hyperbolic flim-flams has been found before, chances are you'll find it in here, with references. The

Encyclopedia is also accessible by electronic mail. An email to

sequences@research.att.com

consisting of one or more lines of the form

lookup 1 1 2 7 37 269 2535 29738 421790 7076459

will elicit a reply giving details of your sequences (if known).

SYMMETRIC FUNCTIONS.

This subject connects with enumeration on one side and with representation on the other. R. P. Stanley's *Ordered Structures and Partitions* (1972) is on the combinatorial side; I. G. Macdonald's *Symmetric Functions and Hall Polynomials* (1979) is condensed but clear.

FAMILIES OF SETS.

I. Anderson, *Combinatorics of Finite Sets* (1987) and B. Bollobás, *Combinatorics* (1986) are recommended.

TRANSVERSAL THEORY (SDRs).

L. Mirsky, *Transversal Theory* (1971); L. Lovász and M. D. Plummer, *Matching Theory* (1986).

RAMSEY THEORY.

R. L. Graham, B. L. Rothschild and J. Spencer, *Ramsey Theory* (1990), is wide-ranging and readable. P. Erdős *et al.*, *Combinatorial Set Theory* (1977), is slanted towards the infinite.

LATIN SQUARES.

See J. Dénes and A. D. Keedwell, *Latin Squares and their Applications* (1974).

DESIGN THEORY.

Two books with the title *Design Theory*, both appearing in 1985, are the textbook by D. R. Hughes and F. C. Piper, and the tome by T. Beth, D. Jungnickel and H. Lenz (which gives details of many recursive constructions).

GEOMETRY.

The classics are É. Artin, *Geometric Algebra* (1957), and J. Dieudonné, *La Géometrie des Groupes Classiques* (1955). A more recent account is in my lecture notes, *Projective and Polar Spaces* (1992), available from the School of Mathematical Sciences, Queen Mary and Westfield College. For finite projective geometries, the three-volume series by J. W. P. Hirschfeld, *Projective Geometries over Finite Fields* (1979, 1985, 1991 – the last with J. A. Thas), is definitive.

PERMUTATION GROUPS.

H. Wielandt, *Finite Permutation Groups* (1964), and D. S. Passman, *Permutation Groups* (1968). T. Tsuzuku, *Finite Groups and Finite Geometries* (1982) and N. L. Biggs and A. T. White, *Permutation Groups and Combinatorial Structures* (1979) deal particularly with the relations to combinatorics. No satisfactory account of the situation post-Classification (of finite simple groups) has appeared.

A class of infinite permutation groups with particular links with combinatorics are discussed in my book *Oligomorphic Permutation Groups* (1990).

CODES.

Start with R. Hill, *A First Course in Coding Theory* (1986) or J. H. van Lint, *Introduction to Coding Theory* (1982). The encyclopædia is F. J. MacWilliams and N. J. A. Sloane, *The Theory of Error-Correcting Codes* (1977). For the relation with information theory, see C. M. Goldie and R. G. E. Pinch, *Communication Theory* (1991); for cryptography, see D. J. A. Welsh, *Codes and Cryptography* (1988). The title of P. J. Cameron and J. H. van Lint's *Designs, Graphs, Codes and their Links* is self-explanatory.

ORDERS.

B. A. Davey and H. A. Priestley, *Introduction to Lattices and Order* (1990).

MATROIDS.

V. W. Bryant and H. Perfect, *Independence Theory in Combinatorics* (1980) sets the scene for the comprehensive treatment by D. J. A. Welsh, *Matroid Theory* (1976).

GRAPH THEORY.

There is a wide choice of books at many levels. R. J. Wilson's *Introduction to Graph Theory*, is just that. For more specialised topics, see B. Bollobás, *Extremal Graph Theory* (1978); B. Bollobás, *Random Graphs* (1985); A. E. Brouwer, A. M. Cohen, and A. Neumaier, *Distance-Regular Graphs* (1989); or several of the books referred to earlier.

Four volumes of surveys edited by L. W. Beineke and R. J. Wilson, three on *Selected Topics in Graph Theory* (1978, 1983, 1987) and one on *Applications of Graph Theory* (1979), give a wide coverage of the subject.

INFINITE COMBINATORICS.

No text-book is devoted exclusively to this. Most books include it in varying proportions. D. König's early classic *Theorie der endlichen und unendlichen Graphen* (1950 reprint) is not prejudiced towards the finite. A recent conference proceedings edited by R. Diestel, *Directions in Infinite Graph Theory and Combinatorics* (1991), contains a number of valuable surveys.

ALGEBRAIC COMBINATORICS.

This growing area includes topics from many parts of combinatorics. C. D. Godsil, *Algebraic Combinatorics*, is recommended.

AND FINALLY ....

A lot of information about combinatorics is available on the World Wide Web. You might try the World Combinatorial Exchange, at

`http://ejc.math.gatech.edu:8080/Journal/ejc-wce.html`

for an electronic journal and many pointers to other sites.

Answers to selected exercises

CHAPTER 2, EXERCISE 5. There are 80 unlabelled families.

CHAPTER 2, EXERCISE 12. (i) Let $m = a_0 + 2a_1 + \ldots + 2^{d-1}a_{d-1}$. Numbering the first row as zero, the two entries in the ith row are $a_i + 2a_{i+1} + \ldots + 2^{d-1-i}a_{d-1}$ and $2^i n$. The first of these is odd if and only if $a_i = 1$; so we add the values $2^i n$ for which $a_i = 1$, that is, we calculate $\sum a_i 2^i n = mn$.

(ii) If n is written in base 2, then doubling has the effect of shifting it one place to the left; so the terms added are exactly those occurring in the standard long multiplication done in base 2.

(iii) With these modifications, we replace $2^i n$ by n^{2^i}, so that the final result is indeed n^m. The method requires $2\lfloor \log_2 m \rfloor$ multiplications at most, viz., $\lfloor \log_2 m \rfloor$ squarings and then at most this number of multiplications in the last step, since m has $1 + \lfloor \log_2 m \rfloor$ digits in base 2.

CHAPTER 3, EXERCISE 4(e). HINT: $(1+t)^n(1-t)^n = (1-t^2)^n$.

CHAPTER 3, EXERCISE 11. 523 words can be made with these letters.

CHAPTER 3, EXERCISE 16. We get a 1-factor by writing $n/2$ boxes each with room for two entries, and filling them with the elements $1, \ldots, n$. These can be written in the boxes in $n!$ ways. However, permuting the boxes (in $k!$ ways), or the elements within the boxes (in 2^k ways) doesn't change the 1-factor. The product of these numbers is $2.4.6 \ldots (2k)$; dividing, we obtain $1.3.5 \ldots (2k-1) = (2k-1)!!$ for the number of 1-factors.

(b) Suppose that the k-set A is exchanged with its complement B by a permutation. Then elements of A and B alternate around each cycle, which thus has even length. Conversely, if all cycles have even length, we may colour the elements in each cycle alternately red and blue; the red and blue sets are then exchanged.

From a permutation with all cycles even we obtain a pair of 1-factors as follows. A 2-cycle is assigned to both 1-factors; in a longer cycle, the consecutive pairs are assigned alternately to the two 1-factors. The process isn't unique, since the starting point isn't specified; indeed, a permutation gives rise to 2^d ordered pairs of 1-factors, where d is the number of cycles of length greater than 2.

Conversely, let a pair of 1-factors be given. Their union is a graph with all vertices of valency 2, thus a disjoint union of circuits, all of even length (since the 1-factors alternate around a circuit). We take these circuits to be the cycles of a permutation. In fact, for a circuit of length greater than 2, there are two choices for the direction of traversal. So the number of permutations obtained is 2^d, where d is as before.

(c) The proportion is

$$((2k-1)!!)^2/(2k)! = \prod_{i=1}^{k}(1 - 1/2i)$$

$$\leq \prod_{i=1}^{k} e^{-1/2i}$$

$$= e^{-\sum_{i=1}^{k}(1/2i)}$$

$$= e^{-\log k/2 + O(1)} = O(k^{-1/2}).$$

CHAPTER 3, EXERCISE 16. 2^{n^2}. (a) $2^{n(n-1)}$, (b) $2^{n(n+1)/2}$, (c) $2^{n(n-1)/2}$, (d) $3^{n(n-1)/2}$.

CHAPTER 3, EXERCISE 19. The numbers are (a) 29, (b) 13, (c) 19, (d) 6. (The numbers of unlabelled structures are 9, 4, 5, 1 respectively.)

CHAPTER 4, EXERCISE 10.

$$f(n) = \begin{cases} (2^{n+1} - 1)/3, & n \text{ even}, \\ (2^{n+1} - 2)/3, & n \text{ odd}. \end{cases}$$

CHAPTER 4, EXERCISE 11. (b) The relationship is $u(n) - u(n-1) = s(n)/2$ for $n \geq 2$.

CHAPTER 4, EXERCISE 16. Imagine that the clown wears a diving suit, and continues to draw balls even after he gets wet. There are $\binom{2n}{n}$ ways in which the balls could be drawn. The clown stays dry if and only if the number of red balls never exceeds the number of blue ones; according to the voting interpretation, this is C_{n+1}. The ratio is $1/(n+1)$.

For a harder exercise, find a direct proof of this exercise, and reverse the above argument to deduce the formula for the Catalan numbers.

CHAPTER 4, EXERCISE 19. The recurrence can be written as

$$\sum_{k=1}^{n+1} \binom{n+1}{k} B(k) = B(n+1) \qquad \text{for } n \geq 2.$$

Multiplying by $t^{n+1}/(n+1)!$ and summing over n, the two sides are $f(t)\exp(t)$ and $f(t)$ with the constant and linear terms omitted. Thus $f(t)\exp(t) - 1 - t + \frac{1}{2}t = f(t) - 1 + \frac{1}{2}t$, whence $f(t) = t/(\exp(t) - 1)$, as required.

Now $f(t) + \frac{1}{2}t = \frac{1}{2}t(\exp(\frac{1}{2}t) + \exp(-\frac{1}{2}t))/(\exp(\frac{1}{2}t) - \exp(-\frac{1}{2}t)) = \frac{1}{2}t \coth \frac{1}{2}t$, an even function. The last recurrence obviously has the solution $b(k) = (-1)^k$.

CHAPTER 5, EXERCISE 1. The pollster is either dishonest or incompetent.

CHAPTER 5, EXERCISE 11. (a) The identity has n cycles, a single cycle just one, and multiplying by a transposition changes the number by 1. So at least $n - 1$ transpositions are required. The proof of (5.5.2) shows that this number is achieved if and only if each transposition involves points lying in different cycles of the product so far. This is equivalent to saying that the transpositions are edges of a connected graph without cycles, a tree (compare the discussion in Section 11.3). There are n^{n-2} trees on $\{1, \ldots, n\}$, and $(n-1)!$ orders to choose the $n-1$ edges of such a tree. So there are $n^{n-2}(n-1)!$ tree-cycle pairs. Each of the $(n-1)!$ cycles occurs equally often (they are all conjugate), necessarily n^{n-2} times.

CHAPTER 6, EXERCISE 2. We look for properties of the group multiplication tables unaltered by row and column permutations. There are two groups of order 4, the cyclic group and the Klein group. The second, but not the first, has the property that given any two rows and any column, there is a (unique) second column so that the entries in these rows and columns form a Latin subsquare of order 2. This property is preserved by row and column permutations. So the two multiplication tables are inequivalent. Since there are only two inequivalent Latin squares, both are group multiplication tables.

There is only one group of order 5, the cyclic group; its multiplication table has the property that any two rows 'differ' by a cyclic permutation. It is not hard to construct a Latin square of order 5 which does not have this property.

CHAPTER 6, EXERCISE 4. (a) $\{\{1, 2, 3\}, \{1, 2\}, \{1, 3\}\}$.
    (b) 24.

CHAPTER 7, EXERCISE 4. HINT: Use Lucas' Theorem (Section 3.4).

CHAPTER 7, EXERCISE 8. (a) The minimal sets of any family form a Sperner family.
    (b) If Z is minimal with respect to meeting every set $F' \setminus F$, for $F' \in \mathcal{F}$, $F' \neq F$, then $Y = \{y\} \cup Z$ has this property. [Why does such a Z exist?]
    (c) If $F \in \mathcal{F}$, then F meets every set of $B(\mathcal{F})$; by (b), if we omit a point of F, this is no longer true, so F is minimal. Thus $\mathcal{F} \subseteq b(b(\mathcal{F}))$. To prove the reverse inclusion, it suffices to show that a

set meeting every member of $b(\mathcal{F})$ contains a member of $\mathcal{F}$, or equivalently, a set Z containing no member of $\mathcal{F}$ is disjoint from some member of $b(\mathcal{F})$. But if Z is such a set, then $X \setminus Z$ meets every member of $\mathcal{F}$, so $X \setminus Z$ contains a member of $b(\mathcal{F})$, as required.

(d) The first part is clear, since an $(n+1-k)$-set meets every k-set but an $(n-k)$-set is disjoint from some k-set.

$\mathcal{F}_0 = \{\emptyset\}$, and no set meets $\emptyset$, so $b(\mathcal{F}_0) = \emptyset$, the empty family. [Note that any set has the property that it meets every set in $\emptyset$, and the unique minimal set is $\emptyset$; so $b(\emptyset) = \{\emptyset\}$, in accordance with (c).]

CHAPTER 8, EXERCISE 2. $1+z$ is a primitive element; its fourth power is $1-z$. So coset representatives are 1, $1+z$, $(1+z)^2 = 3z$, and $(1+z)^3 = 2+z$.

CHAPTER 8, EXERCISE 11. Let z be a point outside Y. Then z lies in $(n-1)/2$ triples. But a triple containing z and a point of Y contains only one point of Y (since a triple with two points in Y is contained in Y); there are m triples of this form. Thus, $(n-1)/2 \geq m$.

Equality holds if and only if every triple through *any* point outside Y meets Y, i.e., no triple is disjoint from Y.

CHAPTER 9, EXERCISE 2. There are q^{n^2} matrices; of these, $\prod_{i=0}^{n-1}(q^n - q^i)$ are non-singular; so the probability is

$$\prod_{k=1}^{n}\left(1 - \frac{1}{q^k}\right)$$

(putting $k = n - 1 - i$). This is a decreasing function of n, so tends to a limit $c(q)$, which is clearly less than 1. Now

$$\log c(q) = \sum \log\left(1 - \frac{1}{q^k}\right) \geq \sum q^{-(k-1)} \log\left(\frac{q-1}{q}\right) = \frac{q}{q-1}\log\left(\frac{q-1}{q}\right),$$

since the curve $y = \log(1-x)$ lies above the line segment from $(0,0)$ to $(1/q, \log((q-1)/q))$. So $c(q) \geq ((q-1)/q)^{q/(q-1)} > 0$.

CHAPTER 9, EXERCISE 11. Let $p_1, \ldots, p_5$ be five points, no three collinear. Then non-zero vectors spanning the first three points form a basis, and the other two points have all their coordinates non-zero relative to this basis. (If p_4 had its third coordinate zero, then p_1, p_2, p_4 would be dependent.) Multiplying the basis vectors by suitable scalars, we can assume that $p_4 = [1, 1, 1]$. Then $p_5 = [1, \alpha, \beta]$, where $1 \neq \alpha \neq \beta \neq 1$ (to ensure the independence of p_i, p_4, p_5 for $i = 1, 2, 3$).

Now the general second degree equation is

$$ax^2 + by^2 + cz^2 + fyz + gzx + hxy = 0.$$

If p_1, p_2, p_3 lie on this curve, then $a = b = c = 0$. Substituting the coordinates of p_4 and p_5 gives

$$f + g + h = 0,$$
$$f + g/\alpha + h/\beta = 0.$$

These equations are independent, so the solution is unique up to scalar multiple (and so defines a unique conic).

Counting arguments show that the number of choices of five points in $\mathrm{PG}(2, q)$ with no three collinear is

$$(q^2 + q + 1).(q^2 + q).q^2.(q-1)^2.(q-2)(q-3).$$

A conic has $q + 1$ points with no three collinear; five points can be chosen in

$$(q+1).q.(q-1).(q-2).(q-3)$$

ways. Division gives the result.

CHAPTER 10, EXERCISE 1. Let a_{ij} and b_{ij} be the heights of the soldiers in row i and column j after the first and second rearrangements. Then $a_{ij} > a_{i+1j}$. Also, $a_{ij} = b_{kj}$ if and only if a_{ij} is the kth largest number in column j. Suppose that $b_{ij} < b_{i+1j}$, and let x be a number lying between these two values. Then fewer than i soldiers in column j have heights exceeding x, but i or more soldiers in column $j + 1$ have heights exceeding x. But this is a contradiction, since each soldier in column j (before the second rearrangement) is taller than his neighbour in column $j + 1$.

CHAPTER 10, EXERCISE 6. (a) It is important to justify the 'by symmetry ...' in the Hint. The colouring is unchanged if we add a fixed residue mod 17 to everything, or if we multiply everything by $\pm 1, \pm 2, \pm 4$ or ± 8 (these are all the quadratic residues mod 17). Suppose that there is a red 4-set $\{a, b, c, d\}$. By adding $-a$, we can assume that $a = 0$; by multiplying by $1/b$, we can assume that $b = 1$. Now the red neighbours of 0 are 1, 2, 4, 8, 9, 13, 15, 16; the red neighbours of 1 are 0, 2, 3, 5, 9, 10, 14, 16. So c and d are chosen from 2, 9 and 16. But all edges between these three points are blue.

Furthermore, multiplication by any fixed quadratic non-residue maps red edges to blue ones and *vice versa*. So, if there were a blue 4-set, there would be a red one as well.

(b) Take the points joined to 0 by red edges in the solution to (a).

(c) By symmetry, we can assume that $x = \emptyset$. (The map on X given by $x \mapsto x \triangle a$, for fixed $a \in X$, preserves the colouring.) Now examine the points joined to $\emptyset$ by edges of each colour.

CHAPTER 11, EXERCISE 1. (a) 13; (b) 10; (c) 3; (d) 4; (e) 8; (f) 13.

CHAPTER 11, EXERCISE 2. In Fig. 11.4, a Hamiltonian cycle must use two or four of the five edges connecting the outer and inner 5-cycles (the pentagon and the pentagram). If two of these 'crossing' edges are used, their ends in the outer cycle must be adjacent, since the remainder of the cycle must be traversed between them. Similarly, their ends in the inner cycle must be adjacent. But this is impossible. Suppose that four crossing edges are used. There is only one possible configuration for these four edges. Then there are two ways to link them in the outer cycle, and two in the inner cycle. Of these four possibilities, three fail to have ten edges altogether, and the remaining one consists of two 5-cycles.

A Hamiltonian path is easily found by traversing the outer cycle, crossing to the inner cycle, and traversing the inner cycle.

CHAPTER 11, EXERCISE 13. Let A_i be the set of neighbours of i. By assumption, $|A_i \cap A_j| = 1$ for $i \neq j$. If $|A_1| = n - 1$, then 1 is joined to all other vertices. Now the remaining vertices are paired up, since any friend of i (other than 1) is a common friend of 1 and i; the graph is a windmill.

If this doesn't happen, then $|A_i| = k$ for all i, by the De Bruijn–Erdős theorem. Now the adjacency matrix A satisfies $A^2 = (k - 1)I + J$, where J is the all-one matrix. It has the eigenvalue k with multiplicity 1, corresponding to the all-1 eigenvector; any other eigenvalue α satisfies $\alpha^2 = k - 1$, so the eigenvalues are $\pm\sqrt{k - 1}$, with multiplicities f and g, say. The trace of A is zero, so $f - g = -k/\sqrt{k - 1}$.

From the last equation, it is impossible that $f = g$. So $k = u^2 + 1$ for some integer u, and $f - g = -(u^2 + 1)/u$. Since this is an integer, we have $u = 1$, whence $k = 2$ and the graph is a triangle (which is a special case of a windmill).

Bibliography

S. N. Afriat, *The Ring of Linked Rings*, Duckworth & Co., London, 1982.

I. Anderson, *A First Course in Combinatorial Mathematics*, Oxford University Press, Oxford, 2nd edition 1989.

I. Anderson, *Combinatorics of Finite Sets*, Oxford University Press, Oxford, 1987.

V. I. Arnol'd, *Huygens and Barrow, Newton and Hooke*, Birkhäuser, Basel, 1990.

É. Artin, *Geometric Algebra*, Wiley, New York, 1957.

N. Baxter, E. Dubinsky, and G. Levin, *Learning Discrete Mathematics with ISETL*, Springer-Verlag, New York, 1988.

L. W. Beineke and R. J. Wilson, *Selected Topics in Graph Theory* (3 volumes), Academic Press, London, 1978/1983/1987.

T. Beth, D. Jungnickel and H. Lenz, *Design Theory*, Cambridge University Press, Cambridge, 1986.

N. L. Biggs, *Discrete Mathematics*, Oxford University Press, Oxford, 1985.

N. L. Biggs, E. K. Lloyd, and R. J. Wilson, *Graph Theory 1736–1936*, Oxford University Press, Oxford, 1976. (Paperback edition 1986.)

N. L. Biggs and A. T. White, *Permutation Groups and Combinatorial Structures*, Cambridge University Press, Cambridge, 1979.

B. Bollobás, *Extremal Graph Theory*, Academic Press, London, 1978.

B. Bollobás, *Random Graphs*, Academic Press, London, 1985.

B. Bollobás, *Combinatorics*, Cambridge University Press, Cambridge, 1986.

J. L. Borges, *Labyrinths*, New Directions, New York, 1964.

R. Brautigan, *The Hawkline Monster: A Gothic Western*, Jonathan Cape, London, 1975.

N. Braybrooke (ed.), *T. S. Eliot: A Symposium for his Seventieth Birthday*, Hart-Davies, London, 1958.

A. E. Brouwer, A. M. Cohen, and A. Neumaier, *Distance-Regular Graphs*, Springer, Berlin, 1989.

V. W. Bryant, *Aspects of Combinatorics*, Cambridge University Press, Cambridge, 1993.

V. W. Bryant and H. Perfect, *Independence Theory in Combinatorics*, Chapman and Hall, London, 1980.

P. J. Cameron, *Oligomorphic Permutation Groups*, Cambridge University Press, Cambridge, 1990.

P. J. Cameron, *Projective and Polar Spaces*, Queen Mary and Westfield College, London, 1992.

P. J. Cameron and J. H. van Lint, *Designs, Graphs, Codes, and their Links*, Cambridge University Press, Cambridge, 1991.

L. Comtet, *Advanced Combinatorics*, D. Reidel, Dordrecht, 1974.

T. Dantzig, *Number: the Language of Science*, Macmillan, New York, 1930.

B. A. Davey and H. A. Priestley, *Introduction to Lattices and Order*, Oxford University Press, Oxford, 1990.

J. Dénes and A. D. Keedwell, *Latin Squares and their Applications*, English Universities Press, London, 1974.

K. J. Devlin, *Fundamentals of Contemporary Set Theory*, Springer-Verlag, New York, 1979.

R. Diestel (ed.), *Directions in Infinite Graph Theory and Combinatorics*, Elsevier, Amsterdam, 1991.

J. Dieudonné, *La Géometrie des Groupes Classiques*, Springer-Verlag, Berlin, 1955.

J. Dieudonné, *A Panorama of Pure Mathematics: As seen by N. Bourbaki*, Academic Press, New York, 1982.

T. S. Eliot, *Four Quartets*, Faber & Faber, London, 1944.

P. Erdős, *The Art of Counting: Selected Writings* (ed. J. Spencer), MIT Press, Cambridge, Mass., 1973.

P. Erdős, A. Hajnal, A. Maté, and R. Rado, *Combinatorial Set Theory: Partition Relations for Cardinals*, North-Holland, Amsterdam, 1977.

E. M. Forster, *Howards End*, Edward Arnold, London, 1910.

M. R. Garey and D. S. Johnson, *Computers and Intractibility: A Guide to the Theory of NP-Completeness*, W. H. Freeman, San Francisco, 1979.

C. D. Godsil, *Algebraic Combinatorics*, Chapman and Hall, New York, 1993.

C. M. Goldie and R. G. E. Pinch, *Communication Theory*, Cambridge University Press, Cambridge, 1991.

S. J. Gould, *Ever Since Darwin: Reflections in Natural History*, Norton, New York, 1977.

I. P. Goulden and D. M. Jackson, *Combinatorial Enumeration*, John Wiley & Sons, New York, 1986.

R. L. Graham, M. Grötschel and L. Lovász (eds.), *Handbook of Combinatorics*, Elsevier, Amsterdam, to appear.

R. L. Graham, B. L. Rothschild and J. Spencer, *Ramsey Theory*, Wiley, New York, 1990.

M. Hall Jr., *Combinatorial Theory* (second edition), Wiley, New York, 1986.

F. Harary and E. M. Palmer, *Graphical Enumeration*, Academic Press, New York, 1973.

G. H. Hardy and E. M. Wright, *An Introduction to the Theory of Numbers*, (5th edition), Oxford University Press, Oxford, 1979.

R. Hill, *A First Course in Coding Theory*, Oxford University Press, Oxford, 1986.

J. W. P. Hirschfeld, *Projective Geometries over Finite Fields*, Oxford University Press, Oxford, 1979.

J. W. P. Hirschfeld, *Projective Spaces of Three Dimensions*, Oxford University Press, Oxford, 1985.

J. W. P. Hirschfeld and J. A. Thas, *General Galois Geometries*, Oxford University Press, Oxford, 1991.

R. Hoban, *Turtle Diary*, Jonathan Cape, London, 1975.

R. Hoban, *Riddley Walker*, Jonathan Cape, London, 1980.

R. Hoban, *Pilgermann*, Jonathan Cape, London, 1983.

D. A. Holton and J. Sheehan, *The Petersen Graph*, Cambridge University Press, Cambridge, 1993.

D. R. Hughes and F. C. Piper, *Design Theory*, Cambridge University Press, Cambridge, 1985.

G. Ifrah, *From One to Zero: A Universal History of Numbers*, Viking Penguin, New York, 1985.

D. König, *Theorie der endlichen und unendlichen Graphen: kombinatorische Topologie der Streckenkomplexe*, Chelsea, New York, 1950.

I. Lakatos, *Proofs and Refutations: The Logic of Mathematical Discovery*, Cambridge University Press, Cambridge, 1976.

U. K. LeGuin, *The Left Hand of Darkness*, MacDonald & Co., London, 1969.

S. Lem, *His Master's Voice*, Martin Secker & Warburg, London, 1983.

J. H. van Lint, *Introduction to Coding Theory*, Springer-Verlag, Berlin, 1982.

J. H. van Lint and R. M. Wilson, *A Course on Combinatorics*, Cambridge University Press, Cambridge, 1992.

L. Lovász, *Combinatorial Problems and Exercises*, North-Holland, Amsterdam, 1978.

L. Lovász and M. D. Plummer, *Matching Theory*, North-Holland, Amsterdam, 1986.

I. G. Macdonald, *Symmetric Functions and Hall Polynomials*, Oxford University Press, Oxford, 2nd edition 1995.

F. J. MacWilliams and N. J. A. Sloane, *The Theory of Error-Correcting Codes*, North-Holland, Amsterdam, 1977.

P. Matthews (ed.), *The Guinness Book of Records/The Guinness Multimedia Disc of Records*, Guinness Publ./Grolier Electronic Publ., New York, 1993.

L. Mirsky, *Transversal Theory*, Academic Press, New York, 1987.

J. von Neumann and O. Morgenstern, *Theory of Games and Economic Behavior*, Princeton University Press, Princeton, 1944.

J. C. Oxtoby, *Measure and Category*, Springer-Verlag, New York, 1980.

D. S. Passman, *Permutation Groups*, Benjamin, New York, 1968.

J. Riordan, *An Introduction to Combinatorial Analysis*, Wiley, New York, 1958.

H. J. Ryser, *Combinatorial Mathematics*, Wiley, New York, 1963.

K. Singer, *Mirror, Sword and Jewel: The Geometry of Japanese Life*, Croom Helm, London, 1973.

N. J. A. Sloane and S. Plouffe, *he Encyclopedia of Integer Sequences*, Academic Press, New York, 1995.

A. Slomson, *Introduction to Combinatorics*, Chapman and Hall, London, 1991.

R. Stanley, *Ordered Structures and Partitions*, Amer. Math. Soc., Providence, 1972.

I. N. Stewart and D. O. Tall, *Algebraic Number Theory*, Chapman and Hall, London, 1987.

R. J. Stewart, *Where is Saint George? Pagan Imagery in English Folksong*, Blandford Press, London, 1977.

T. Tsuzuku, *Finite Groups and Finite Geometries*, Cambridge University Press, Cambridge, 1982.

N. Ya. Vilenkin, *Stories about Sets*, Academic Press, New York, 1968.

A. W. Watts, *The Book: On the Taboo against Knowing Who You Are*, Random House, New York, 1972.

D. J. A. Welsh, *Matroid Theory*, Academic Press, London, 1976.

D. J. A. Welsh, *Codes and Cryptography*, Oxford University Press, Oxford, 1988.

H. Wielandt, *Finite Permutation Groups*, Academic Press, New York, 1964.

R. J. Wilson, *Introduction to Graph Theory*, Longmans, Harlow, 3rd edition 1985. (4th edition due 1996.)

R. J. Wilson and L. W. Beineke, *Applications of Graph Theory*, Academic Press, London, 1979.

R. J. Wilson and J. J. Watkins, *Graphs: An Introductory Approach*, Wiley, New York, 1990.

W. Wharton, *Birdy*, Alfred A. Knopf, New York, 1979.

Index